Biological Trace Element Research

Biological Trace Element Research
Multidisciplinary Perspectives

K.S. Subramanian, EDITOR
Health & Welfare, Canada
G.V. Iyengar, EDITOR
National Institute of Standards and Technology, U.S.
K. Okamoto, EDITOR
National Institute of Environmental Studies, Japan

Developed from a symposium sponsored
by the International Chemical Congress
of Pacific Basin Societies,
Honolulu, Hawaii,
December 17–22, 1989

American Chemical Society, Washington, DC 1991

Library of Congress Cataloging-in-Publication Data

Biological trace element research: multidisciplinary perspectives/
K. S. Subramanian, editor G. V. Iyengar, editor K. Okamoto, editor

Developed from a symposium sponsored by the International Chemical Congress of Pacific Basin Societies, Honolulu, Hawaii, December 17–22, 1989.

p.   cm.—(ACS Symposium Series, 0097–6156; 445).

Includes bibliographical references and indexes.

ISBN 0–8412–1888–9
1. Trace elements—Research—Methodology—Congresses.

I. Subramanian, K. S., 1944– . II. Iyengar, G. V. (Govindaraja V.)
III. Okamoto, Kensaku. IV. International Chemical Congress of Pacific
Basin Societies (1989: Honolulu, Hawaii) V. Series

QP534.B563   1991
612'.01524—dc20   90–20772
                                          CIP

The paper used in this publication meets the minimum requirements of American National Standard for Information Sciences—Permanence of Paper for Printed Library Materials, ANSI Z39.48–1984.   ∞

ACS Symposium Series

QP534
B563
1991
Chem

M. Joan Comstock, *Series Editor*

1991 ACS Books Advisory Board

Foreword

THE ACS SYMPOSIUM SERIES was founded in 1974 to provide a medium for publishing symposia quickly in book form. The format of the Series parallels that of the continuing ADVANCES IN CHEMISTRY SERIES except that, in order to save time, the papers are not typeset, but are reproduced as they are submitted by the authors in camera-ready form. Papers are reviewed under the supervision of the editors with the assistance of the Advisory Board and are selected to maintain the integrity of the symposia. Both reviews and reports of research are acceptable, because symposia may embrace both types of presentation. However, verbatim reproductions of previously published papers are not accepted.

Contents

DETERMINATION

INDEXES

Preface

CHEMICAL ELEMENTS AS ESSENTIAL CONSTITUENTS sustain life processes, but they can also be toxic components that hinder normal metabolic events. The study of these roles has generated a large body of analytical data. However, as many of us have now realized, a significant portion of the existing analytical results is of questionable value and in some cases is totally worthless. Fortunately, careful research has identified some of the sources of analytical errors, especially those arising from contamination of the samples and improper use of analytical techniques.

Many laboratories have recognized the indispensable role of reference materials as quality markers of analytical results. Through the use of reference materials and good laboratory practices, these laboratories have been trying to improve their analytical performance. These measures, which addressed the problem predominantly at the postsampling stage, offered only a partial remedy to the overall situation. A comprehensive improvement in the quality of the analytical result required recognition of not only the analytical component, but also the biological component (especially presampling factors).

A well-planned biological trace element study requires total quality control, including experimental design, collection of valid samples (biological and analytical integrity), chemical analysis, data evaluation, and interpretation. In other words, we need a multidisciplinary approach. Dr. Iyengar has cogently handled this part in Chapter 1 by cautioning, "The lack of a multidisciplinary approach has been the Achilles heel of biological trace element research."

Both analysts and life scientists recognize that the key to the development of a reliable data base for the elemental content and speciation of biological systems lies in an interdisciplinary strategy involving: (1) an understanding of the biological basis of the problem, (2) developments in bioanalytical concepts, (3) a controlled-environment approach for metabolic studies, (4) preparation of well-defined protocols, (5) procurement of valid samples, (6) use of appropriate analytical techniques and procedures, and (7) analytical quality assurance.

In the symposium upon which this book is based, we sought to address these concerns by bringing together experts from the fields of analytical chemistry, biochemistry, biophysics, biostatistics, medicine, and nutrition. The speakers came from Asia, Canada, Europe, and the United States. Although our intention was to achieve a comprehensive multidisciplinary gathering, we feel that we made only a modest beginning. However, the mix of review papers, original research manuscripts, and reports of new work presented in this book best reflect the current state of the subject. The opportunity to publish such diverse material in a single place is particularly welcome, for much would have been lost if the information presented here had been broken into sections appropriate for specialized journals. The interdisciplinary nature of this book should make it valuable to scientists and clinicians in the areas of analytical chemistry, biochemistry, clinical chemistry, medicine, and nutrition.

Much remains to be done. The multidisciplinary approach to solving biological trace element problems is exciting as well as challenging. We hope that the readers will find this book interesting and useful in their many and varied quests involving trace element analytical chemistry of biological systems.

K. S. SUBRAMANIAN
Environmental Health Centre
Health & Welfare Canada
Tunney's Pasture, Ottawa, Ontario K1A 0L2, Canada

G. V. IYENGAR
Center for Analytical Chemistry
National Institute of Standards and Technology
Gaithersburg, MD 20899

K. OKAMOTO
National Institute for Environmental Studies
Environmental Agency of Japan
Onogawa 16–2, Tsukuba, Ibaraki, 305, Japan

August 29, 1990

Chapter 1

The Need for Multidisciplinary Approaches in Biological Trace Element Research

G. V. Iyengar

National Institute of Standards and Technology, Reactor Building 235, B125, Gaithersburg, MD 20899

Biological Trace Element Research (BTER) is a multidisciplinary science and directing research in this field requires a combination of both biological insight and analytical awareness. Quite frequently, the complexities involved here necessitate the use of a variety of scientific talents and a combination of several analytical techniques. For example, some of the special difficulties can be visualized in meeting the requirements of a "total" quality control in the overall context of a BTER investigation. These include experimental design, collection of biologically and analytically "valid" samples, the ability to carry out accurate analytical measurements on those specimens, and a proper evaluation of the analytical data (including data interpretation). Therefore, it follows that a team approach ensuring the required multidisciplinary interactions is a crucial component in any BTER endeavor.

"The lack of a multidisciplinary approach has been the Achilles heel of biological trace element research" (1).

Although a large body of data exists for most trace elements in biological media, in many cases they are of limited use because of inherent inaccuracies (1,2). This has generated a lot of skepticism among biological trace element researchers. Primarily, the variations in analytical findings that still are prevalent can be linked to several pitfalls that have failed to receive adequate attention by trace element investigators. Therefore, first

of all, there is a need to reestablish credibility in trace
element analytical research by generating well controlled
quality data to dispel fears in the minds of many users. In
this context, it may be emphasized that development of a
reliable data base for the elemental composition of
biological systems is a multidisciplinary task involving an
understanding of the biological basis of the problem,
preparation of well defined protocols, adequately tested
methods for sampling and handling biomaterials and, use of
appropriate analytical procedures. In other words, it is a
team effort requiring cumulative knowledge from several
disciplines (3).

In earlier investigations, especially until about the
early seventies, a great proportion of the analytical
results obtained for various biological materials were
mainly intended to demonstrate the powerful capabilities of
the newly emerging methodologies, e.g., multielement
techniques. Unfortunately, in the wake of analyst's
enthusiasm to prove the effectiveness of these
extraordinary technical achievements, this "black box"
attitude of using multielement analyses assumed a sort of
"cure all" posture and little or no consideration was given
to incorporate the biologic basis for the investigations.
To give an example, practically countless number of studies
have been performed on hair due to the fact that this
specimen was the easiest to procure, and a great majority
of these investigations has merely resulted in a trail of
pointless publications. The position is no better for a
number of other tissues and body fluids, since in most
cases, the medical and other health sciences professionals
who were involved in supplying the samples for analysis,
were unaware of the spurious analytical implications of
uncontrolled specimen collection and, the analysts for
their part who found access to the samples on their own,
did not have either the insight to assess the biological
integrity or possess the exacting training required to deal
with real world biological collections.

It is not my intention to summarily disqualify all the
earlier BTER investigations. For example, reliable results
have been reported for elements that did not pose
insuperable analytical difficulties. Similarly, in some
cases, the conclusions drawn were based on relative
measurements to estimate the differences, and there were
grounds to accept these findings as valid. In addition to
these, there were instances where an investigation was
carried out with meticulous care to understand the problems
confronted. However, it was not always easy to identify
such cases so that the handful of good results were also
drowned in the ocean of confusion. It is now generally
conceded that a great majority of the earlier BTER
experiments suffered from methodological inadequacies.

Developments in Bioanalytical Concepts

"A prudent combination of analytical awareness and biological insight is crucial for success in biomedical trace element research studies" (1).

The developments which contributed to the emergence of a viable bioanalytical concept capable of addressing BTER problems are the result of several aspects such as a) technological developments in metrology, b) analytical quality assurance and quality standard, c) controlled-environment concept for metabolic studies, d) recognition of BTER as a multidisciplinary science, and e) trace element speciation and bioavailability. These aspects will be discussed briefly in the following paragraphs.

Technological Developments in Metrology. Metrology is the science of measurements, devoted to all aspects of perfecting an analytical measurement to generate as accurate a result as possible. This branch of science has undergone tremendous changes since the days when the laboratories contained simple equipment that were conveniently termed "apparatus". The progress in analytical instrumentation and the concept of a modern analytical laboratory has brought about a bewildering array of analytical instruments whose performance is not optimally explored without the use of yet another laboratory tool namely, computer in analytical chemistry! Several analytical techniques namely atomic absorption (flame and flameless), atomic emission (direct current and inductively coupled plasma), chemical and electroanalytical methods, gas and liquid chromatography, mass spectrometry in different modes, nuclear activation techniques and X-ray fluorescence, which offer sufficiently low detection limits applicable to a wide variety of biomatrices. Furthermore, many of these methodologies possess simultaneous multielement capabilities and a gradual understanding of the analytical problems with biomatrices has significantly revolutionized BTER efforts. Thus, many interesting findings have surfaced unexpectedly as a result of these studies. From an analytical point of view, the non-destructive modes offer possibilities for generating simultaneous data for several elements for comparison, thus acting as internal quality control agents so that unusual situations involving any specific element can be evaluated. Moreover, in a carefully designed study, multielement assays can provide very useful information on a large number of elements at relatively low costs.

However, even as we have entered the era of the 90s, I must sadly pointout that very few laboratories in the world carry out reliable trace element determinations, while a large proportion of the laboratories working with biomaterials find it difficult to achieve a consistent

capability to maintain even a 10-20 % accuracy and
precision. This is an indication that high detection
capability and sensitivity alone are not the solution to
the problem of accuracy and precision, and that unawareness
of various interferences (e.g., matrix related problems),
flaws in sample and standard preparation and inadequate
calibration procedures are evident.

Analytical Quality Assurance and Quality Standard.
Frequent analysis of reference standards is the key for
overcoming procedural inconsistencies in establishing
quality control in any analytical laboratory. Major
advances have been made in the area of development of
reference materials with diverse matrix properties (1,4-
8). In this context, it should be recognized that the
standards set for the quality of an analytical result is an
important factor and is of course dependent upon the end
use of the results. Thus, for example, if the aim is
restricted merely to scan different biological matrices as
a provisional step to establish the relative levels of
elemental concentration profiles in them, it is obvious
that a reasonable degree of quality standard is sufficient;
whereas, for meeting the requirements of a typical BTER
laboratory aiming to use the results for medical diagnostic
purposes and legal regulatory processes, depending upon the
problem, results with 5 to 10 % total uncertainty may be
acceptable. On the other hand, an exceptionally high
quality standard (< 1% total uncertainty) is mandatory in
some cases, e.g., certification of reference materials. If
the tolerance limits are set narrower than the
investigation really requires or not feasible under
practical laboratory conditions, it can cause unnecessary
expense and loss of time (9).

Controlled-Environment Concept for Metabolic Studies. The
realization that several trace elements are required at
very minute concentrations and that under routine
laboratory conditions the ambient levels of the metals
could mask the effects of an investigation led to the
development of controlled-environment approach or the trace
element-free-isolater system. This is basically an all-
plastic assembly, with filtered air, free of dust
particles down to fraction of a μm. Further, continuous
monitoring of all the items in the system is maintained for
elements of interest. Under these conditions, the diet is
the main source of trace element supply, and by using a
chemically well-defined diet, reliable BTER investigations
can be carried out. Thus, the introduction of high purity
diets on one hand, and the ability to exercise strict
analytical quality control on the other, have led to rapid
advances in BTER investigations (10-13).

BTER- a Multidisciplinary Science. Biological trace
element research is a multidisciplinary science. Therefore,

it is necessary for researchers in this field to strive for a reasonable degree of biological insight and analytical awareness to the problems involved. This in turn will enable them to design meaningful BTER investigations. Quite frequently, the complexities involved here necessitate the use of a variety of scientific talents and a combination of several analytical techniques. Some of the special difficulties are seen in, for example, providing a "total" quality control in the overall context of an investigation. These include: experimental design; collection of biologically and analytically "valid" samples; understanding the implications of presampling factors such as biological variations, post mortem changes and intrinsic errors (14,15) in context of overall accuracy of an analytical result as reflected in Figure 1; and finally, the ability to carry out accurate analytical measurements on those specimens, data evaluation and a meaningful interpretation of the findings (16). For example, "status at sampling" can be an important source of variation of analytical results (Table I) reflecting the differences in residence times of trace elements and their compounds in the blood stream. To understand this, some knowledge of gut absorption of chemicals and nutrients from foods is necessary. Therefore, it follows that team work is an integral component of any BTER investigation.

In a carefully planned BTER investigation, several parameters deserve equal consideration. The actual process of an analytical measurement is merely one critical point that is preceded by a series of bioanalytical precautions and followed by a meaningful data interpretation. The parameters involved often cross boundaries of different disciplines, a signal that it is prudent to incorporate a team approach to sustain the comprehensiveness in such investigations. It should be recognized that sample quality as applied to biomedical specimens demands both the routine analytical precautions such as prevention of extraneous contamination or loss of an analyte, as well as retention of biological integrity of the specimen. For example, analysis of a specimen representing the cells of an organ (e.g., blood or liver cells) implies that the results obtained were finger prints of that cellular system attributable to a viable status. Any impairment to such a status would vitiate the results of an analysis, depending upon the question posed in utilizing that analytical finding. If the question is to address the net elemental contents of that cellular system, then the requirement is simply no extraneous contamination and no loss of any of the cellular fragments (and therefore loss of the analyte) during analysis. On the other hand, if the question is to address the biochemical moiety of a metal species in a specific compartment of the chosen cellular system, depending upon the situation, a series of biochemical measures will be required and for these results to be valid, a total quality control concept is obligatory. An

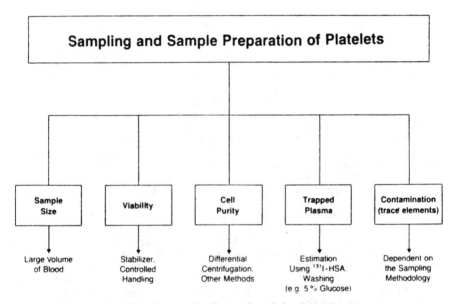

Figure 1. Sources of errors in elemental analysis of biological systems.
Reprinted from ref. 15. Copyright 1982 American Chemical Society.

Table I. Influence of "status at sampling" on the concentration
          levels of selected elements: some examples

Sample	Element	Condition	Influence	Remarks
Serum	Zn	Fasting	Elevated	Even overnight fasting
		After normal food	Lowered	During first few hours
		After high Zn food,	Elevated	e.g., Oysters, liver, etc.
		Low Zn intake	Lowered	Rather sudden effect
		Stress	Elevated	e.g., Standing posture
		Pregnancy	Lowered	Progressive decline
	Cu	Pregnancy	Elevated	Progressive increase
	F	Fasting	Normal	Overnight fasting
		After normal food	Variable	If sampled immediately
		Tea consumption	Elevated	If sampled immediately
		Low F intake	Lowered	Over a few days
	I	High intake e.g. sea foods, iodinated salts etc.,	Elevated	Uptake is rapid but decline is slow since biological half-life is 2.5 weeks
		Low I intake	Lowered	Over a few days
	As	Fish intake	Elevated	Well absorbed and slowly excreted
Blood	Cd	Tobacco smoking	Elevated	Remains chronically high in smokers
	Pb	Alcohol consumption	Elevated	Especially wine drinkers
Milk	Mn	High input	Elevated	Mn rich cereals
	Fe	Fore-milk Hind-milk	Elevated Elevated	Variation due to high fat content
	I	Low I intake High I intake e.g., dietary algae	Lowered Elevated	Over a few days Extremely high levels of I excreted within a short period
Urine	Several elements	Low or high intakes	Variations	Sensitive to fluid intake, requiring 24-h collections.

excellent example of total quality control concept is provided by the study that was designed to obtain platelets from human blood for trace element investigations. The steps involved namely, the problems of sample size, cell viability, cell purity, trapped plasma and extraneous contamination (Figure 2), required the expertise of several disciplines to obtain valid samples (17).

Trace Element Speciation and Bioavailability. A great majority of conventionally conducted BTER studies even now generate data only on the total amount of trace elements. For example in dietary intake studies it is common to carryout elemental analysis and use the results to draw comparisons and conclusions in context of recommended dietary allowances (RDA). It is being recognized now that the biologically available fraction of a trace element differs among different foods and therefore, bioavailability of various elements from single or mixed foods should be determined for an accurate estimation of the intake from diets. However, metal speciation in complex biological and dietary media is a difficult task due to detection limit considerations and the danger of inducing changes in the speciation states during analysis. Many biochemical methods for isolation of metallocomplexes such as ion-exchange chromatography, affinity chromatography, immunopre-cipitation, electrophoresis and isoelectricfocussing are in use, but are faced with problems of loss and contamination of the metal and its species. Radioactive labeling of organo-arsenic compounds has been described for animal studies (18). A two-stage in vitro system to simulate gastric and intestinal digestion of food in combination with Inductively Coupled Plasma-Mass Spectrometry has been successfully used in recent investigations (19). Several techniques based on gas chromatography and high pressure liquid chromatography and anodic stripping voltametry have been applied for speciation chemistry. Many analytical methods are useful for speciation chemistry work provided that their limitations are recognized (20). Concerning bioavailability, use of stable isotopes in mineral nutrition research is well established. Stable isotopes as isotopic tracers have no radiation risk and therefore, are widely accepted. Use of various versions of Mass Spectrometry and Neutron activation analysis for in vitro specimens following stable isotope application have resulted in considerable progress in mineral metabolism research.

Experimental Design and Data Interpretation

"The ultimate purpose of an analytical result is to address the problem, not merely generating numbers for intercomparisons" (1).

A RELIABLE CONCLUSION
DEPENDS ON THE QUALITY
OF THE ANALYTICAL RESULT

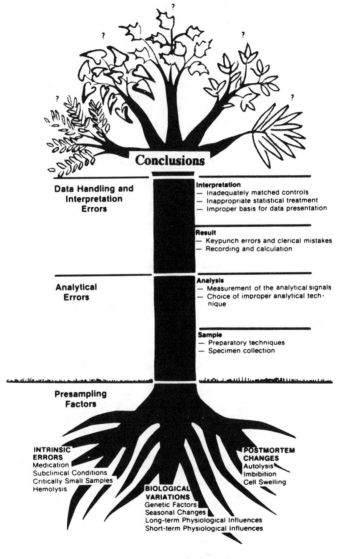

Figure 2. Multidisciplinary aspects of sampling human blood platelets for trace element analysis. (Reprinted from ref. 17. Copyright 1979 American Chemical Society.)

Studies on the elemental composition studies in biological systems can be divided into 4 stages: experimental design, collection of valid samples, chemical analysis and data evaluation (and interpretation). Collection of valid samples and chemical analysis have been touched upon in previous sections. Experimental design and data interpretation will be briefly commented upon here.

Experimental Design. The designing of a BTER experiment is possibly the most important stage at which multidisciplinary approach is crucial. A lack of this approach has led to either a total failure of an investigation or has generated false conclusions. To understand this situation one may consider the following four situations in resolving a problem related to lead in the environment:
Case 1: Analytically questionable and biologically unsound- e.g., earlier studies on hair as a monitoring specimen, and Pb in blood plasma.
Case 2: Analytically questionable and biologically sound- e.g., determination of Pb in liver.
Case 3: Analytically sound and biologically questionable- e.g., increase of Pb in whole blood related to smoking.
Case 4: Analytically reliable and biologically sound- e.g., Pb in whole blood (and also Cd in blood of smokers and non-smokers).
Case 1 illustrates a worst situation of choosing a questionable or unsuitable biological specimen presented with unreliable results. There are many problems in measuring the low Pb levels in plasma and more than 95 % of the whole blood Pb is associated with the erythrocytes and therefore, blood plasma is not an ideal indicator to reflect changes in Pb concentrations. Case 2 is a reflection of a sound biological specimen to study, yet prone to analytical pitfalls, especially at very low concentration levels. Case 3 is a typical situation where extreme care is exerted to solve a problem of wrong biological perception. Smokers are usually consumers of alcohol too, and it has been demonstrated that the elevated blood lead was traceable to wine consumption (21). Obviously, based on these findings if smoking had been banned as a measure to control blood lead levels, the effort would have been a futile move! Case 4 illustrates an ideal combination of biological relevance and analytical soundness since the validity of this approach has been documented through several studies (22).

Data Presentation and Interpretation. It is important to present results from BTER investigations in an unambiguous way and it is prudent to choose simple methods of data presentation and analysis over complicated procedures. As

expressed by Cherry, the numbers generated by trace element
studies are often very large; therefore, an appreciable
fraction of resources must be devoted for data acquisition
and processing (23). Unfortunately, relating the elemental
concentration data to meaningful base has not received
adequate attention. This can lead to wrong interpretation
of the results even though the analysis has been carried
out properly. This is due to the fact that biological
materials consist of multiple components with different
element content, and the ratio of these components may
differ even in the same type of samples (24). Several
examples, recommendations concerning parameters for
evaluation of elemental analysis data and a checklist of
essential features related to elemental analysis of
biological systems are presented (16).

Conclusions

"Cells in the body do not function in isolation. Perhaps
the trace element researchers may wish to emulate them"
(1).

    Obtaining analytically meaningful and biologically
interpretable data for trace elements in biomedical
investigations is a tedious task and requires dedicated
efforts through a multidisciplinary approach by analytical
chemists and health sciences investigators. Examples such
as the RDAs (25), the permissible limits for exposure to
various trace elements (26), composition of Reference Man
(27) and evaluation of reference elemental concentration
values in human milk (28) and in other clinical specimens
(29-31) for monitoring purposes, demonstrate the public
health benefits of multidisciplinary approaches in BTER
studies. The problems accounted for in the preceding
sections reemphasize the fact that good analytical
measurements are not only crucial for success with future
biological trace element research studies, but are
mandatory for considering the analytical findings for
diagnostic and related evaluations. In this context Mertz's
statement "that an analytical chemist should be more than
procurer of data and a life scientist more than their
interpreter" (32) cannot be emphasized sufficiently.

Literature Cited

1. Iyengar G.V. Elemental Analysis of Biological Systems.
   Vol 1, CRC Press, Boca Raton, FL, 1989; pages 1, 19,
   71.
2. Versieck, J. Trace Ele. Med. 1984, 1, 2-6.
3. Iyengar, G.V. Sci. Total Environ. 1981, 19, 105-109.
4. Third International Symposium on Biological Reference
   Materials, Fres. Zschr. Analyt. Chem. 1988, 332, No.6.

5. Ihnat, M. In _Quantitative Analysis of Biological Materials_, McKenzie, H.A.; Smythe, L.E., Ed.; Elsevier, Amsterdam, New York, 1988; Chapter 19, p 331.
6. Okamoto, K.; Fuwa, K. _Anal. Lett._ 1985, _1_, 206-213.
7. Muramatsu, Y.; Parr, R.M. Survey of currently available Reference Materials for use in connection with the determination of trace elements in biological and environmental materials. IAEA Report IAEA/RL/128, 1985.
8. Horwitz, W. In this ACS Symposium Proceedings, 1990.
9. Parr, R. M. _Nut. Res. Suppl._ 1985, _I_, 5-11.
10. Schwarz, K. In _Proceedings of Nuclear Activation Techniques in the Life Sciences_, 1972, IAEA, Vienna, p 3.
11. Underwood, E.J. 1977. _Trace Elements in Human and Animal Nutrition._ 4th Edition, Academic Press, New York.
12. Smith, J.C. In _Trace Elements in Human and Animal Nutrition_, Mertz, W., Ed.; Academic Press, New York, 1988, p 21.
13. Mertz, W. _Trace Elements in Human and Animal Nutrition_, Volume 1 and 2, Academic Press, New York. 1987/1988.
14. Iyengar, G.V. _J. Path._ 1981, _134_, 173-181.
15. Iyengar, G.V. _Anal. Chem._ 1982, _54_, 554A-558A.
16. Iyengar, G.V.; Behne D. _J. Res. Nat. Bur. Stds._ 1988, _93_, 326-327.
17. Iyengar, G.V.; Borberg, H.; Kasperek, K.; Kiem, J.; Siegers, M.; Feinendegen, L.E.; Gross, _R. Clin. Chem._ 1979, _25_, 699-704.
18. Sabbioni, E.; Edel, J.; Goetz L. _Nut. Res. Suppl._ 1985, _I_, 32-43.
19. Crews, H.M.; Massey, R.; McWeeny, D.J. _J. Res. Nat. Bur. Stds._ 1988, _93_, 349-350.
20. Toelg, G. In _Trace Element Analytical Chemistry in Medicine and Biology_, Braetter, P.; Schramel P. Eds.; de Gruyter, Berlin, 1988, p 1.
21. Grandjean, P. 1983. In _Chemical Toxicology and Clinical Chemistry of Metals_, Brown, S.S.; Savory, J. Eds.; p 99.
22. Iyengar, G.V. _Biol. Trace Ele. Res._ 1987, _12_, 263-295.
23. Cherry, W.H. _Sci. Total Environ._ 1983, _34_, 199-200.
24. Behne, D. _Clin. Chem. Clin. Biochem._ 1981, _19_, 115.
25. _Recommended Dietary Allowances_, National Research Council, Washington D.C. 1989, 10th Ed.
26. Joint FAO/WHO Expert Committee on Food Additives: Technical Report Series 751, WHO, Geneva, 1987.

27. International Commission on Radiological Protection (ICRP). Report of the Task Group on Reference Man, ICRP Publication 23, Pergamon Press, Oxford, 1975, (Revision in progress).
28. Minor and Trace Elements in Breast Milk: World Health Organization Report of a Joint WHO/IAEA Collaborative Study, WHO, Geneva, 1989.
29. Hamilton, E.I. The Chemical Elements and Man. Charles Thomas, Spring Fields, IL, 1979.
30. Heydorn, K. Neutron Activation Analysis in Clinical Trace Element Research, CRC Press, Boca Raton, FL, 1984.
31. Iyengar, G.V.; Woittiez J.R.W. Clin. Chem. 1988, 34, 474-481.
32. Mertz, W. Clin. Chem. 1975, 21, 468-473.

RECEIVED July 16, 1990

PLANNING   CONSIDERATIONS

Chapter 2

Bioavailability Considerations in Planning Dietary Intake Studies

Bo Lönnerdal

Departments of Nutrition and Internal Medicine, University of California, Davis, CA 95616

Dietary intake studies yield data on the total amount of trace elements taken in from various food items. Frequently these intake data are compared to RDA's and the diet may be considered adequate or inadequate with regard to trace elements. In the case of many trace elements, however, bioavailability is quite different from one dietary source to another. Thus, the ability of various diets to meet the biological need for a particular element may be over- or underestimated. Information on the bioavailability of trace elements from single dietary items and combined meals in humans is therefore required to properly assess the adequacy of trace elements provided in the diet. When such data are available, dietary intake studies will result in more appropriate information and subsequent recommendations.

Dietary intake studies are commonly used to assess intake of individual nutrients in populations at risk and are subsequently utilized to formulate recommendations. Depending on the situation, these recommendations may be for supplemental foods, changes in dietary habits/food combinations or fortification of food items with individual nutrients. It is therefore important that the data obtained are as reliable as possible. For many nutrients, assessment of the quantities consumed will give a reasonable estimate of how well they will meet the requirements of certain groups or populations. A condition that needs to be fulfilled, however, is that the effects of processing, storage and cooking on the particular nutrient are accounted for; i.e., that the nutrient content of the food as consumed is used. For several nutrients, however, the constituents of the food will have marked effects on the absorption and utilization of the nutrient. This is particularly relevant for minerals and trace elements such as calcium, iron, zinc and copper. For example, iron as non-heme iron in a cereal-based meal may be absorbed to only 2% while heme iron from meat may be absorbed to 30% (1). Thus, iron bioavailability will be strongly affected by the composition of the meal/diet

0097–6156/91/0445–0016$06.00/0
© 1991 American Chemical Society

in which the iron is consumed, and the total amount of iron in the diet will not be the dominant determinant of whether the requirement will be adequately covered or not. While some consideration of bioavailability was made when the Recommended Dietary Allowances (RDA's) were established, it is unusual for bioavailability to be taken into account when dietary intakes are assessed. As a consequence, evaluation of whether or not a diet will provide adequate quantities of nutrients (often compared to the RDA's) rests on shaky ground. An example of this is given in Table I, which demonstrates the absorbed amount of zinc from four typical diets for young children. If we assume that the RDA for children, 5 mg, is ingested, these diets would provide 1.5, 1.0, 1.0 or 0.7 mg available zinc, respectively.

Table I. Zinc Absorption from Different Diets

Diet	% Absorbed[a]	Amount Ingested	Amount Absorbed
Milk, homogenized	30	5 mg	1.5 mg
Milk, cereal mixed formula	20	5 mg	1.0 mg
Meat-based composite meal	20	5 mg	1.0 mg
Soy formula	14	5 mg	0.7 mg

[a]Data from references 2 and 3.

It is obvious that it would be of considerable value if some factors that influence nutrient bioavailability could be quantified in dietary intake studies. With a factorial approach, absorbable (bioavailable) nutrient intake could be estimated. In this paper, I have attempted to describe some of the major dietary factors that affect trace element bioavailability. By compiling such information, it should be possible to establish better dietary assessments.

Inhibitors of Trace Element Absorption

Phytate. The highly negatively charged phytic acid, or inositol hexaphosphate, which is present in many plants such as cereals and legumes, is known to form insoluble complexes, phytates, with minerals and trace elements (4). Since the human gastrointestinal tract is unable to hydrolyze phytate, these mineral-phytate complexes will remain through the passage of the gut and consequently have a negative effect on mineral and trace element bioavailability.

A negative effect of phytate on the absorption of zinc (5,6) and iron (7,8) has been demonstrated in several animal models. Isotope studies in humans have shown that when phytic acid is added to a liquid diet, zinc absorption is reduced by 50% (9). Similarly, addition of phytic acid to cow's milk formula at a level similar to that of soy formula reduced zinc absorption by 50% and made it equal to that for soy formula (10). Addition of phytate to white bread has been shown to reduce both zinc (11) and iron absorption (12). The effect of phytate appears to be dose-dependent (12) and no adaptation to high intakes seems to occur in humans (13). While it could be argued that added purified phytic acid may not exert the same affect as phytate present in foods, several studies have shown that removal of phytate from a particular food or meal has a positive effect on trace element absorption. Dephytinization by either precipitation methods or by phytase treatment resulted in a significant increase in zinc absorption in infant rhesus monkeys (14). Similarly, removal of phytate from bran had an enhancing effect on iron absorption in humans (15).

A negative correlation between the phytate content of the diet and trace element absorption was suggested already by McCance et al. in 1943 (16). Since that time, many studies have demonstrated trace element deficiencies in populations with high dietary intake of phytate. It has been suggested that there is a threshold effect, i.e. that the effect of phytate on trace element absorption is not exerted until a particular [phytate]/[trace element] ratio is exceeded. This ratio has been suggested to be 12.5:1 or 15:1 based on results from experimental animals (17,18). However, Sandstrom (19) has shown that the effect of phytate on zinc absorption in humans can be expressed as a curvilinear function and that ratios of phytate to zinc between 5:1 and 10:1 also can inhibit zinc absorption. This will be dependent on the diet used; for example, phytate in a formula diet inhibited zinc absorption at a molar ratio of phytate:zinc of 4:1 (10). In addition, the study of Hallberg et al. (12), in which the dose-dependency of phytic acid inhibition of iron absorption was investigated, demonstrated that even low amounts of phytate had a significant negative effect. Thus, it appears that quantitation of phytate in dietary intake studies would be helpful when it comes to assessing trace element bioavailability. Unfortunately, the situation is complicated by the fact that abundance of one cation may have a negative effect on the bioavailability of another one, so that not only dietary phytate but also the ratio of some minerals and trace elements to phytate must be considered. For example, a high level of dietary calcium may precipitate more of a calcium-zinc-phytate complex (4) and therefore reduce zinc bioavailability to a greater extent than anticipated when phytate only is taken into account. It has therefore been proposed that zinc bioavailability can be predicted by a [Ca] x [phytate]/[Zn] ratio (20). However, while a high ratio has been shown to impair zinc absorption in experimental animals, there is little support for this occurring in human subjects (10,19).

The total content of phytate in the diet may, however, not be directly proportional to the inhibitory effect. Different types of food preparation, such as baking, fermentation and extrusion cooking, can actually cause a dephosphorylation of phytate, so that inositols with fewer phosphate groups are formed (21). It has been shown that inositols with four or fewer phosphate groups do not inhibit mineral or trace element absorption (22). Thus, the use of methods that analyze "total" phytate, which includes these forms of phytates, will over-estimate the inhibitory effect, provided that they constitute a significant proportion of the total content. This emphasizes that both the content of inhibitory components and the method of food preparation should be taken into account when assessing "bioavailable" trace elements from the diet.

Minerals and Trace Elements

Calcium was previously mentioned to possibly affect trace element bioavailability negatively by facilitating formation of calcium-phytate-protein complexes, thereby co-precipitating trace elements such as zinc and iron. It has recently been shown that calcium can also have a direct inhibitory effect on iron absorption. Barton et al. (23) showed that addition of calcium to human milk had a marked negative effect on iron absorption in the rat and that the previously high bioavailability of iron from human milk became similar to that from cow's milk, which is higher in calcium. Subsequent studies in humans have confirmed this inhibitory effect. Dawson-Hughes et al. (24) showed that a calcium supplement of 500 mg decreased iron absorption to 43-46% of the absorption value for the meal without calcium supplement. Hallberg et al. (25) added calcium to the bread in a breakfast meal and found a reduction in iron absorption of 59% when 165 mg calcium was added. It is therefore possible that a high calcium intake, particularly when taken as a supplement, could have a negative influence on iron bioavailability. Further studies are needed to quantify this inhibitory effect and to study the possible mechanism(s). A similar negative effect is not observed for zinc absorption; studies in human subjects given high doses of calcium (500 mg) did not show any effect on zinc absorption (24).

Trace elements have the capacity to competitively interact with each other at the level of absorption. The reason for this is that several trace elements form coordination complexes in aqueous solution that are structurally similar and therefore can compete for absorptive pathways (26). Examples of such interactions are copper-zinc and iron-manganese. Another type of interaction is when trace elements compete for carrier ligands within the mucosal cell, as is likely to be the case for iron and zinc. In both these types of interactions, an inordinately high level of one trace element may inhibit the absorption of another. This may be of particular relevance for food items that are fortified with trace elements, such as iron. It has been shown that high levels of iron can have a negative effect on zinc absorption (27,28), although the effect appears less pronounced when these elements are present in or given with a meal than what it is in aqueous solution (28). For example, infants fed iron-fortified formula were described to have lower plasma zinc levels than infants fed the same formula without iron (29). It has recently been shown that high levels of dietary manganese inhibit iron absorption in humans in a dose-dependent manner (30). Thus, some consideration should be taken to the ratio of trace elements in the diet when considering bioavailability, particularly when supplementation or fortification is used.

Phenolic Compounds

Tannic acid,which is present in tea, coffee, red wine and some vegetables, is one example from a diverse group of phenolic substances that form complexes with trace elements, particularly iron. Most of this complexes are insoluble and will have a negative influence on trace element bioavailability; however, there are also soluble complexes. The effect of such phenolic compounds on iron absorption has recently been studied systematically in human subjects. Brune et al. (31) found that tannic acid strongly inhibits iron absorption in a dose-dependent manner. Gallic acid had the same effect per galloyl groups (tannic acid has 10 galloyl groups), while chlorogenic acid had less of an effect and catechin no effect on iron absorption. These observations are likely to explain the negative effect of tea (32), coffee (33) and some red wines (high in tannates) (34) on iron bioavailability.

Coffee

It is well-known that coffee exerts a negative effect on iron absorption in humans (33,35). When coffee was added to a breakfast meal, iron absorption was reduced by 40% as compared to when no coffee was given. Brune et al.

(*31*) showed that the negative effect of coffee on iron bioavailability was primarily due to its content of tannic acid, but in part also to chlorogenic acid and phytate. In a study in Costa Rica, women drinking several cups of coffee per day during pregnancy and lactation were shown to have a higher incidence of anemia and lower milk iron than women not drinking coffee (*36*). A study in rats showed similar results and also that decaffeinated coffee produced the same effect (*37*), suggesting that caffeine was not the inhibitory component.

It should be noted that iron absorption did not appear to be impaired by the coffee, but that liver iron was increased. This suggests that a component(s) of coffee interfered with iron metabolism in that iron mobilization from liver was reduced, resulting in low hemoglobin values. Thus, the role of coffee in iron absorption and metabolism needs to be evaluated further. Zinc metabolism does not appear to be affected by coffee consumption, at least in experimental animals (*37*).

Dietary Fiber

It is often assumed that dietary fiber has a negative effect on trace element bioavailability. While several studies support this opinion (*38,39*), it should be emphasized that the type of dietary fiber needs to be defined and examined more closely. Recent studies on the effect of different types of fiber on iron absorption in human subjects strongly suggest that it is the phytate content of fiber that is the major inhibitory factor (*15,38*). Several types of dietary fiber which do not contain phytate had no negative influence on iron bioavailability (*40,41*). In addition, dephytinization of bran reduced the negative effect of bran on iron absorption significantly (*15*). Similar results have been obtained for zinc (*11*). Several types of fiber that do not contain phytate, such as cellulose (*9*), beet pulp fiber (*42*) and pectin (*43*) have been shown to have no effect on zinc absorption. Thus, phytate content would be a more relevant dietary factor to assess than fiber, as it concerns trace element bioavailability.

Enhancers of Trace Element Absorption

Protein.   Dietary protein can affect trace element bioavailability in two principal ways; either specifically such that one protein source has a more (or less) positive effect on absorption than another or generally in that the total protein level in a meal affects trace element absorption. One example of the first category is meat - it is well-known that a protein component of meat, the so-called "meat-factor" enhances the absorption of non-heme iron (*1*). Several studies have attempted to define this component and it appears that cysteine (or cysteine-containing peptides) formed during digestion is largely responsible for this effect (*44*). Free cysteine may not have the same effect as it easily becomes oxidized, while peptide-bound cysteine may be protected until released in the gut, where it could chelate iron and facilitate its absorption (*44*). Meat has also been shown to have a positive effect on zinc absorption, but for this element it seems more likely that this is the result of a high zinc content of meat rather than by a specific component enhancing zinc absorption (*45*).

Individual proteins have been shown to affect trace element bioavailability to varying extent. Using bovine serum albumin as a reference protein, Hurrell et al. (*46*) found that casein, whey protein and soy protein resulted in lower iron absorption than from serum albumin. Subsequent studies have suggested that other components of the protein sources, such as organic phosphate may have been responsible for most of the negative effects observed (47). Thus, not only the protein structure itself, such as its content of cysteine, will affect trace element absorption, but also other organic and inorganic constituents associated with protein. While there appears to be some effect of protein source on zinc absorption, it does not affect it as much as iron absorption. Casein in cow's milk has been shown to have a negative effect on zinc absorption (10), possibly by formation of casein phosphopeptides that inhibit zinc absorption or by inorganic constituents of the casein micelle, such as colloidal calcium phosphate (48). It is evident that further work is needed on the effects of various proteins on trace element absorption before any factorial approach can be used in dietary intake studies.

A general positive effect of protein level of a meal on zinc absorption has been demonstrated (19). It should be cautioned that this could also be explained by an increase in zinc level, as zinc is bound to dietary proteins. It is possible, however, that amino acids and small peptides formed during digestion will chelate zinc and aid in its absorption (49). Thus, a higher level of protein may increase chelation and subsequent absorption of zinc. While this has not been systematically studied for iron, a similar effect would be expected to occur.

Organic Acids

Ascorbic acid is known to enhance iron absorption (1), both by reducing ferric iron to absorbable ferrous iron and by chelating this reduced iron. This stimulatory effect on iron bioavailability is so strong that it can overcome the negative effect of phytate (12). There is some limit to this effect, however. Gillooly et al. (50) showed that ascorbic acid addition to cow's milk formula increased iron absorption in humans in a dose-dependent manner, while similar additions to soy formula had an enhancing effect on iron absorption, but to a much more modest extent. Thus, the ascorbic acid intake should be assessed in dietary intake studies and be used as a factor when assessing iron bioavailability. In contrast, ascorbic acid has been shown to have no effect at all on zinc absorption (51,52).

Other organic acids, such as citrate, malate and fumarate, can have a positive effect on trace element bioavailability (53). This effect is likely to occur by chelation of the trace element via negatively charged carboxylate groups so that the element is kept in solution and subsequently absorbed. It is also possible that these anions can affect ion fluxes across the mucosal barrier and thus affect trace element absorption (54). This could possibly explain the higher concentrations of citrate needed for an enhancing effect; while 15 mg of ascorbate stimulated iron absorption, more than 35 mg of citrate was needed to achieve a similar effect. It is also obvious that organic acids, such as citrate, are less capable of counteracting the negative effect of phytate on trace element bioavailability (10,50).

Conclusion

Quantitation of the dietary intakes of phytate, ascorbic acid, calcium and tannates is likely to be valuable when establishing "bioavailable nutrients" in the diet. Other factors such as protein, meat, organic acids and competing trace elements would also be of value, although their quantitative importance is less pronounced. By taking both negative and positive factors into account, more accurate assessments should be feasible, particularly in areas such as India where iron intake sometimes can be reasonable but iron status is poor

22

BIOLOGICAL TRACE ELEMENT RESEARCH

due to a low intake of ascorbic acid. Factorial estimates should be followed by experiments in human subjects to validate the approach. This can be combined with studies of "typical" meals in different regions to obtain data for the factors in the relevant dietary context. Such studies have been performed for "Western" (55), Southeast Asian (56) and Latin American diets (57,58). With proper validation, more accurate estimates of the capacity of various diets to meet nutrient requirements will be obtained.

Literature Cited

1. Hallberg, L. *Ann. Rev. Nutr.* **1981**, *1*, 123-47.
2. Sandström, B.; Arvidsson, B.; Cederblad, Å.; Björn-Rasmussen, E. *Am. J. Clin. Nutr.* **1980**, *33*, 739-745.
3. Sandström, B.; Cederblad, Å.; Lönnerdal, B. *Am. J. Dis. Child*, **1983**, *137*, 726-29.
4. Oberleas, D.; Muhrer, M. E.; O'Dell, B. L. *J. Nutr.* **1966**, *90*, 56-62.
5. O'Dell, B.L.; Savage, J.E. *Proc. Soc. Exp. Biol. Med.* **1960**, *103*, 304-09.
6. Morris, E.R.; Ellis, R. *J. Nutr.* **1980**, *110* 1037-45.
7. Davies, N.T.; Nightingale, R. *Br. J. Nutr.* **1975**, *34*, 243-58.
8. Morris, E.R.; Ellis, R. *J. Nutr.* **1980**, *110*, 2000-10.
9. Turnlund, J. R.; King, J. C.; Keyes, W. R.; Gong, B.; Michel, M. C. *Am. J. Clin. Nutr.* **1984**, *40*, 1071-77.
10. Lönnerdal, B.; Cederblad, Å.; Davidsson, L.; Sandström, B. *Am. J. Clin. Nutr.* **1984**, *40*, 1064-1070.
11. Nvert, B.; Sandström, B.; Cederblad, Å. *Br. J. Nutr.* **1985**, *53*, 47-53.
12. Hallberg, L.; Brune, L.; Rossander, L. *Am. J. Clin. Nutr.* **1989**, *49*, 140-44.
13. Brune, M.; Rossander, L.; Hallberg, L. *Am. J. Clin. Nutr.* **1989**, *49*, 542-45.
14. Lönnerdal, B.; Bell, J. G.; Hendrickx, A. G.; Burns, R. A.; Keen, C. L. *Am. J. Clin. Nutr.* **1988**, *48*, 1301-06.
15. Hallberg, L.; Rossander, L.; SkÅnberg, A-B. *Am. J. Clin. Nutr.* **1987**, *45*, 988-96.
16. McCance, R. A.; Edgecombe, C. N., Widdowson, E. M. *Lancet* **1943**, *2*, 126-28.
17. Davies, N.T.; Olpin, S.E. *Br. J. Nutr.* **1979**, *41*, 591-603.
18. Lo, G. S.; Settle, S. L.; Steinke, F. H.; Hopkins, D. T. *J. Nutr.* **1981**, *111*, 2223-25.
19. Sandström, B.; Lönnerdal, B. In: *Zinc in Human Biology*; Mills, C.F., Ed.; Springer-Verlag: Berlin, 1989; pp. 57-78.
20. Davies, N. T.; Carswell, A. J. P.; Mills, C. F. In: *Trace Elements in Man and Animals*; Mills, C. F.; Bremner, I., Chesters, J. K., Eds.; Commenwealth Agricultural Bureaux, Slough, 1985; p 440.
21. Sandberg, A.-S.; Anderson, H.; Carlsson, N.-G.; Sandström, B. *J. Nutr.* **1987**, *117*, 2061-65.
22. Lönnerdal, B.; Sandberg, A-S.; Sandström, B.; Kunz, C. *J. Nutr.* **1989**, *119*, 211-14.
23. Barton, J. C.; Conrad, M. E.; Parmley, R. T. *Gastroenterology* **1983**, *84*, 90-101.
24. Dawson-Hughes, B.; Seligson, F. M.; Hughes, V. A. *Am. J. Clin. Nutr.* **1986**, *44*, 83-88.
25. Hallberg, L.; Brune, M.; Rossander, L. *Am. J. Clin. Nutr.* (in press).
26. Hill, C. H.; Matrone, G. *Fed. Proc.* **1970**, *29*, 1474-88.
27. Solomons, N. W.; Jacob, R. A. *Am. J. Clin. Nutr.* **1981**, *34*, 475-82.

28. Sandström, B.; Davidsson, L., Cederblad, Å.; Lönnerdal, B. *J. Nutr.* **1985**, *115*, 411-14.
29. Craig, W.J.; Balbach, L.; Harris, S.; Vyhmeister, N. *J. Am. Coll. Nutr.* **1984**, *3*, 183-86.
30. Hallberg, L.; Rossander, L.; Brune, M.; Sandström, B.; Lönnerdal, B. In: *Trace Elements in Man and Animals - 6*; Hurley, L.S., Keen, C.L., Lönnerdal, B., Rucker, R.B., Eds.; Plenum Press: New York, **1988**, pp. 233-4.
31. Brune, M.; Rossander, L.; Hallberg, L. *Eur. J. Clin. Nutr.* (in press).
32. Disler, P.B.; Lynch, S.R.; Carlton, R.W.; Torrance, J.D.; Bothwell, T.H.; Walker, R.B.; Mayet, F. *Gut* **1975**, *16*, 193-200.
33. Morck, T. A.; Lynch, S. R.; Cook, J. D. *Am. J. Clin. Nutr.* **1983**, *37*, 416-20.
34. Bezwoda, J. R.; Torrance, J. D., Bothwell, T. H.; MacPhail, A. P.; Graham, B.; Mills, W. *Scand. J. Haematol.* **1985**, *34*, 121-27.
35. Hallberg, L.; Rossander, L. *Hum. Nutr.: Appl. Nutr.* **1982**, *36A*, 116-23.
36. Munoz, L. M.; Lönnerdal, B.; Keen, C. L.; Dewey, K. G. *Am. J. Clin. Nutr.* **1988**, *48*, 645-51.
37. Munoz, L.; Keen, C. L.; Lönnerdal, B.; Dewey, K. G. *J. Nutr.* **1986**, *116*, 1326-33.
38. Simpson, K. M.; Morris, E. R.; Cook, J. D. *Am. J. Clin. Nutr.* **1981**, *34*, 469-78.
39. Cook, J.D.; Noble, N.L.; Morck, T.A.; Lynch, S.R.; Petersburg, S.J. *Gastroenterology* **1983**, *85*, 1354-58.
40. Gillooly, M.; Bothwell, T.H.; Charlton, R.W.; Torrance, J.D.; Bezwoda, W.R.; MacPhail, A.P.; Derman, D.P.; Novelli, L.; Morrall, P.; Mayet, F. *Br. J. Nutr.* **1984**, *51*, 37-46.
41. Rossander, L. *Scand. J. Gastroenterol.* **1987**, *22*, (suppl. 12a) 68-72.
42. Sandström, B.; Davidsson, L.; Kivistö, B.; Hasselblad, C. *Br. J. Nutr.* **1987**, *58*, 49-57.
43. Sandberg, A-S.; Ahderinne, R.; Andersson, H.; Hallgren, B.; Hulten, L. *Hum Nutr: Clin. Nutr.* **1983**, *37C*, 171-83.
44. Taylor, P.G.; Martinez-Torres, C.; Romano, E.L.; Layrisse, M. *Am. J. Clin. Nutr.* **1986**, *43*, 68-71.
45. Sandström, B.; Cederblad, Å. *Am. J. Clin. Nutr.* **1980**; *33*, 1778-83.
46. Hurrell, R.F.; Lynch, S.R. Trinidad, T.P.; Dassenko, S.A.; Cook, J.D. *Am. J. Clin. Nutr.* **1989**, *49*, 546-52.
47. Hurrell, R.F. (personal communication).
48. Kiely, J.; Flynn, A.; Singh, H.; Fox, P.F. In: *Trace Elements in Man and Animals - 6*, Hurley, L.S., Keen, C.L., Lönnerdal, B., Rucker, R.B., Eds. Plenum Press: New York, **1988**, pp. 499-500.
49. Wapnir, R.A.; Khani, D.E.; Bayne, M.A.; Lifshitz, JF. *J. Nutr.* **1983**, *113*, 1346-54.
50. Gillooly, M.; Torrance, J.D.; Bothwell, T.H.; MacPhail, A.P.; Derman, D.; Mills, W.; Mayet, F. *Am. J. Clin. Nutr.* **1984**, *40*, 522-27.
51. Solomons, N. W.; Jacob, R. A.; Pineda, O.; Viteri, F. E. *Am. J. Clin. Nutr* **1979**, *32*, 2495-99.
52. Sandström, B.; Cederblad, Å. *Int. J. Vitam. Nutr. Res.* **1987**, *57*, 87-90.
53. Gillooly, M.; Bothwell, T. H.; Torrance, J. D.; MacPhail, A. P.; Derman, D. P.; Bezwoda, W. R.; Mills, W.; Charlton, R. W.; Mayet, F. *Br. J. Nutr.* **1983**, *49*, 331-42.
54. Rolston, D.D.K.; Moriarty, J.J.; Farthing, J.J.G.; Kelly, J.J.; Clark, M.; Dawson, A.M. *Digestion* **1986**, *34*, 101-04.
55. Hallberg, L.; Rossander, L. *Am. J. Clin. Nutr.* **1982**, *35*, 502-09.

56.   Hallberg, L.; Björn-Rasmussen, E.; Rossander, L.; Suwanik, R.; Pleehachinda, R.; Tuntawiroon, M. *Am. J. Clin. Nutr.* **1983**, *37*, 272-77.
57.   Hallberg, L.; Rossander, L. *Am. J. Clin. Nutr.* **1984**, *39*, 577-83.
58.   Pabon, M.L.; Munevar, E.; Ramirez, R.; Martinez-Torres, C.; Layrisse, M.; *Nutr. Res.* **1990**, *10*, 15-22.

RECEIVED July 16, 1990

Chapter 3

Planning In Vivo Body Composition Studies in Humans

Kenneth J. Ellis

Children's Nutrition Research Center, Department of Pediatrics, U.S. Department of Agriculture, Agricultural Research Service, Baylor College of Medicine, 1100 Bates Street, Houston, TX 77030

As multi-elemental analyses of body fluids and tissues become increasingly more sensitive, their interpretation may require additional knowledge about the relationship of the fluids and tissues to the composition of either the total body or specific organs. Until recently, only biopsy or autopsy data could be used to estimate total body composition in humans. Nuclear-based techniques, however, have been developed in a few centers for direct in vivo measurements of total body calcium, sodium, chlorine, nitrogen, phosphorus, potassium, hydrogen, carbon, and oxygen in man. Analyses of selected trace elements, cadmium, lead, and aluminum have also been developed for specific clinical or toxicological studies. Interest in the availability of these measurements, which often require specialized facilities, is increasing throughout the world. Activation and counting techniques routinely implemented for in vivo analyses in humans will be emphasized in this report. Techniques will be described that have levels of accuracy and precision of at least ± 3 to 5%. Institutional requirements and preferred nuclear techniques will be presented to assist those who are considering in vivo body composition studies in humans. Also included is information on the planning of these studies.

Neutron activation is a well-established analytical technique for the in vitro analysis of biological samples. For in vivo analyses, however, neutron activation is limited by the radiation dose and exposure time, as well as by the irregular geometry of the body and its elemental composition. The long exposure times required for effective activation of many

elements in the body significantly reduce the range of elements
available for in vivo analysis. Bulk elements (Ca, P, H, N) and
selected trace elements (Cd, Hg, Fe, Mn, I), i.e., those with
large thermal neutron cross-sections, have been examined in
vivo (1-4). Nevertheless, the elements that have been measured
in vivo have provided useful clinical, toxicological, and basic
physiological information which otherwise would not be
obtainable in the living human (1-4).
      A number of in vivo techniques have been investigated,
including neutron activation (delayed and prompt procedures),
nuclear resonant scattering of gamma-rays, inelastic
scattering, photonuclear production, and the collection of
exhaled radioactive gases in the breath after exposure to
neutrons. Although the latter four techniques are of
considerable interest, they will not be addressed in any detail
in this paper, because experience with their use on humans is
not sufficient to enable an adequate assessment of their
potential. Several review articles have been published on in
vivo neutron activation analysis techniques (1-4). The focus of
this paper will be the techniques 'routinely' implemented for
in vivo analyses in humans, and where possible, will suggest
which method is preferred. In each application, the magnitude
of the radiation dosage has to be balanced against the medical
value of the in vivo analysis data. My primary aim is to assist
those who are considering in vivo body composition studies in
humans and to provide information helpful to the planning of
these studies.

IN VIVO NEUTRON ACTIVATION ANALYSIS (IVNAA)

Neutron Reactions in the Body. The basic principle of in vivo
elemental analysis is the detection of gamma-rays externally
emitted from the body which are present during or after
exposure to neutrons. Although the principle is quite simple,
the physical and experimental factors which determine whether
or not an element can be measured accurately and with precision
are frequently complex. A number of physical factors such as
neutron reaction cross-section, complexity of the gamma-ray
spectra, interference peaks, amount of element in the body,
neutron flux, and exposure and counting times governs the
interactions.
      The general composition and geometry of the human body are
beyond the investigator's control; the only variables which the
experimenter may determine are the neutron source, irradiation
geometry, and counting technique. Thermal neutron reactions are
used most commonly for in vivo analyses. These neutrons are
produced mainly in the body by scattering fast neutrons on body
hydrogen. The characteristics of the neutron reactions
considered for IVNAA are listed in Table I for the elements
normally measured in vivo. The values in the last column give
only a sense of the relative sensitivity for each element; the
complexity of the gamma spectra, background levels, or possible
interference gamma-rays are not considered.

Table I.   Main Reactions Considered for In Vivo Neutron Activation Analysis

Element	Amount in Body[a] (g)	Target Isotope	Neutron Reaction	Gamma Energy (MeV)	Emission Rate	Half-Life	Relative Detection Sensitivity[d]
Hydrogen	7000	^{1}H	^{1}H(n,γ)^{2}H	2.223	1.00	Prompt	1550
Nitrogen	1800	^{14}N	^{14}N(n,γ)^{15}N	10.83	0.15	Prompt	0.44
		^{14}N	^{14}N(n,2n)^{13}N	0.511	2.00	9.96 min	0.61
Calcium	1000	^{48}Ca	^{48}Ca(n,γ)^{49}Ca	3.084	0.92	8.72 min	0.022[c]
Potassium	140	^{40}K	Natural activity	1.461	0.11	1.277×10^9 g	
Sodium	100	^{23}Na	^{23}Na(n,γ)^{24}Na	2.754	1.00	15.02 h	0.15
Chlorine	95[b]	^{37}Cl	^{37}Cl(n,γ)^{38}Cl	2.168	0.42	37.3 min	0.10
Cadmium	0.024[b]	^{113}Cd	^{113}Cd(n,γ)^{114}Cd	0.559	0.40	Prompt	0.55
Oxygen	43000	^{16}O	^{16}O(n,p)^{16}N	6.134	0.69	7.2 sec	7.20
Phosphorus	780	^{31}P	^{31}P(n,α)^{28}Al	1.779	1.00	2.24 min	1.57

[a] ICRP 23 - Reference Man
[b] Liver cadmium (3 μg/g) and kidney cadmium (3.0 mg) are assumed to represent 50-75% of total body content.
[c] < 20 cpm per g K
[d] Complexity of spectra, background levels, and interference peaks are not considered.

Selection of Neutron Source. The three types of neutron
sources currently used for in vivo activation in man are listed
in Table II. The radioactive sources are ^{238}Pu/Be, ^{241}Am/Be,
and ^{252}Cf. The Pu and Am sources have the advantage of long
half-lives (87.5 yr and 458 yr, respectively), thus providing a
relatively constant neutron output with time. Cf has a much
shorter half-life (2.65 yr), and requires replacement at
regular intervals, which adds to the total cost of the system.
The neutron generator has the unique advantage of no neutron
beam (thus no major health physics considerations) except when
the unit is operating. It does deliver higher energy neutrons,
so that the dose per neutron is increased. These systems
usually require more maintenance than do radioactive sources,
their output decreases with use, and tube replacement requires
that the system be completely recalibrated. The ranges of total
source output, flux (n/cm^2 x sec) incident on the body, and
irradiation and counting times are also given in Table II. The
predominant reactions for clinical applications have been the
thermal (n,γ) reactions listed in Table I. If one is not
planning long-term studies that include in vivo measurements or
examination of only a specific organ or element, ^{252}Cf should
be the first source to consider for this limited application.

Activation and Counting Geometries. The most basic requirement
is that the whole body irradiation and detector geometries
subtend a solid angle that is as great as possible. Classic
examples of this approach are the 2π irradiation chamber and
whole-body counter at Baylor College of Medicine (5). The
irradiator uses 56 ^{241}Am/Be sources (15.5 Ci each) distributed
above and below the subject. The whole body counter comprises
32 NaI detectors (crystal size of 10 cm x 10 cm x 46 cm each)
arranged in two banks above and below the subject. The counting
response is relatively invariant to the distribution of
activity within the body, thus enabling absolute measurements.
Each detector can be operated as a single unit or in
combinations to optimize the total signal to background ratio.
This system enables ^{40}K counting in infants through adults
(Ellis, K.J. and Shypailo, R.J., Phys. Med. Biol., in press). A
more typical whole body counter arrangement is that at the
Leeds General Infirmary (6). This counter has 8 NaI detectors
(circular crystals of 15 cm diameter x 10 cm thickness), 4
above and 4 below the subject. Another approach, selected
mainly for economical reasons, is the 'shadow shield' counter
which usually comprises two large NaI crystals, typically 30 cm
diameter by 10 cm thickness, that are shielded by a lead
covering. In this case, the subject must be scanned between the
detectors at a rate determined by the half-life of the
radioisotope and the desired level of precision.

Background and Detector Shielding. A shielded whole body
counter is needed to detect low levels of induced activity. It
should be located reasonably close to the irradiation chamber,
but at a sufficient distance to avoid activating the crystals
or increasing the background signal. The Baylor 1000 Ci
^{241}Am/Be irradiator is within 100 ft of the low background

Table II.   Three Types of Neutron Sources Currently in Use for In Vivo Neutron Activation Analysis of Man.

Type of Source Characteristics	$^{238}Pu/Be$ $^{241}Am/Be$	^{252}Cf	Neutron Generator
Total source output (n/sec)	$2.0x10^7-1.1x10^9$	$2.2x10^9-1.8x10^{10}$	Up to $3x10^{10}$
Incident neutron flux density in body (n/cm²-sec)	$1x10^3-5x10^4$	$3x10^5-10^6$	$3x10^3-5x10^5$
Irradiation time	2-30 min	10-30 min	10 sec-5 min
Counting time	2-40 min	10-40 min	
Applications			
Delayed Gamma Spectra	Hand, torso, whole body Ca, P, Cl, Na	Hand, forearm, spine Ca	Total body N, Ca, Na, Cl, P
Prompt Gamma Spectra	Liver and kidney Cd Total and partial body N	Liver and kidney Cd Total and partial N, Ca	Total Body C (pulsed)
Advantages	Reliable, constant output, long half-life, portable, moderate cost	Reliable, constant output, portable very small source Highest thermal flux per unit dose	Beam on only when required. Good energy for thick sections. No dose when not in use.
Disadvantages	Continuously active, some regulatory difficulties in disposal of source may be encountered.	Continuously active short half-life replacement, disposal, calibration at 5-yr intervals	More maintenance required than for radionuclide sources. Cost of replacement tubes adds to expense. High dose per unit neutron

whole body counter. The shielding required for prompt gamma
studies, although smaller in physical size because of the lower
source activity, is more complex because the detectors must
monitor the signal emitted from the body simultaneously with
the neutron irradiation. One obvious constraint is that the
detectors are not placed in full view of the neutron beam.
Nevertheless, it is estimated that 1-5% of the neutron flux is
scattered towards the detectors from the subject's body. The
number of neutrons reaching the detectors is reduced by
surrounding the detectors with polyethylene or wax impregnated
with lithium or boron.

NaI and Ge Detector Systems. The major consideration governing
the choice of detectors is that counting efficiency be
maximized while an acceptable energy resolution is maintained
at a reasonable cost. Normally, the first choice is NaI because
of its higher detection efficiency at the lower cost. If the
spectrum is complex, the higher energy resolution of Ge
detectors may be needed. This improvement in energy resolution
is obtained only at the expense of detection efficiency and at
a major increase in the cost of the system. The NaI crystals
typically used for IVNAA have efficiencies that are 10 to 20
times greater than the larger (>25% efficiency) Ge detectors
and at one-tenth the cost. Whether specific organs, regions of
the body, or the whole body is to be measured will determine
the physical arrangement of the detector system. The geometry
of the counter and its detection efficiency are interdependent,
again governed mainly by cost. Multidetector arrays require
computer-based operations for data collection and analyses.
For the measurement of induced activities of calcium, sodium,
chlorine, and phosphorus, NaI detectors, housed in
low-background rooms, are used. For the prompt gamma
measurements of body nitrogen, hydrogen, and carbon,
neutron-shielded large volume (15 cm diameter x 15 cm
thickness) NaI crystals have proven acceptable. Recent
investigations by Ryde et al. (7) have shown that body calcium
can be detected in vivo by the prompt gamma technique, if Ge
detectors are used.

Dosimetry. Any neutron exposure, no matter how small, can be
assigned some level of risk. Therefore, it is essential to
weigh the benefits of such measurements against the risks. The
World Health Organization (WHO) has examined the risks
associated with different levels of radiation and has proposed
the three dosage categories listed in Table III (8). The

Table III. Dose (mSv) Guidelines for In Vivo
Activation Analysis

Body Regions Exposed	WHO Category		
	I	II	III
	Within variations of natural background	Within dose limits for general public	With dose limits for persons occupationally exposed to radiation
Order of magnitude of dose (mSv)[a]	0.1	1.0	10
Range of whole body doses (mSv)	< 0.5	0.5–5.0	5.0–50
Dose to single organ (mSv):			
Thyroid	< 17	17–170	170–1700
Liver or kidney	< 8.3	8.3–83	83–830
Gonads	< 2.0	2.0–20	20–200

[a] 1mSv = 100 mrem

categories are helpful in identifying the relative exposure
levels used for body composition measurements. The WHO report
indicates that category I dosages neither require particular
radiation protection nor pose radiobiological problems; i.e.,
any resultant risk would be considered negligible. Most of the
delayed IVNAA procedures are within the range of category II.
These WHO guidelines are observed in the UK for in vivo neutron
activation techniques. Most investigations use dosages that
are well within the WHO categories II and III. Whole-body
calcium measurements, for example, can be made at a precision
of $\pm$ 1 % with dosages as low as 2.5 mSv (250 mrem). Whole body
nitrogen measurements are routinely performed at dosages below
0.3 mSv (30 mrem).

RESEARCH APPLICATIONS AND ESTABLISHED PROCEDURES

Body Calcium and Bone Mineral Mass. Total or partial body
calcium has been the most widely studied element. The majority
of the studies have reported on the various disease conditions
in which calcium is lost; the dominant investigations are those
of post-menopausal osteoporosis (2). Body calcium levels have
served as an index of the progressive nature of diseases that
affect the skeleton and also as a monitor of the efficacy of
treatment. To quantify the degree of demineralization in the
patient groups, reference values must be established on the

basis of age, body size, race, and sex for the general
population. Bone calcium has been determined for the whole
body, the trunk, the spine, and the appendicular skeleton by
neutron activation analysis (1-4). Bone mineral content can
also be estimated by computerized axial tomography (CT
scanning) at a higher exposure dosage or in body regions by
photon absorptiometry (2). It is important to note, however,
that when available, the measurement preferred is that of total
body calcium. It has served as the 'reference standard' for the
calibration of these two alternate techniques.

Body Nitrogen and Protein Mass. Measurement of body nitrogen
by prompt gamma neutron activation is currently the most direct
method by which to determine body protein mass. In new
installations where nitrogen is the only element to be
determined, the facility should employ radioactive neutron
sources. Such facilities are reliable, relatively inexpensive,
require minimum floor space (< 3 m x 8 m), and can use
commercial electronics and software. In addition, gammas from
body sodium, chlorine, phosphorus, and calcium, although weak,
can be observed in the prompt spectra. Body calcium is already
being measured sequentially using Ge detectors (7). The higher
dose, loss of absolute measurements, and reduced precision must
undergo additional improvements before this approach can be
recommended for the measurement of calcium. One can expect,
however, that development in this direction will be continued.
If the cost of Ge detectors can be reduced, the dosage
maintained at the level currently used for nitrogen, and the
precision improved to 3% or better, the prompt-gamma technique
could replace the counting of the delayed activity in the body.
If one had to choose only one type of system, prompt or delayed
gamma counting, the choice will depend on the importance placed
on the immediate need for multi-elemental capability.
     Both cross-sectional and longitudinal studies have
provided estimates of protein mass and nitrogen balance
information in healthy and diseased humans. Clinical
investigations have included the examination of body nitrogen
after major surgery, changes in patients receiving total
parenteral nutrition, dietary support for cancer patients, and
the effects of protein-sparing diets for weight reduction in
obesity (9-13).

Other Elements and Additional In Vivo Techniques. The in vivo
prompt gamma measurements of kidney and liver cadmium levels in
humans are currently the only direct methods for accurate
assessments of cumulative exposure (14). Examination of
industrially-exposed populations have led to a redefinition of
the critical concentration of cadmium for the kidney, an
evaluation of the relationship of cadmium level in the kidney
to levels in blood and urine, and of air monitoring data
(15-17). In vivo X-ray fluorescence (IVXRF) techniques have
also been developed, mainly when neutron activation is not
practical (low sensitivity or high dose). These techniques are
similar to those of partial body prompt neutron activation; the
interested reader is referred to several reviews (18-20). IVXRF

techniques are used to measure iodine content and its distribution in the thyroid; lead and strontium in teeth and bones; mercury in the skull and brain; and cadmium concentration in the renal cortex (18,21).

   Measurements of body sodium, chlorine, and phosphorus are also routinely obtained, but have not been utilized to any major extent in clinical research protocols. In vivo methods for measuring other elements, among them, carbon, iron, aluminum, selenium, copper and silicon, have been developed. An assessment of their contribution to health care or a recommendation that they be considered in the initial planning of in vivo elemental studies would be premature (1-4,22,23).

Absolute or Sequential Measurements.   One of the strengths of the IVNAA measurements has been their use in long term-balance studies. Previous reports have shown that the classical balance techniques incorrectly estimate the changes that occur in vivo (24). The time interval between two in vivo measurements will determine the sensitivity of in vivo data for balance studies (Table IV). In planning an in vivo system, absolute calibration

Table IV. Sensitivity of In Vivo Balance Studies
in an Individual[a]

Time Interval between Measurements	Average Balance Sensitivities		
	Ca ($\pm$ mg/d)	N ($\pm$ g/d)	K ($\pm$ mmol/d)
2 wk	–	–	7.10
1 mo	–	1.78	3.53
3 mo	–	0.59	1.18
6 mo	110	0.30	0.59
1 yr	57	0.15	0.29
3 yr	18	0.05	0.10
5 yr	11	0.03	0.06
Estimates of Daily Intake	1000	16	84.4
Physiological Balance[b] (M)	-8.7	-0.019	-0.035
(F)	-9.5	-0.014	-0.023

[a]Significant change between two measurements at the $p < 0.01$ level
[b]Estimated on the basis of total body change between ages 25 and 65, assuming a constant rate of loss due to aging.

may not be necessary if only sequential measurements are to be
made. The precision for repeated studies on the same individual
is usually greater than that for an absolute calibration. The
absolute value is needed, however, if it is important to relate
an observed increase in the body to replace a deficiency, or an
increment above an already normal level. Optimum use of the
IVNAA techniques will be achieved when absolute measurements
are made. Their full diagnostic potential can best be realized
when a healthy reference population is also examined. Body
composition data are more fully utilized when combined with
other clinical or biochemical indices for the assessing the
health or disease status of the individual. Nitrogen
measurements, for example, are more useful when used in
conjunction with measurements of body potassium and water. The
combination enables a more comprehensive view of the combined
changes that can occur in the lean tissue mass of different
body compartments (25).

PLANNING OF IN VIVO STUDIES

Institutional Infrastructure. A more widespread interest in
noninvasive techniques will emerge as the clinical community
continues to recognize their potential to provide unique
information on human body composition. In vivo neutron
activation is sufficiently advanced to warrant consideration of
the medical economics in planning the next phases of
development. Although each system built to date has been custom
designed, the integrated laboratory concept used at the newest
facilities at Baylor College of Medicine in Houston, Texas,
should help reduce some costs for other centers (5). It seems
best to avoid the proliferation of lower cost equipments which
will considerably compromise the accuracy and precision of
procedures that can be obtained with larger state-of-the-art
systems. The following suggestions, therefore, are aimed
primarily at institutions planning to establish fixed
installations with a strong emphasis on multi-elemental
analysis. For this application, the irradiation facility,
including the neutron sources, should be located within
reasonable proximity (~ 30 m)to the counting equipment,
although at a sufficient distance or with adequate shielding
(2 m concrete for 1000 Ci ^{241}Am/Be source) to protect the
detectors from neutron damage. The data collection and pulse
height analysis techniques will require a computer-based
system.
     Several considerations emphasize the advantage of locating
these facilities at major clinical centers where the best
possible equipment, in terms of sensitivity, low dose, and
technical expertise, would be available. The requirements for
clinical and research use would probably result in high
productivity on a 24-h basis. Although some patients would be
inconvenienced by having to travel to a regional center, it is
less expensive and more productive than to establish smaller
facilities at a number of hospitals. The regional approach also
has an advantage in that a variety of other sophisticated
procedures and equipment available for diagnosis and therapy,

for example, CT scanning, nuclear magnetic imaging, nuclear
medicine, and ultrasound imaging, could be used in conjunction
with the IVNAA analyses. These procedures are generally
complementary rather than competitive. The patient is offered
the best possible service when a variety of techniques is
available.

One important exception to the regional approach is the
use of smaller prompt gamma systems to assess body protein
status. Successful systems have been established worldwide in
clinical settings (Toronto, Canada; Brookhaven, New York;
Houston, Texas; Auckland, New Zealand; Swansea, Wales;
Edinburgh, Scotland; and Menai, Australia) (5,7,26-30).
Additional systems are being planned for New York, Galveston,
Boston, and San Francisco in the US alone (Ellis KJ, Baylor
College of Medicine, personal communication, 1989). This
procedure can be expected to be dispersed even into general
hospitals once the equipment becomes reliable; is shown to
require minimum maintenance; and is packaged into a 'turn-key'
instrument reducing technical support requirements.

Location, Design, and Accommodation Requirements. The IVNAA
facility should have easy access for patients from the hospital
floors and for outpatients. The accommodation should be
designed so that operating personnel are not exposed to
excessive radiation. The heavy shielding of the irradiation and
counting equipment for the induced activity measurements will
require floor space on ground level or below. This requirement
is recommended although not necessary for the prompt gamma
system. Waiting areas, changing rooms, and a clinical
examination area for patients should be part of the total
facility. Although an electronics workshop is recommended,
adequate maintenance can usually be provided by the central
services of the institution. There are no special electrical
power requirements. If Ge detectors are used, a regular supply
of liquid nitrogen is needed. Routine hospital health physics
support is adequate once installation is completed.

Staffing for Routine Measurements. A professionally qualified
scientist or clinician is required to supervise the work and
ensure quality control. The measurements can be performed by
nuclear medicine technicians. Their training can include
routine maintenance and minor repair of the instruments. One
or two persons can perform most of these tasks, but generally
for a full work load, three or more persons are required.

Clinical, Diagnostic, and Research Protocols. A summary of the
various types of protocols that have used IVNAA techniques are
given in Table V. Continued interest in these applications is
expected as more research centers establish their own
facilities. In each case, the WHO recommendations (8) should be
followed if local and state guidelines governing the use of
ionizing radiation are not available. A most important aspect
of any facility is to keep the radiation doses at the lowest
levels without loss of adequate diagnostic or research value.
Pregnant women and children under 18 years of age should not be

Table V. Clinical, Diagnostic, and Research Applications of In Vivo Neutron
                              Activation Analysis

Total Body Calcium and Phosphorus

Population Group	Clinical Condition and Research Protocols
Normal subjects	Reference range for age, sex, race and body size
Aging Osteopenia Osteoporosis	Postmenopausal Effects,Osteogenesis imperfecta, Drug induced bone loss, Vitamin D deficiency states, Renal osteodystrophy
Endocrine disorders	Cushing's syndrome, Acromegaly, Parathyroid disorders, Thyroid disorders, Hypogonadism
Other Disorders	Myotonic dystrophy, Thalassemia, Alcoholic cirrhosis, Cadmium-induced bone loss

Total Body Nitrogen and Potassium

Population Group	Clinical Condition and Research Protocols
Normal subjects	Reference range for age, sex, race and body size
Athletes/exercise therapy	Effects of exercise and/or anabolic agents
Obese subjects[a]	Weight reduction programs: caloric restriction, intestinal bypass surgery, lypo-suction
Surgical patients[a]	Evaluation of nutritional management following major abdominal,cardiac or transplant surgery
Protein malnutrition	Malnutrition, Effects of diet and total parenteral nutrition
Cancer	Maintance of Protein Mass –effects of cancer therapy and TPN
Renal failure	Changes of body composition of patients on dialysis and with endstage renal failure
Cardiovascular disease[a]	Post-MI monitoring of body composition of patients
Thyroid and Parathyroid	Body composition changes with treatment and advance disease
Patients with ascites	Cirrhosis with ascites and the effect of an arteriovenous shunt

[a]Areas where future investigation appear most promising for the study of body chlorine and sodium in obesity with or without hypertension, oral contraception, and urolithiasis.

exposed unless there are clear indications for medical reasons.
A signed informed consent should be obtained for each procedure
after protocol review and approval by the appropriate
institutional review board.

CONCLUSIONS

In vivo measurements of whole body calcium and nitrogen, and
kidney and liver burdens of cadmium are likely to continue in
regular use. These measurements can be applied to a variety of
clinical and toxicological problems; such information is
virtually unobtainable by other means in the living human. The
risks involved with each procedure are minimal and pose no
significant hazard for sequential observations in the same
subjects and for individuals in the general population. The
radiation dosages are generally low compared with many common
radiological examinations and diagnostic tests (31). When a new
activation facility is planned, one should expect to start with
measurements of at least one of these three elements. Although
the preferred neutron source is probably ^{241}Am/Be, ^{252}Cf may be
more accessible. Pu sources should not be considered unless
already available to the user; these sources require extensive
regulatory control and security. The preferred method of
measuring nitrogen is undoubtedly to use its neutron capture
gammas, whereas a multi-elemental analysis of body sodium,
chlorine, phosphorus, and calcium is best carried out at
present using the delayed spectra. Areas of further development
are likely to focus on body carbon and oxygen as measures of
body fat and water, respectively.
    At present, NaI detectors should be used, because of the
prohibitive cost associated with the number of Ge detectors
that would be needed for these measurements. The NaI crystals
should be relatively large (>2000 cc volume) to have reasonable
efficiency for detecting high energy gammas (1-11 Mev).
Commercially available electronics for gamma spectroscopy and
computer-based data collection systems are more than adequate
for the tasks associated with IVNAA. The commercial analytical
software programs are generally adequate, although some
additional modifications are usually needed to conform to the
geometrical design of the system, number of detectors, and the
specific in vivo application.
    When planning IVNAA studies of body composition, it is
important to consider other 'noninvasive' techniques which may
also be available. For example, there are alternative means by
which to measure bone mineral, such as radiography, photon
absorptiometry, Compton scattering, and CT scanning. None of
these methods including IVNAA is likely to displace all others.
Total body calcium measurements, however, continue to provide
unique and valuable data that have been the 'reference
standard' for comparison with these alternate techniques.
Nuclear magnetic resonance spectroscopy of body nitrogen,
sodium, and phosphorus remains a consideration; its
development, however, will continue to be delayed until proton
mapping techniques for imaging are fully developed.

ACKNOWLEDGMENTS

The author thanks E.R. Klein, J.D. Eastman, and S. Smith for
editorial assistance and preparation of the manuscript. This
work is a publication of the USDA/ARS Children's Nutrition
Research Center, Department of Pediatrics, Baylor College of
Medicine and Texas Children's Hospital, Houston, TX. This
project has been funded with federal funds from the USDA/ARS
under Cooperative Agreement No. 58-7MN1-6-100. The contents of
this publication do not necessarily reflect the views or
polices of the USDA, nor does mention of the trade names,
commercial products, or organizations imply endorsement by the
US Government.

Literature Cited

1. Chettle, D.R.; Fremlin, J.H. Phys. Med. Biol. 1984, 29,
   1011-43.
2. Cohn, S.H.; Parr, R.M. Clin. Phys. Physiol. Meas. 1985, 6,
   275-301.
3. East, B.W. Trends Anal. Chem. 1982, 1, 179-93.
4. Boddy, K. Biolography on In Vivo Activation Anlaysis;
   Scottish Universities and Reactor Center, SURRC-65/78,
   1978.
5. Ellis, K.J.; Shypailo, R.J. Med. Phys. 1988, 15, 438-47.
6. Oxby, C.B.; Appleby, D.B.; Brooks, K.; Burkinshaw L.;
   Krupowicz, D.W.; McCarthy, I.D.; Oldroyd, B.; Ellis, R.E.;
   Collins, J.P.; Hill, G.L. Int. J. Appl. Radiat. Isot. 1978,
   29, 205.
7. Ryde, S.J.S.; Morgan, W.D.; Sivyer, A.; Evans, C.J.;
   Dutton, J. Phys. Med. Biol. 1987, 32, 1257-63.
8. Use of ionizing radiation and radionuclides on human
   benings for medical research, training and non-medical
   purposes, World Health Organization Technical Report
   Series, No. 611, WHO, Geneva, 1977.
9. Hill, G.L.; McCarthy, I.D.; Collins, J.P.; Smith A.H. Br.
   J. Surg. 1978, 65, 732-35.
10. McNeill, K.G.; Harrison, J.E.; Mernagh, J.; Jeejeebhoy,
    K.N. J. Enter. Parenter. Nitrit. 1982, 6, 106-08.
11. Cohn. S.H.; Gartenhaus, W.; Sawaitsky, A.; Rai,K.;
    Zanzi,I.; Vaswani, A.; Ellis, K.J.; Yasumura, S.; Cortes,
    E.; Vartsky, D. Metabolism 1981, 30, 222-29.
12. Archibald, E.H.; Harrison, J.E.; Pencharz, P.B. Am. J. Dis.
    Child. 1983, 137, 658-62.
13. Vaswani, A.N.; Vartsky, D.; Ellis, K.J.; Yasumura, S.;
    Cohn, S.H. Metabolism 1983, 32, 185-88.
14. Ellis, K.J.; Morgan, W.D.; Zanzi, I.; Yasumura, S.;
    Vartsky, D.; Cohn S.H. J. Toxicol. Environ. Health 1981, 7,
    691-703.
15. Ellis, K.J.; Morgan, W.D.; Zanzi, I.; Yasumura, S.;
    Vartsky, D.; Cohn, S.H. Am. J. Ind. Med. 1980, 1, 339-48.
16. Ellis. K.J.; Vartsky, D.; Cohn, S.H. Neutrotoxicology 1983,
    4, 164-68.
17. Ellis, K.J.; Cohn, S.H.; Smith T.J. J. Toxicol. Environ.
    Health 1985, 15, 173-87.

18. Scott, M.C.; Chettle, D.R. Scand. J. Work Environ. Health 1986, 12, 81–96.
19. Ellis, K.J. In Methods for Biological Monitoring; Kneip, T.J.; Crable, J.V. Eds., American Public Health Association, Washington, D.C.,1988; p. 65.
20. Ellis, K.J. In Biological Monitoring of Toxic Metals; Clarkson, T.W.; Friberg, L.; Nordberg, G.F.; Sager, P.R. Eds.; Plenum Press, New York, 1988; p. 499.
21. Palmer, D.W.; Kaufman, L.; Deconinck, F. In Medical Applications of Florescent Excitation Analysis; Kaufman, L.; Price. D. Eds.; CRC Press, Boca Raton, FL, 1979; p. 139.
22. Ellis, K.J.; Kelleher, S.; Raciti, A.; Savory, J.; Wills, M. J. Radioanal. Nucl. Chem. 1988, 124, 85–95.
23. Kehayias, J.J.; Ellis, K.J.; Cohn, S.H.; Yasumura, S.; Weinlein, J.H. In In Vivo Body Composition Studies; Ellis, K.J.; Yasumura, S.; Morgan W.D. Eds.; Institute of Physical Sciences in Medicine, London, 1987; p. 427.
24. Forbes, G.B. Am. J. Clin. Nutr. 1983, 38, 347–48.
25. Cohn, S.H.; Vartsky, D.; Yasumura, S.; Vaswani, A.N.; Ellis, K.J. Am. J. Physiol. 1983, 244, E305–10.
26. Meragh, J.R.; Harrison, J.E.; McNeill, K.G. Phys. Med. Biol. 1977, 22, 831–39.
27. Vartsky, D.; Ellis, K.J.; Vaswani, A.N.; Yasumura, S.; Cohn, S.H. Phys. Med. Biol., 1984, 29, 209–18.
28. Beddoe, A.H.; Streat, S.J.; Hill, G.L. Phys. Med. Biol. 1987, 32, 191–201.
29. Allen B.J.; Blagojevic, N.; McGregor, B.J.; Parsons, D.E.; Gaskin, K.; Soutter, V.; Waters, D.; Allman, M.; Stewart, P.; Tiller, D. In In Vivo Body Composition Studies, Ellis, K.J.; Yasumura, S.; Morgan W.D. Eds.; Institute of Physical Sciences in Medicine, London, 1987; p. 77.
30. Mackie, A.; Hannan, W.J.; Smith, M.A.; Tothill, P. J. Med. Eng. Technol. 1988, 12, 152–57.
31. Faulkner, K.; Wall, B.F. (Editors) Are X-rays Safe Enough? Patient Doses and Risks in Diagnostic Radiology; Institute of Physical Sciences in Medicine, London, 1988, p. 32–53.

RECEIVED August 14, 1990

Chapter 4

In Vitro Investigations in Biological Trace Element Analysis
Positive and Negative Aspects

Nicholas M. Spyrou

Department of Physics, University of Surrey, Guildford, Surrey, GU2 5XH, England

'In vitro' and 'in vivo' analytical methods for trace element studies in biological systems play an increasingly important role in assessing environmental impact on health, in diagnosis of disease and in monitoring treatment and it is not possible to identify the merits and limitations of 'in vitro' methods without discussion of the capabilities of the 'in vivo' techniques. Problems in the selection of representative samples and in sample inhomogeneity are common to both. However, neither makes use of computerised tomography and of advances in tomographic techniques to help in resolving these, despite claims that the field is multidisciplinary. Difficulties in 'in vitro' investigations are illustrated by examples whose errors occur in the analytical technique, in the choice of representative sample and in sample inhomogeneity.

In the last twenty years, trace element studies related to human health have continued to grow in a variety of fields, be it to assess environmental impact, to determine daily intake, to diagnose disease or to monitor the course of treatment. However, in order to carry out useful investigations in determining elemental concentrations, it is important to select samples of tissues and body fluids for analysis which will provide meaningful results. Selection should, therefore, generally be focussed on biological specimens which reflect elemental status and can be obtained from subjects preferably with ease and without trauma. The problems then arise in establishing whether these samples are representative and how elemental levels are being influenced by, for example, temporal changes due to metabolism and physiological processes, diet, medication or in general, the lifestyle of the subjects. Otherwise erroneous interpretation may result however good the analytical precision and accuracy.

In other situations, when a pathological specimen is excised from the body, interpretation of results may depend on the comparison of the elemental concentrations found in this sample with those found in normal tissue adjacent to it, or in the absence of the latter, with baseline values or ranges reported in the literature for the particular type of tissue, if they exist. It is therefore highly desirable to establish for populations normal levels in tissues and body fluids so that changes during disease conditions can be detected (1). Biopsy tissue samples, i.e. those that require surgical intervention, would require to be representative of the

0097–6156/91/0445–0040$06.00/0

tissue or organ from which the specimen has been removed and present the additional analytical challenges of being unique and of small mass, typically less than a few mg.

Autopsy samples have been useful in determining elemental distributions in whole organs e.g. in liver (2), kidney and lung (3), despite the problems associated with postmortal changes in elemental concentration and content, as a function of time after death, which can be severe (4). These elemental distribution studies are useful in indicating the extent of variation within an organ and that found between the same organ from different individuals. They may also suggest that sampling and analysing a specific region of an organ (eg kidney cortex or medulla) may be more fruitful than determining elemental levels of the entire organ. However, there are medical, ethical and legal restrictions to contend with when acquiring postmortem specimens. Acquistion of suitable material will become even more difficult (and more expensive) as the considerable progress made in organ transplantation makes ever increasing demands in obtaining healthy organs.

'IN VIVO' METHODS

In order to discuss the advantages and disadvantages of 'in vitro' measurements with regards to elemental analysis of biological systems it is important to assess what 'in vivo' methods have to offer in this respect. At present, there are twenty-three elements for which nuclear-based techniques exist or are being developed for determination of their concentration 'in vivo' in the human body (5). The concentration of elements 'routinely' determined are Ca, P, Na, Cl, K, N, O and Cd, whereas those of H,C,I and Pb can also be measured. There is an almost equal number of elements, Li, Be, Mg, Al, S, Si, Fe, Cu, Se, Ag and Hg for which techniques have been developed and detection capabilities demonstrated. All are based on the detection of products resulting from nuclear interactions, except for Pb which is measured by X-ray fluorescence and K which depends on the detection of the characteristic gamma-rays emitted by the naturally occuring radionuclide, K-40. In general the methods are specific to a single element, although a group of elements may be detected simultaneously e.g. Ca, P, Na, Cl and K in bone studies.

Whole body or partial body 'in vivo' neutron activation analysis provides information about elemental composition for diagnosis and for monitoring therapy, which in many instances would otherwise not be available. Other 'in vivo' methods are also being explored and developed for human compositon studies to provide information, not necessarily in the form of elemental concentrations upon which medical diagnosis can be based (6). Techniques like magnetic resonance imaging (MRI) which can define the relative amounts of bound and free water in a biological system or total body electrical conductivity measurements which determine total lean mass or total body fat (7) are proving useful in such body composition studies and are not only non-invasive but also ionising radiation free. In addition other methods, for example, employing computerised X-ray transmission tomography and photon densitometry can provide specific information about bone mineral content, in terms of a weighted (or average) atomic number and the density of the bone matrix, which may be sufficient to serve as an index for diagnosis of osteoporosis (or other bone disease) and for monitoring its treatment (8,9). This is achieved for about the same radiation dose as (or less than) neutron activation analysis which determines Ca and P bone concentrations in the same region of interest.

THE ROLE OF TOMOGRAPHY

The advantages of 'in vivo' methods of analysis are that measurements are carried out at the site of interest and therefore no variations in elemental concentrations arise from removing the tissue from the body and the possibility of contaminating the sample is avoided. But, unless the 'in vivo' measurements are carried out for total body analysis or total organ analysis (the degree of localisation may be difficult to achieve anyway), the problem of how representative is the region selected for the investigation of the total organ or tissue, as in 'in vitro' analysis, still remains. However, if the technique has tomographic capabilities or can be combined with tomography i.e it can represent the distribution of a physical property in a plane or slice through the region of interest, without interference from overlying and underlying planes or slices (10), then variations within the 'volume sampled' may be detected and taken into account. When photon transmission tomography is used, for example, for determination of bone mineral content, the distribution of the photon linear attenuation coefficient (a function of the atomic number and the physical density) in a selected slice through the body, be it the wrist, the ankle or the spinal column, can represent the amounts of cortical and trabecular bone, as well as soft tissues, marrow, fat blood, muscle etc., that may be present. This can be achieved with a spatial resolution (expressed in terms of a pixel or picture element) as good as 1mm by 1mm and slice thickness of 10mm (or less, if desired). It depends on the photon energy, photon source intensity and time of irradiation i.e the radiation dose.

X-ray transmission computerised tomography is now established worldwide as a method of diagnosis, made possible with the introduction of the first commercial scanner by Hounsfield in 1972. There has been a demand over the years for more accurate, quantitative data regarding the photon linear attenuation coefficients of tissues and organs (and hence their composition) in the body, so that not only accurate determinations of the size of organs and tissues can be made but also for normal variations in the values can be established and compared with pathological states. Some of the techniques developed, especially those using discrete photon energies, can be sensitive to changes in the attenuation coefficients of soft tissue (water is usually used as the comparator) of 1 in 1000. This represents a detectable mass fraction (11) for higher atomic number elements (above, say, atomic number 30) of about 10 μg/g, with a 5 percent precision, in water.

If an organ or a region of the body is to be analysed by an 'in vivo' method, in order to determine its elemental concentrations, or a tissue sample is to be removed from the body for 'in vitro' trace element analysis, selection of the appropriate site in terms of its homogeneity or specific heterogeneity can be made more accurate by obtaining tomographic information. Radiation dose must be taken into consideration for 'in vivo' analysis and sampling but if an organ or tissue is removed from the body, for example, for specimen banking, then tomography can be performed and variations measured without restriction. The measurements can also be repeated, in order to monitor changes in the stored tissue or organ with time.

We first used tomography in conjunction with neutron activation analysis in an environmental study where the elemental composition of tree rings was determined and ring to ring variations found (12). Use of photon transmission tomography of a tree trunk section rapidly identified the areas where elemental variations occured and therefore made possible the selection of specific samples for detailed trace element anlaysis. In another investigation the changes in the linear photon attenuation coefficient in the cortical and trabecular compartments across and along the length of a human tibia, which had been removed from a subject, were correlated with the results of elemental levels obtained from neutron activation analysis of a number of thin transverse sections cut from the bone along its length (13). Tomographic information about the relative size of the cortical and trabecular compartments, which could not be easily separated by microtome sectioning of the samples for trace element analysis, allowed calculations of relative weighting and estimates of elemental compositon of each compartment to be carried out.

In carrying out 'in vivo' neutron activation analysis of liver for determination of Cd and Se concentrations in the organ, an X-ray transmission tomographic image through the region of interest was found to be useful in calculating the volume of interaction between the neutron beam and the tissues and also in estimating the volume of irradiated tissues emitting radiations 'seen' by the gamma-ray detector (14). In addition, since the sizes, the distribution and the value of the photon linear attenuation coefficients of the tissues were known, the attenuation of the gamma-rays, resulting from neutron interactions, on reaching the detector could also be calculated. More recently, the techniques of neutron transmission tomography and neutron induced emission tomography have been developed (15, 16) which depend on the neutron linear attenuation coefficient (the macroscopic neutron cross-section) and the neutron capture cross-section, respectively. The latter technique combines the principles of tomography with neutron activation analysis, thus providing a method which can determine the concentration of elements in a selected plane or section, through a sample, nondestructively. The technique, which is multielemental, has at present a spatial resolution of 0.5 mm for relatively small samples of about 50 mm diameter and because of radiation dose is only applied to 'in vitro' analysis.

'IN VITRO' PROBLEMS

There are many examples cited in the literature, concerning problems in the elemental analysis of biological systems and sources of error in sampling and sample preparation (e.g. 17, 18). Here, a small number of examples from recent work using instrumental nuclear-based methods of in vitro trace element analysis are briefly discussed.

First an example where the analytical technique as applied may lead to errors. In a collaborative project with the European Community Laboratories at Ispra, Italy, a preliminary investigation to determine the fluorine concentration in bone biopsy samples removed from patients with osteoporosis and on renal dialysis treatment, was carried out (Spyrou N.M., Altaf W.J., Gill B.S., Jeynes C., Nicolaou G.E., Pietra R., Sabbioni E and Surian M., J. Biol. Trace Analysis Research, in press). Concern has been expressed about the uptake of certain elements, particularly aluminium, in bone and other tissues of patients receiving kidney dialysis treatment. It has also been suggested that incidence of osteoporosis and bone disease where aluminium may be playing a part is reduced in the presence of fluorine. Therefore one of the objectives was to find out whether a correlation between the two elements existed. Fresh biopsy samples from the iliac crest of the subjects were obtained, each had an average diameter of 3 mm with masses varying between 10 and 140 mg.

Two methods were employed for the determination of fluorine concentrations. Cyclic neutron activation analysis (CNAA) to determine fluorine and other elemental levels in the whole bone sample and proton induced gamma-ray emission (PIGE) analysis to measure variations of the element along the length and across the surface of the bone sample. Determination of the trace element concentration using PIGE depends on the range of penetration of the proton beam into the sample and thus on its major element composition, so that the mass of the sample excited by the protons can be calculated. If there are variations in the range of penetration of protons and the gross elemental composition between comparator and sample, or sample to sample or within or along a sample, then these must be corrected, otherwise errors in interpretation occur since these are not "true variations' in the concentration of the trace element of interest. The major element composition (C, O, P and Ca) of the bone samples and bone standard was obtained using Rutherford Backscattering (RBS), from which the mass excited by the proton beam was calculated and the fluorine concentration in the bone accurately determined. If this correction was not applied, the differences in the mass excited, amounting to 30 percent, would have been attributed entirely to variations in the fluorine composition of the specimens.

The second example is from an on-going Se supplementation project where elemental changes in blood components are monitored (19); it relates to which component of blood should be analysed, if not whole blood, and also emphasises that conditions under which blood samples are collected, should be stated as fully as possible. Healthy, young adults of both sexes, partaking in a gastric motility study, in which the rate of stomach emptying of a liquid meal was being measured non-invasively, were given as a supplement a commerical product containing the element together with vitamins A, C and E. Blood samples were obtained on a one-day-a-week basis from each individual over a period of about 10 weeks. The subject fasted overnight, then in the morning was given to drink 500 ml of water flavoured with orange squash and stomach emptying measurements were made over an hour or so or until emptying was complete and there was a return to a baseline value. Blood was collected prior to the drink and then again at the half-emptying stage, as well as at the end of the experiment. This was repeated in the afternoon, after four hours, with no food or drink taken in between. Blood samples were separated into erythrocytes and plasma and a whole blood sample was also retained for trace element analysis. Results indicated that the level of selenium in whole blood and its components, erythrocytes and plasma, increased with dose and time and concentration of the element was maintained for at least six weeks following cessation of supplementation. However, the influence of the supplement on some of the electrolytes, simultaneously determined with Se in the samples, such as Na, Cl, Br and Rb could only be detected clearly in the erythrocyte and plasma and not in the whole blood. Further, when the variations of Se concentrations during the morning and afternoon sessions of the same day were examined, it was interesting to note a measureable decrease in the concentration for both plasma and erythrocytes during the day (20). One may have expected, from reading the literature, that the plasma levels would vary for the element during the day but not those for erythrocytes, which we are told may reflect a longer term intake and accumulation of the element. It is obvious, as with other elements, that more work requires to be done in determining the uptake (and washout) of Se in blood, under controlled conditions of food intake.

The third example is related to the above but underlines the importance of selecting the appropriate tissue. Public consumption of multivitamin and mineral supplements in the UK has grown significantly in the last decade, following trends set earlier in the USA. A reliable and preferably inexpensive method for monitoring the levels of these supplements in humans is sought in order to ensure that toxicity does not arise from prolonged usage. This is expecially true for trace elements where 'windows' of intake for good health can be very narrow and synergistic effects between elements may alter body burdens significantly. The possibility of using fingernails as a biomonitor for Se and other elements is being investigated (21), as blood samples become more difficult to obtain because of the rigorous 'screening' procedures required before analysis. The ease of collection, storage, preparation and handling of nail tissue has always been an attraction. The degree to which nail stores different elements normally depends on the supply of these elements during nail growth and the presence of 42 elements in the normal nail implies that the tissue may be a suitable candidate for monitoring. However, what are the elemental variations from finger to finger and from hand to hand? Does one select, as many do, the thumb nail from the right or left hand (whichever is used more) because it provides the largest mass ?

Fingernail clippings were collected from two healthy male adults, aged 45 and 28 years, over a period of 6 months. The first subject (A) was on supplements; Se (100μg/day) with vitamins A, C and E and Zn (5mg/day) provided by commercial products, whereas the second subject (B) was not given any supplement. Nail-to-nail variations and hand-to-hand variations were obtained for both subjects and Table I, shows the Se concentrations over the whole six month period; deviations for the same finger over this period were less than 20%. It can be seen that the highest concentrations for Se occur in both subjects in the nails of the little fingers of both hands and that variations exist between the fingers

Table I: Average Se concentrations (μg/g ± 20%) over six month period of six sets of fingernail samples from both hands of subjects A (supplemented) and B (unsupplemented).

Subject	little finger	ring finger	middle	forefinger	thumb
A(left)	17	N/A	5	N/A	10
(right)	17	N/A	6	N/A	10
B(left)	37	13	16	17	19
(right)	48	15	20	19	13
N/A	all samples not available				

of one hand. It may also be worth noting that Se levels in the fingernails of subject A did not exhibit the same degree of variation as those found in subject B (not supplemented) and that supplementation may have acted as a regulating mechanism in the excretion of the element from the body and its distribution through the fingernails. The profile of elemental concentrations in the fingernails may therefore give us additional information about the regular intake or otherwise, of trace elements by the subject.

Finally, it was decided to determine the homogeneity of the nail tissue and estimate the minimum mass required for a representative sample to be collected from the unsupplemented individual (subject B), who exhibited large variations in the nail-to-nail and hand-to-hand elemental concentrations. This was done by employing the concept of the sampling factor (22, 23) for the elements detected and using all 140 fingernail samples collected from subject B over the six month period. The values for the sampling factors, defined at being homogeneous to within 1 percent, are given in Table II. These values are very high but should not be considered unusual. Fingernails must be accepted as inhomogeneous materials for analysis, providing an integrated value of elemental concentration only for the specific period of growth for which the sample has been collected.

Table II: The sampling factor, K, at one percent inhomogeneity, for 12 elements detected in 140 fingernail samples from both hands of subject B (unsupplemented), collected over 6 months

Element	Au	Br	Cl	K	Mg	Mn
K(g)	53.0	5.6	1.1	8.5	1.4	3.1

Element	Na	Rb	S	Se	V	Zn
K(g)	18.1	0.08	4.5	191	1.9	4.9

CONCLUSIONS

The advantages and disadvantages of 'in vitro' trace element studies are better illustrated in the context of 'in vivo' methods of analysis. Both 'in vitro' and 'in vivo' methods have common problems when it comes to selecting a sample or site which is representative of the particular biological system under investigation and in assessing sample inhomogeneity. The use of tomographic techniques and of the vast amount of data already accumulated in hospital departments from computerised tomography scanning of body tissues, fluids and organs, should serve as aids in resolving some of these problems and in providing information about homogeneity and size of organs and compartments in terms of major element composition.

It is advocated that analytical techniques will benefit significantly by applying whenever possible the principles of tomography in order to obtain information about elemental distribution in a sample, non-destructively. In addition a number of non-invasive, radiation free methods are becoming available for composition studies. They define body compartments e.g. fat, water and lean mass, which can be useful in normalising elemental concentrations. It should then be possible, by employing the different information supplied from a combination of techniques, to reduce the variations in composition generally attributed to 'individual physiology and anatomy' and therefore allow the more subtle differences that trace element analysis may reveal to be interpreted correctly. Availability, acquisition and interpretation of such information is only possible through a multidisciplinary approach in such studies.

LITERATURE CITED

1. Iyengar, G.V. Clin. Nutr. 1987, 6, 4, 154-58.
2. Lievens, P.; Versieck, J.; Cornelis, R.; Hoste, J. 1977, J. Radioanal. Chem. 1977, 7, 483-96.
3. Vanoeteren, C.; Cornelis, R.; Sabbioni, E. Critical Evaluation of Normal Levels of Major and Trace Elements in Human Lung Tissue, Report EUR 10440 En, Commission of the European Communities, Luxembourg, 1986.
4. Iyengar, G.V. J. Path. 1981, 134, 173-80.
5. Spyrou, N.M. J. Trace and Microprobe Tech. 1988, 6, 4, 603-20.
6. In Vivo Body Composition Studies, Ellis K.J.; Yasumura, S.; Morgan, W. D. Eds., The Institute of Physical Sciences in Medicine: London, 1987.
7. Proceedings Symposium: Body Composition Methods, Roche, A.F. Ed., 1987, Human Biology 59, 209-335.
8. Non-Invasive Bone Measurements and Their Clinical Application, Cohn, S.H. Ed., CRC Press: Florida, 1981.
9. Banks, L.M.; Stephenson, J.C. J. Comput. Assist. Tomogr. 1986, 10, 463-67.
10. Kouris, K.; Spyrou, N.M.; Jackson, D.F. Imaging with Ionising Radiations, Surrey University Press/Blackie and Son Ltd: Glasgow and London, 1982, Vol. 1.
11. Kouris, K.; Spyrou, N.M. Nucl. Inst. & Meth. 1978, 153, 477-87.
12. Spyrou, N.M.; Neofotistou, V.; Kouris, K. In Proceedings Third International Conference on Nuclear Methods in Environmental and Energy Research, 1977, Columbia, Missouri, Technical Information Centre, U.S. Department of Energy, CONF-771072, 126-135.
13. Kidd, P.M.; Nicolaou, G.E.; Spyrou, N.M., J. Radioanal. Chem. 1982, 71, 489-507.
14. Nicolaou, G.E.; Matthews, I.P.; Stephens-Newsham, L.G.; Othman, I.; Spyrou, N.M. J. Radioanal. Chem. 1982, 71, 519-27.
15. Spyrou, N.M.. J. Radioanal. Nucl. Chem. 1987, 110, 641-53.
16. Spyrou, N.M.; Kusminarto, Nicolaou, G.E. J. Radioanal. Nucl. Chem. 1987, 112, 57-64.
17. Behne, D. J. Clin. Chem. Clin. Biochem. 1981, 19, 115-120.
18. Iyengar, G.V. In Trace Element Metabolism in Man and Animals, McHowell, J.; Gawthorn, J.M.; White, L. Eds., Australian Academy of Sciences: Canberra, 1981, 667-73.
19. Damyanova, A.A.; Akanle, O.A.; Spyrou, N.M. J. Radioanal. and Nucl. Chem. 1987, 113, 431-36.
20. Spyrou, N.M.; Akanle, O.A.; Damyanova, A.A. Am. Nucl. Soc. Transact. 1986, 53, 187-89.
21. Spyrou, N.M.; Altaf, W.J. Am. Nucl. Soc. Transact. 1989, 60, 28-29.
22. Heydorn, K.; Damsgaard, E.; Rietz, B. Anal. Chem. 1980, 52, 1045-53.
23. Spyrou, N.M.; Al-Mugrabi, M.A. J. Trace and Microprobe Techn. 1988, 6, 425-35.

RECEIVED August 6, 1990

QUALITY ASSURANCE

Chapter 5

Biologically Related National Institute of Standards and Technology Standard Reference Materials

Variability in Concentration Estimates

William Horwitz and Richard Albert

U.S. Food and Drug Administration, HFF-7, Washington, DC 20204

The extensive compilation of data on SRMs by Gladney et al. (Standard Reference Materials: Compilation of Elemental Concentration Data for NBS Clinical, Biological, Geological, and Environmental Standard Reference Materials, NBS Special Publication 260-111 (1987)) provides a remarkable opportunity to check the statistical distribution of analytical results for SRMs reported under unusually favorable circumstances. The precision estimates obtained for 11 biologically related SRMs by the compilers, who exercised subjective, critical analytical judgment to eliminate data "clearly beyond the limit of acceptability" by removing values beyond 2 standard deviations from the mean, provided published consensus values that are compared to the results predicted by the Horwitz equation: relative standard deviation between-laboratories (RSD_R, %) = 2 exp (1 - 0.5 log C), where C is the concentration expressed as a decimal fraction. The compilers removed 14% of the reported values from the 117 certified analyte-matrix combinations, when at least 8 values were available, for 28 elements over a certified concentration range from 5 ng/g to 40 mg/g. Only 2 of the resulting RSD_R values are outside the acceptable maximum of twice the RSD_R value calculated from the Horwitz equation. However, removal of 14% of the data as nonconforming is regarded as excessive. This finding suggests that many laboratories lack a quality assurance program to investigate the occurrence of gross deviations from certified values.

Very little information exists in the literature as to the magnitude of the underlying error structure of analytical chemistry. One of the best and most understandable expositions is that of Currie (1). One particularly practical statement by Currie is that the

variability should not exceed about 10% for application to the trace
concentration level. A number of professional and commodity-
oriented organizations conduct programs to control potential sources
of analytical error. For example, the Association of Official
Analytical Chemists (AOAC) controls the method variable by providing
a mechanism to validate methods of analysis applied to specific
commodities but on a case-by-case basis. Similarly, the American
Oil Chemists' Society (AOCS) and the American Association of Cereal
Chemists (AACS) not only validate methods but also furnish
laboratory (analyst) proficiency evaluation services for oilseeds,
grains, and related commodities. The American Association of Feed
Control Officials (AAFCO) provides guidelines for allowances for
sampling and analytical error in constituent declarations placed on
feed labels. In recent years, many international standards
organizations, such as International Organization for
Standardization (ISO), International Union for Pure and Applied
Chemistry (IUPAC), and International Dairy Federation (IDF), have
initiated routine programs for obtaining accuracy and precision
parameters from method-performance (collaborative) studies. The
National Institute of Standards and Technology (NIST, formerly
National Bureau of Standards) and numerous other intergovernmental,
governmental, and private institutions provide certified reference
materials against which accuracy or "trueness" (the currently
ISO-recommended (2) substitute term for accuracy) can be evaluated.
An extensive compilation of sources for biological reference
materials can be found in the recent book by Iyengar (3).

As a result of reviewing numerous AOAC commodity-oriented collaborative studies, Horwitz and his collaborators (4) elaborated an
equation which characterizes the between-laboratories precision of
chemical analysis fairly well. This equation relates relative
standard deviation between-laboratories in percent (RSD_R) to concentration expressed as a decimal fraction (e.g., for 1%, C = 0.01):

$$RSD_R \ (\%) = 2 \exp (1 - 0.5 \log_{10} C). \tag{1}$$

This equation states that beginning with 2% for pure materials
(C = 1), RSD_R doubles for every 0.01-fold decrease in concentration,
as shown in Table I, which provides some calculated RSD_R values for
the range of concentrations of interest here.

Table I. Relative Standard Deviation Between-Laboratories,
RSD_R (%), for Some Benchmark Concentrations, C,
as Calculated from Equation (1)

Concentration			RSD_R, %
Fractional	Log C	%	
1.0	0	100	2
0.01	-2	1	4
0.0001	-4	0.01	8
0.000001 (1 ppm)	-6	0.0001	16
0.00000001 (10 ppb)	-8	0.000001	32

In general, RSD_R appears to be a function of concentration
only, independent of analyte, matrix, and method. This

approximation has been tested for numerous analytes and matrices of regulatory interest, such as pharmaceuticals, pesticide formulations and residues, contaminants, mycotoxins, trace elements, fertilizers, feeds, dairy products, and other foods, over a concentration range of $C = 1$ to 10^{-9}. With some methods, i.e., those in which the analyte concentration is commonly adjusted to the point corresponding to maximum sensitivity of the measuring instrument, e.g., those using spectrophotometry, S_R (absolute standard deviation between-laboratories) is a constant, even independent of concentration, since an attempt is made always to measure at about the same concentration.

The availability of the extensive compilation of published data on Standard Reference Materials (SRMs) by Gladney et al. (5) provided a remarkable opportunity to check the statistical distribution of analytical results reported under unusually favorable circumstances. This compilation included original data on SRMs from 67 major journals in analytical chemistry, geology, geochemistry, and environmental science, as well as books and institutional reports for the period 1972-1985. Although many of the papers are research reports of methods development and application, quality control data also may be included. For the present purpose, the original documents were not consulted to classify the source or quality of the data. It is assumed that the compilation contains practical data resulting from a conscious decision on the part of those members of the analytical chemistry community who were willing to exhibit their performance on analysis of homogeneous materials of known or assigned composition to public scrutiny. Under such conditions, it would be expected that this collection of published values would show the best performance that the analytical profession has to offer.

The present paper compares these data with the between-laboratories variability produced under method-performance (collaborative) study conditions. A method-performance study subjects a number of homogeneous test samples to analysis as unknowns by a number of laboratories, all utilizing the same method. Analysts generally work very carefully during the analysis of such test samples. The compilation by Gladney et al. (5) contains data from what can be considered supplementary-to-certification (material-performance) studies of specific, homogeneous test materials by any valid method that provide substantiated or assigned values with stated uncertainties against which other measured values can be compared. Analysts who provided these literature values were aware of the certified, information, or published concentration value associated with the analyte of interest. Interpreters of analytical data generally expect better precision when analysts know the value expected than when the analyses are conducted "blind" (6, 7). The compilers of the SRM document (5) recognized the many uses to which their collection would be put and invited recalculations to reflect specific purposes.

Method

To limit the work to a manageable extent, this initial examination was confined to the 11 biological SRMs numbered 1566 to 1577a. It was further restricted to those analytes with 8 or more quantitative entries. The data were entered into an APL computer language workspace, and outliers were examined by the IUPAC-1987 protocol (8), which requires a minimum of 8 laboratories (entries). For further simplification, each published entry was considered as a single value with a uniform weight. Therefore, the Cochran test for replicate outliers could not be applied. Only the Grubbs and the paired Grubbs outlier tests were applied to the data, and aberrant individual values were removed only if they exceeded the very conservative critical value equivalent to a probability (P) of false rejection = 0.01. The readily available Grubbs tables for a single terminal outlier extended only to 40 values. Since as many as 212 values were present in a data set of this database, the critical value table for a single extreme value was expanded to 325 values by simulation and smoothing. For the paired Grubbs tests (high-high, low-low, and high-low, simultaneously), we adapted the values given by Kelly (9). The extended tables for the single Grubbs test and for the paired Grubbs tests were added to the APL program that executed the IUPAC-1987 protocol for the statistical analysis.

The IUPAC-1987 protocol is silent with respect to recycling outlier tests. Statisticians hesitate to reapply outlier tests after an initial removal of a value because "the probabilities change." However, with some of the data sets, it was obvious that outlier tests had to be applied consecutively, if only for consistency in removal of values in a tight cluster some distance from the main cluster. The outlier removal operation used here was that called for by the IUPAC-1987 protocol as interpreted by us for multiple outliers. The procedure consisted of first testing for a single Grubbs outlier; if present, the value was removed and the set was retested for a further single Grubbs outlier; when no single Grubbs outlier was removed during a cycle, the set of values was tested for a paired Grubbs outlier. If found, the pair was removed and the set was recycled to begin again with the single and then the paired Grubbs tests until no further values were removed.

The IUPAC-1987 outlier procedure blindly applied to the original data retained considerably more values than were kept by the compilers (5) who rightly exhibited subjective, critical analytical chemical judgment "to eliminate data on either end of the concentration spectrum that we judged to be clearly beyond the limit of acceptability." This initial screening by the data compilers usually removed less than 1% of the total data. Then they calculated a preliminary mean and standard deviation, S, and removed points outside 2S from the mean. Then a recalculated mean and an S were reported as the "consensus value and associated standard deviation." Unquantitated values reported as "less than" or "greater than" were not considered in the screening or calculations. In only 1 case of the 206 data sets (Mn in Pine Needles) did the IUPAC-1987 rules, as elaborated here, remove more outliers than did the compilers.

For the present paper, the published data under each SRM for each element, if 8 or more values were tabulated, were reanalyzed by the IUPAC-1987 protocol, using the appropriate outlier critical values, and the results were presented as relative standard deviations between-laboratories (RSD_R). The use of this parameter

permits consistent comparison of precision across materials,
analytes, and concentrations, unadjusted for analytical chemistry
judgment, which may differ from reviewer to reviewer. These RSD_R
values were then used as numerators in ratios containing, as
denominators, the RSD_R values based on concentration, C, as a
decimal fraction (e.g., C = 0.002 for a 0.2% level), as calculated
from the Horwitz equation (4). The resulting ratio, designated as
the HORWITZ RATIO or HORRAT in the tables and graphs, has been found
to be informative as to the acceptability of the underlying data.
Collaborative assays that give a HORRAT greater than 2.0 have been
considered as suspect in method performance studies (10).

Results

The NIST values, the consensus results from the compilation by
Gladney et al. (5), and the data from the present IUPAC
recalculation are given in the computer printout, Table II,
arranged alphabetically by analyte. The first column contains a
consecutive number assigned by the computer. Columns 2 and 3 give
the analyte and the NIST SRM matrix numbered 1566 to 1577a. Two
SRMs exist for Bovine Liver. The original SRM 1577 is designated
"Liver1-Bovine" and the second one, "Liver2-Bovine." The next 3
columns refer to NIST values: The fourth (NIST VALUE) is the
certified or informational value; the fifth (UN) is units in parts
per 100 (%) as H, in parts per million (ug/g) as M, and in parts per
billion (ng/g) as B; and the sixth column (NIST RSDR) refers to the
NIST uncertainty values, when given, recalculated to RSD_R (= S_R x
100/NIST value). The NIST uncertainty value in the compilation may
not necessarily be given as a standard deviation, S, but it is
treated as such. As a matter of fact, the certificates of different
SRMs give different definitions for the NIST uncertainty. This
point is not important in the present context. If the NIST value is
"for information" (i.e., no uncertainty is indicated), the NIST RSDR
column is blank; if no NIST value is available, both columns 4 and 6
are blank.
     Columns 7-10 refer to the compilers' consensus values given in
the publication, compiled to 3/1/1986 (5). Column 7 is the mean,
column 8 is its S recalculated to an RSD_R; column 9 gives the range
of values included in the consensus data; and column 10 indicates
the number of values (considered as laboratories) removed by the
compilers/total number listed with quantitative values.
     The last 4 columns refer to values calculated by the IUPAC-1987
rules (8) as extended here. The IUPAC VALUE is the mean; IUPAC RSD_R
is the S of the concentration estimates obtained by the IUPAC rules,
recalculated to an RSD_R; the IUPAC OMIT/LABS column, which follows,
contains the number of laboratories removed as outliers by the
IUPAC-1987 rules/total number listed. The final column gives the
HORRAT value as calculated from the IUPAC RSD_R value used as the
numerator in a ratio containing, as the denominator, the RSD_R
calculated from the NIST certified or information value, if given;
otherwise the consensus mean was used as the denominator in the
calculation.
     The NIST SRM values are given in 3 categories: (a) a certified
value with an associated uncertainty; (b) a value without an

uncertainty for information (uncertified); and (c) no value
provided. In the compilation, some of the individual published
values for a specific analyte for a given SRM also had an associated
within-laboratory uncertainty. These values were not included here
not only because the data were incomplete but also because of the
complication of interpretation that would have been introduced by
any kind of weighting. We do not believe that this simplification
affected the conclusions reached. Although the raw data were also
categorized by method, that information was not incorporated into
the database. The method type and references provided in the
compilation with each value were used only incidentally to determine
data patterns consistently responsible for extreme values across
analytes and matrices. For the analytes and matrices we have
examined to date (feeds, fertilizers, foods, mycotoxins and
contaminants, nutrients, pesticide residues and formulations,
pharmaceuticals, trace elements), we have found that generally
precision is independent of method. In this compilation, however,
some studies or methods, such as X-ray fluorescence and optical
emission spectroscopy, seem to consistently provide extreme values.
A specific study of the relationship of "removed values" of this
compilation to analyte, matrix, method, and concentration may
provide interesting correlations.

Table III presents the differences between the consensus values
and the NIST values and between the IUPAC values and the NIST values
for the 117 certified concentrations only. Other columns give the
percent outliers removed by both protocols and the corresponding
HORWITZ RATIOS. The "N" of this table is the original number of
quantitative values reported by the compilers before outlier
removal, for use in calculating confidence limits or standard
errors. In this table the original Bovine Liver SRM-1577 is
designated as "Bovn Li"; the new lot SRM-1577a, "Bovn LiA".

Results by Element

The identification information and the certification, consensus, and
IUPAC-recalculated data for 206 analyte-matrix combinations
available in the compilation (5) are accumulated in the database
(Table II). The certified analyte component contains 117 items; the
informational component, 37; and no NIST value, 52. To keep the
discussion within reasonable bounds, further remarks are confined to
the certified values.

Table III presents the relative differences, with sign ignored,
of the consensus values from the NIST certified values, the
corresponding differences for the IUPAC-1987 protocol means, and the
number of extreme values that were removed to achieve the recorded
values. These differences provide a measure of systematic error.
The NIST uncertainties and the RSD_R values from consensus and
IUPAC-1987 calculations transformed to Horwitz ratios are given
under the HORRAT column head. The numbers of HORRAT values
exceeding 2.0 are NIST, 4; consensus, 2; and IUPAC-1987, 30. These
data are further summarized in Table IV for those elements certified
in at least 3 matrices. The average difference of the consensus
value from the NIST certified value is 3.5%, with a minimum of 0.5%
for zinc and a maximum of 8.7% for chromium. The average HORRAT

Table II. Database of elemental composition data of NIST biological Standard Reference Materials

OBS	ANALYTE	REFERENCE MATERIAL	NIST VALUE	UN	NIST RSDR	CONSEN VALUE	CONS RSDR	CONSENSUS RANGE	CONSENSUS OMIT/LABS	IUPAC VALUE	IUPAC RSDR	IUPAC OMIT/LABS	IUPAC HORRAT
1	ALUMINUM	ORCHARD-L	1200.00	M	.	323.00	34.7	123.0-520.0	4/57	312.2	37.8	1/57	5.6
2	ALUMINUM	TOMATO-LE		M		1000.00	30.0	628.0-1300	12/22	646.3	61.6	1/22	11.2
3	ALUMINUM	PINE-NEED	545.00	M	5.5	510.00	11.8	399.0-620.0	5/29	511.8	20.9	1/29	3.4
4	ALUMINUM	LIVER-BO		M		816.00	17.5	609.0-909.6	6/26	893.3	89.3	2/26	8.5
5	ALUMINUM	SPINACH-1	870.00	M	5.7	810.00	11.1	180.0-909.0	6/19	73.3	29.5	0/19	5.1
6	ANTIMONY	ORCHARD-L		B	10.3	2.90	10.6	180.0-220.0	1/85	197.8	11.2	1/85	0.8
7	ANTIMONY	PINE-NEED	200.00	B		198.00	8.6	180.0-220.0	1/13	26.8	8.6	1/13	0.4
8	ANTIMONY	LIVER-BO	5.00	B		9.60	49.0	2.0-26.0	2/31	128.4	128.4	3/31	3.6
9	ANTIMONY	SPINACH-1	40.00	B	14.2	9.60	22.5	5.1-16.3	1/21	36.3	33.5	1/21	1.2
10	ANTIMONY	OYSTER-TI	13.40	B		40.00	9.2	27.0-50.0	2/21	12.8	13.2	1/20	1.3
11	ARSENIC	WHEAT-FLO	6.00	B		13.00	5.3	5.4-16.5	1/21	410.0	5.4	1/11	0.2
12	ARSENIC	RICE-FLOU	410.00	B	12.2	5.70	6.3	370.0-464.0	17/199	7.6	7.6	5/199	0.4
13	ARSENIC	ORCHARD-L	6.00	B	20.0	414.00	12.1	8.0-14.3	1/27	15.7	15.7	1/27	1.9
14	ARSENIC	TOMATO-LE	410.00	B	18.5	410.70	14.2	170.0-310.0	2/26	211.0	18.4	0/24	0.9
15	ARSENIC	PINE-NEED	270.00	B	19.0	253.00	18.7	180.0-240.0	2/20	202.7	11.2	10/64	0.5
16	ARSENIC	LIVER-BO	210.00	B	9.1	207.00	10.9	114.0-70.0	11/64	54.2	13.2	1/19	0.6
17	ARSENIC	SPINACH-1	55.00	M	33.3	55.00	13.1	35.0-52.0	1/19	152.7	13.0	3/55	0.6
18	ARSENIC	ORCHARD-L	150.00	M		153.00	9.3	40.0-69.0	3/55	43.3	16.1	0/10	1.7
19	BARIUM	TOMATO-LE	44.00	M		43.00	15.8	6.0-8.4	10/10	57.1	16.1	1/9	1.8
20	BARIUM	PINE-NEED		M		57.00	33.3	13.7-36.0	2/9	6.7	23.9	0/9	3.0
21	BERYLLIUM	ORCHARD-L	27.00	B	37.0	24.00	11.1	30.0-160.0	2/10	38.1	82.4	0/10	2.5
22	BISMUTH	ORCHARD-L	100.00	B		90.00	44.4	25.5-38.0	4/40	82.7	57.1	1/40	1.3
23	BORON	ORCHARD-L	33.00	M	9.1	33.00	9.1	7.1-25.3	0/18	32.4	12.9	0/18	1.3
24	BORON	TOMATO-LE	30.00	M		17.00	12.1	19.0-25.3	2/62	33.3	13.8	0/18	1.4
25	BORON	PINE-NEED		B		9.50	11.6	7.3-41.5	6/13	16.9	16.0	1/62	1.6
26	BROMINE	ORCHARD-L	10.00	M		21.00	11.5	42.4-55.7	1/52	22.1	13.0	1/13	1.4
27	BROMINE	TOMATO-LE	26.00	M		9.10	8.9	20.0-40.0	6/15	13.0	17.0	1/52	1.5
28	BROMINE	LIVER-BO		M	11.4	48.00	8.3	70.0-190.0	1/18	47.8	8.1	1/15	0.9
29	BROMINE	SPINACH-1	54.00	M	21.8	3.43	16.7	140.0-340.0	1/11	8.1	4.7	1/18	0.4
30	CADMIUM	OYSTER-TI	3.00	M	9.1	30.00	14.8	230.0-337.0	1/98	3.4	23.4	0/11	0.7
31	CADMIUM	WHEAT-FLO	32.00	B		27.00	18.5	120.0-1549	11/31	32.1	19.8	6/98	1.2
32	CADMIUM	RICE-FLOU	29.00	B		119.00	18.0	170.0-208.0	0/21	124.6	26.3	0/31	1.3
33	CADMIUM	ORCHARD-L	3.00	M	14.8	2.50	7.5	135.0-162.0	10/123	227.5	26.3	3/123	1.0
34	CADMIUM	TOMATO-LE		M		281.00	9.8	1.7-2.1	4/34	281.1	16.0	1/135	1.5
35	CADMIUM	PINE-NEED	270.00	B		1.43	5.8	3600-5000	3/17	1392.5	16.2	1/17	2.7
36	CADMIUM	LIVER1-BO	1.50	B	13.3	1400.00	5.4	87.0-151.0	1/19	141.9	14.2	1/19	0.9
37	CALCIUM	SPINACH-1	1500.00	M	15.3	190.00	5.9	111.3-129.7	13/105	146.0	9.6	3/105	0.9
38	CALCIUM	OYSTER-TI	190.00	M	14.3	148.00	8.6	1.7-3.3	5/36	2.9	6.2	3/36	2.5
39	CALCIUM	WHEAT-FLO	140.00	M		2.04	11.5	3600-5000	5/32	4172.2	8.8	3/32	3.2
40	CALCIUM	RICE-FLOU	2.09	H	1.0	3.00	1.5	111.3-129.7	7/66	119.4	10.4	4/66	2.0
41	CALCIUM	ORCHARD-L	3.00	M	4.9	4200.00	6.1	2.4-3.3	4/30	121.3	15.1	4/30	0.6
42	CALCIUM	TOMATO-LE	4100.00	M	4.8	122.00	8.6	0.8-1.3	7/28	37.7	4.5	4/28	4.0
43	CALCIUM	PINE-NEED	124.00	M	5.8	121.00	11.5	20.0-50.0	3/23	57.3	15.4	2/21	0.7
44	CALCIUM	LIVER1-BO	120.00	M	2.2	1.33	6.1	43.0-70.0	4/23	14.7	23.0	3/23	0.6
45	CALCIUM	LIVER2-BO	1.35	H		0.99	23.7	630.0-810.0	1/8	50.00	14.7	4/23	1.7
46	CALCIUM	SPINACH-1		M		38.00	14.2	2410-3000	2/23	713.1	14.00	2/23	1.4
47	CERIUM	ORCHARD-L	40.00	M		57.00	21.7	540.0-750.0	5/42	2673.4	50.00	3/42	2.5
48	CERIUM	ORCHARD-L		B		17.00	42.1		3/38	667.2	14.00	3/38	
49	CESIUM	TOMATO-LE		B		730.00	5.5		5/15	1.7	7.00	0/15	
50	CESIUM	LIVER1-BO	690.00	M	39.1	2680.00	5.5		5/21		24.9	3/21	
51	CHLORINE	ORCHARD-L	2700.00	B		650.00	12.3				35.9		
52	CHLORINE	LIVER1-BO	690.00	B		2.00	13.0						
53	CHROMIUM	OYSTER-TI	690.00	B	2.4								
54	CHROMIUM	BREWERS-Y	2.12	M									

No.	Element	Material
55	CHROMIUM	ORCHARD-L
56	CHROMIUM	TOMATO-LE
57	CHROMIUM	PINE-NEED
58	CHROMIUM	LIVER1-BO
59	CHROMIUM	SPINACH-1
60	CHROMIUM	OYSTER-TI
61	COBALT	ORCHARD-L
62	COBALT	TOMATO-LE
63	COBALT	LIVER1-BO
64	COBALT	SPINACH-1
65	COPPER	OYSTER-TI
66	COPPER	WHEAT-FLO
67	COPPER	RICE-FLOU
68	COPPER	ORCHARD-L
69	COPPER	TOMATO-LE
70	COPPER	PINE-NEED
71	COPPER	LIVER1-BO
72	COPPER	LIVER2-BO
73	COPPER	SPINACH-1
74	EUROPIUM	ORCHARD-L
75	FLUORINE	ORCHARD-L
76	GOLD	ORCHARD-L
77	HAHNIUM	LIVER1-BO
78	IODINE	ORCHARD-L
79	IODINE	ORCHARD-L
80	IODINE	LIVER1-BO
81	IRON	OYSTER-TI
82	IRON	WHEAT-FLO
83	IRON	RICE-FLOU
84	IRON	ORCHARD-L
85	IRON	TOMATO-LE
86	IRON	PINE-NEED
87	IRON	LIVER1-BO
88	IRON	LIVER2-BO
89	IRON	SPINACH-1
90	LANTHANU	ORCHARD-L
91	LANTHANU	LIVER1-BO
92	LEAD	OYSTER-TI
93	LEAD	ORCHARD-L
94	LEAD	TOMATO-LE
95	LEAD	PINE-NEED
96	LEAD	LIVER1-BO
97	LITHIUM	SPINACH-1
98	LUTECIUM	ORCHARD-L
99	MAGNESIU	OYSTER-TI
100	MAGNESIU	WHEAT-FLO
101	MAGNESIU	ORCHARD-L
102	MAGNESIU	TOMATO-LE
103	MAGNESIU	PINE-NEED
104	MAGNESIU	LIVER1-BO
105	MAGNESIU	SPINACH-1
106	MAGNESIU	OYSTER-TI
107	MAGNESIU	WHEAT-FLO
108	MANGANES	WHEAT-FLO

Continued on next page

Table II. Continued

OBS	ANALYTE	REFERENCE MATERIAL	NIST VALUE	UN	NIST RSDR	CONSEN VALUE	CONS RSDR	CONSENSUS RANGE	CONSENSUS OMIT/LABS	IUPAC VALUE	IUPAC RSDR	IUPAC OMIT/LABS	IUPAC HORRAT
109	MANGANES	RICE-FLOU	20.10	M	2.0	20.50	4.9	19.1-22.4	1/17	20.5	4.9	1/17	0.5
110	MANGANES	ORCHARD-L	91.00	M	4.4	89.00	5.6	76.0-103.0	20/162	89.2	8.5	7/162	1.1
111	MANGANES	TOMATO-LE	238.00	M	2.9	224.00	5.8	197.0-252.0	3/50	224.8	6.5	2/50	1.1
112	MANGANES	PINE-NEED	675.00	M	2.2	650.00	10.8	430.0-738.0	12/150	663.4	8.5	6/150	1.1
113	MANGANES	LIVER1-BO	10.30	M	9.7	10.20	6.9	8.4-12.0	12/150	10.2	8.5	6/150	0.8
114	MANGANES	LIVER2-BO	9.90	M	8.1	9.80	4.0	9.1-10.8	13/100	165.4	5.1	1/35	0.7
115	MANGANES	SPINACH-1	165.00	B	3.6	164.00	3.7	155.0-178.0	13/50	156.3	12.5	5/50	0.5
116	MERCURY	ORCHARD-L	155.00	B	9.7	155.00	9.0	122.0-190.0	1/50	16.8	12.4	2/100	0.4
117	MERCURY	LIVER1-BO	16.00	B	12.5	16.40	9.8	13.7-20.0	7/34	404.4	14.9	4/50	0.4
118	MOLYBDEN	WHEAT-FLO	400.00	B		420.00	7.1	380.0-470.0	9/34	768.3	154.7	0/34	8.1
119	MOLYBDEN	ORCHARD-L	300.00	M	33.3	290.00	24.1	200.0-410.0	6/10		127.3	3/12	7.2
120	MOLYBDEN	TOMATO-LE		M		0.53	17.0	0.4-0.6	5/10	1.1	115.3	3/10	5.4
121	MOLYBDEN	PINE-NEED		M		0.15	33.3	0.1-0.2	7/65	1.1	16.3	1/65	1.0
122	MOLYBDEN	LIVER1-BO		M		3.20	12.5	2.3-4.1	3/10	3.0	16.3	0/12	2.2
123	NEODYMIU	ORCHARD-L	3.40	B		510.00	26.5	320.0-765.0	10/70	1.4	21.6	5/17	2.6
124	NICKEL	OYSTER-TI		M	18.4	1.01	8.9	0.9-1.1	3/17		18.9	7/20	1.6
125	NICKEL	ORCHARD-L	1.03	M	15.4	1.30	15.4		4/28	2.6	18.1	0/28	2.3
126	NICKEL	TOMATO-LE	1.30	M		1.30	15.4	1.7-1.8	2/17		35.1	6/17	2.5
127	NICKEL	PINE-NEED		M		0.50	50.0		3/17	368.6	21.8	0/16	2.9
128	NICKEL	LIVER1-BO	3.50	M	1.8	160.00	12.5	50.0-270.0	4/28	96.5	96.6	2/63	1.9
129	NITROGEN	SPINACH-1		H		5.60	10.6	4.-7.5	2/17	5.2	25.6	1/16	3.7
130	PHOSPHOR	ORCHARD-L	6.00	M		7600.00	5.9	6530.-8200	1/16		14.6	2/63	1.9
131	PHOSPHOR	OYSTER-TI	8100.00	M	4.8	5.20	6.9	4.-8.2	1/63	7392.1	8.1	7/31	1.9
132	PHOSPHOR	WHEAT-FLO		H	5.9	1390.00	6.6	1360.-1450	7/31	1392.9	7.7	3/29	2.2
133	PHOSPHOR	ORCHARD-L	2100.00	M	16.7	2000.00	2.2	1560.-2400	4/29	2012.0	10.1	3/25	2.3
134	PHOSPHOR	TOMATO-LE	3400.00	M		3370.00	6.5	2800.-3900	3/25	3313.7	12.5	2/27	4.7
135	PHOSPHOR	PINE-NEED	1200.00	M		1190.00	10.6	1000.-1410	3/27	1220.0	13.6	1/14	3.3
136	PHOSPHOR	LIVER1-BO		H	3.6	5240.00	5.9	4530.-5700	2/14	5211.1	21.6	1/15	1.8
137	PHOSPHOR	SPINACH-1	5500.10	H	0.5	1.13	7.5	0.9-1.3	3/15	0.9	13.2	0/11	0.9
138	POTASSIU	OYSTER-TI	0.97	M	2.9	0.93	8.6	0.8-1.0	2/11	1306.1	5.8	1/97	1.4
139	POTASSIU	WHEAT-FLO	1360.00	M	1.8	1300.00	6.9	1130.-1500	14/87	1093.1	11.6	1/97	0.9
140	POTASSIU	RICE-FLOU	1120.00	H	2.0	1050.00	4.9	900.-1150	8/26		14.2	3/36	1.7
141	POTASSIU	ORCHARD-L	1.47	M	0.7	1.44	5.4	1.3-1.6	13/75	3946.9	9.4	5/26	0.7
142	POTASSIU	TOMATO-LE	4.46	H	5.4	4.44	8.4	3.8-4.8	5/31		18.4	3/75	0.3
143	POTASSIU	PINE-NEED	3700.00	H	0.8	3670.00	6.1	2700.-5100	6/74	3.6	31.8	5/74	0.8
144	POTASSIU	LIVER1-BO	0.97	H	0.5	0.98	4.2	0.8-1.1	6/8	4.5	35.0	6/8	0.7
145	POTASSIU	SPINACH-1	3.56	H	2.0	3.56	11.1	3.5-3.9	1/67	17.4	30.1	1/67	0.7
146	POTASSIU	OYSTER-TI		H	8.3	4.50	10.5	3.8-5.3	2/27	18.4	10.6	4/27	0.6
147	RUBIDIUM	ORCHARD-L	12.00	M	0.6	11.40	14.5	10.0-14.8	0/9	12.3	11.2	0/9	5.0
148	RUBIDIUM	TOMATO-LE	16.50	M	5.5	17.30	15.1	15.1-22.2	0/9	107.2	14.2	3/9	12.0
149	RUBIDIUM	LIVER1-BO	18.40	B	1.7	18.40	10.0	15.2-21.2	1/20	173.4	9.1	1/21	3.2
150	RUBIDIUM	SPINACH-1	12.10	B		11.50	17.0	10.0-12.7	1/21		18.3	0/34	
151	SAMARIUM	ORCHARD-L		B		114.00	15.3	88.0-150.0	0/38	165.8	31.8	0/38	
152	SCANDIUM	TOMATO-LE	130.00	M	23.8	173.00	33.6	138.0-220.0	2/27	2.1	35.0	3/107	
153	SCANDIUM	LIVER1-BO		M	18.2	166.00	9.6	150.0-180.1	0/9	383.4	30.6	9/187	
154	SCANDIUM	SPINACH-1		B	25.5	0.90	10.9		1/20	81.7	10.9	0/12	
155	SCANDIUM	OYSTER-TI		B	12.1	1.03	7.8	0.7-2.4	1/21		19.3	2/22	0.7
156	SELENIUM	SPINACH-1		M	9.1	380.00	12.3	280.0-480.0	1/38	135.9	130.1	15/107	0.3
157	SELENIUM	WHEAT-FLO	160.00	M		81.00	7.3	55.0-110.0	15/107	1110.2	74.1	0/10	0.8
158	SELENIUM	RICE-FLOU	400.00	M		40.00	35.0	24.0-66.0	15/187	81.4	77.2	2/22	0.7
159	SELENIUM	ORCHARD-L	80.00	B		550.00	21.0	475.8-750.0	0/12				0.6
160	SELENIUM	LIVER1-BO	1.10	B		62.00	20.0	40.0-100.0	5/22				5.0
161	SILICON	SPINACH-1		B									12.0
162	SILVER	LIVER1-BO	60.00	B									3.2

No.	Element	Sample	Cert.	Code	CV	Mean	CV	Range	n/N	Consensus	CV	n/N	HR
163	SODIUM	OYSTER-TI	5100.00	M	5.9	4950.00	4.4	4600-5300	2/12	4882.0	6.4	1/12	1.4
164	SODIUM	ORCHARD-L	82.00	M	7.3	89.00	16.9	74.0-140.0	16/65	95.9	33.9	3/65	4.1
165	SODIUM	TOMATO-LE		M		470.00	23.4	326.0-650.0	5/24	549.7	37.2	1/24	5.1
166	SODIUM	PINE-NEED		M		50.00	60.0	18.0-105.0	5/18	49.6	63.2	1/18	7.1
167	SODIUM	LIVER1-BO	2430.00	M	5.3	2395.00	7.0	1940-3010	5/64	2401.4	10.4	2/64	2.1
168	SODIUM	SPINACH-1		M		1.42	6.9	1.2-1.6	1/18	1.1	7.7	1/18	2.3
169	STRONTIU	OYSTER-TI	10.36	M	5.4	10.10	8.6	8.6-11.0	12/65	36.0	13.7	1/65	0.7
170	STRONTIU	ORCHARD-L	37.00	M	2.7	36.00	8.3	28.0-44.2	3/14	44.5	18.2	1/14	1.5
171	STRONTIU	TOMATO-LE	44.90	B	4.2	42.00	11.9	35.6-54.0	4/9	54.1	32.5	1/9	2.2
172	STRONTIU	PINE-NEED	4.80	M		5.00	8.0	4.4-5.5	3/9	4.5	66.6	1/9	2.6
173	STRONTIU	LIVER1-BO	140.00	M	2.3	170.00	41.2	100.0-300.0	0/8	274.3	15.9	0/8	1.8
174	STRONTIU	SPINACH-1	87.00	M		80.00	6.1	72.5-87.0	0/9	80.4	6.7	0/9	0.7
175	SULFUR	WHEAT-FLO		M		1810.00	4.4	1623-1980	5/38	1810.5	8.1	1/38	3.4
176	SULFUR	RICE-FLOU		M		1350.00	4.4	1256-1400	4/11	1347.2	18.1	0/11	1.6
177	SULFUR	ORCHARD-L	1900.00	B	2.2	2040.00	11.8	1660-2600	2/12	1996.1	113.3	1/12	1.5
178	SULFUR	CITRUS-LE	4070.00	B		4080.00	6.5	3822-4400	3/13	4097.1	111.4	1/13	5.3
179	SULFUR	TOMATO-LE		B		6200.00	12.7	1200-1500	4/10	6204.8	44.4	3/10	4.0
180	SULFUR	PINE-NEED		B		1320.00	10.8	5500-9000	3/16	1310.0	12.7	1/16	3.7
181	SULFUR	LIVER1-BO		M		7900.00	20.7	3600-4860	4/23	7541.9	16.4	0/23	1.8
182	SULFUR	SPINACH-1		B		4350.00	20.7	9.0-85.0	5/23	4213.4	33.2	0/23	5.8
183	TERBIUM	ORCHARD-L		B			35.0	40.0-85.0	2/9	21.6	111.1	1/9	4.4
184	THORIUM	ORCHARD-L		B			10.3	180.0-375.0	2/46	53.6	16.9	1/46	0.5
185	TIN	ORCHARD-L		M			7.4	6.6-30.0	9/9	469.1	29.9	3/9	1.6
186	TITANIUM	ORCHARD-L	64.00	B	9.4	58.00	22.0	441.0-490.0	2/14	19.8	14.9	1/14	2.1
187	URANIUM	BREWERS-Y		B		290.00	16.7	2.3-3.1	3/13	28.6		0/13	4.5
188	URANIUM	ORCHARD-L		B		470.00	13.8	25.0-34.3	2/21	2.6	5.2	0/21	1.1
189	VANADIUM	OYSTER-TI	29.00	M	17.2	2.70	13.3	300.0-700.0	3/19	505.6	11.6	2/19	1.1
190	VANADIUM	ORCHARD-L	2.30	B	4.3		2.8	0.9-1.5	1/10	1.1	20.4	1/10	0.5
191	VANADIUM	TOMATO-LE		M		500.00	3.8	248.0-470.0	24/212	354.8		8/212	1.3
192	VANADIUM	PINE-NEED		B		390.00	4.6	33.0-66.2	11/56	167.0		3/56	1.3
193	VANADIUM	LIVER1-BO		B		58.00	13.4	20.0-34.0	19/210	24.9	4.4	5/210	2.4
194	VANADIUM	SPINACH-1		B		25.00	3.3	805.0-887.6	4/47		12.2	1/29	0.6
195	YTTERBIU	ORCHARD-L		M				18.7-11.3	1/8		56.5	1/8	
196	ZINC	OYSTER-TI	852.00	M	1.6	854.00	2.8	59.0-70.0		834.6			1.4
197	ZINC	WHEAT-FLO	10.60	M	9.4	10.60	3.6	19.0-32.0		11.0			3.9
198	ZINC	RICE-FLOU	19.40	M	5.2	19.70	4.6	52.0-71.0					
199	ZINC	BREWERS-Y		M		65.00	8.6	51.0-87.0		65.1			
200	ZINC	ORCHARD-L	25.00	M	12.0	25.00	13.4	112.0-150.0		62.2			
201	ZINC	TOMATO-LE	62.00	M	9.7	67.00	3.3	130.1-30.1		130.1			
202	ZINC	PINE-NEED		M		67.00	8.0	42.0-60.1		121.7			
203	ZINC	LIVER1-BO	130.00	M	10.0	130.00	5.1	0.4-3.8		51.3			
204	ZINC	LIVER2-BO	123.00	M	6.0	122.00	3.3			2.0			
205	ZINC	SPINACH-1	50.00	M	4.0	50.00	4.0						
206	ZIRCONTU	ORCHARD-L		M		2.00	55.0						

SOURCE: Compiled from ref. 5: identification information, concentration estimates, and variability parameters as calculated from the NIST certification data, the consensus procedure, and the IUPAC-1987 protocol. Only analytes with 8 or more quantitative literature values given were used.

Table III. Percent Differences of Consensus (CONS) and IUPAC Mean Values
from Certified Concentrations of Analytes in Biological NIST Standard
Reference Materials; Percent Outliers Removed by Compilers and
by Applying IUPAC Protocol to Original Data; and HORRAT
Values of Relative Standard Deviations for Uncertainties

Analyte	SRM	NIST concn[a]		N[b]	Difference, %		Outliers, %		HORRAT		
					CONS	IUPAC	CONS	IUPAC	NIST	CONS	IUPAC
Aluminum	Spinach	870	M	19	6.9	16	32	0	1.0	1.9	5.1
	Pine Nd	545	M	29	6.4	6.1	17	3	0.9	1.9	3.4
Antimony	Orchard	2.9	M	85	0.0	0.0	11	2	0.7	0.7	0.8
Arsenic	Oyster	13.4	M	21	3.0	4.5	20	0	1.3	0.8	1.2
	Rice Fl	410	M	27	1.0	0.0	11	7	0.7	0.3	0.4
	Spinach	150	M	19	2.0	2.0	5	5	1.6	0.6	0.6
	Orchard	10	M	199	7.0	10	9	3	1.8	1.1	1.4
	Tomato	270	B	26	6.3	7.0	8	0	1.1	0.8	1.0
	Pine Nd	210	B	20	1.4	3.3	8	0	0.9	0.4	0.5
	Bovn Li	55	B	64	0.0	1.8	17	16	0.4	0.4	0.5
Berylium	Orchard	27	B	9	11.0	29	22	0	1.3	1.2	3.0
Boron	Orchard	33	B	40	0.0	1.8	10	2	1.1	1.0	1.3
Cadmium	Oyster	3.50	M	18	2.0	2.0	6	6	0.8	0.4	0.4
	Wheat F	32	B	11	6.2	0.0	9	0	0.8	0.6	0.9
	Rice Fl	29	B	9	6.9	0.0	22	11	0.5	0.5	0.7
	Orchard	110	B	98	8.2	14	11	6	0.4	0.8	1.2
	Bovn Li	270	B	123	4.1	4.4	8	2	0.6	0.3	5.2
Calcium	Oyster	1500	M	17	6.7	7.3	18	6	2.4	1.6	2.7
	Wheat F	190	M	19	0.0	1.1	11	5	0.7	0.8	0.9
	Spinach	1.35	H	28	1.5	2.2	25	4	0.6	1.6	4.0
	Orchard	2.09	H	105	2.4	2.4	12	3	0.4	1.6	2.5
	Tomato	3.00	H	36	5.7	4.3	14	3	0.3	2.4	3.2
	Rice Fl	140	M	16	5.7	4.3	12	6	1.8	0.7	1.0
	Pine Nd	4100	M	32	2.4	1.7	14	3	1.1	1.9	2.5
	Bovn Li	124	M	66	1.6	4.0	11	5	0.6	1.5	1.9
	Bovn LiA	120	M	30	0.8	0.8	13	13	0.8	0.5	0.5
Chromium	Oyster	690	B	15	5.8	3.2	20	7	2.3	0.7	1.5
	Brewr Y	2.12	M	21	5.7	19	24	0	0.2	0.9	2.4
	Spinach	4.6	M	40	6.5	64	38	0	0.5	0.9	7.4
	Orchard	2.6	M	108	0.0	0.0	14	2	0.8	0.8	1.3
	Tomato	4.5	M	25	11	13	24	0	0.9	1.0	1.8
	Pine Nd	2.6	M	21	0.0	1.2	24	0	0.6	0.6	2.0
	Bovn Li	88	B	71	32	364	28	1	0.6	1.9	6.0
Copper	Oyster	63	M	25	0.0	0.8	16	8	0.6	0.4	0.7
	Wheat F	2.0	M	24	2.0	0.5	17	0	1.0	0.4	0.8
	Rice Fl	2.2	M	20	5.5	4.5	10	0	1.0	0.5	0.8
	Spinach	12	M	50	1.7	2.5	8	4	1.5	0.5	0.7
	Orchard	12	M	182	0.0	1.7	8	3	0.8	1.1	1.3

Continued on next page

Table III. Continued

Analyte	SRM	NIST concn [a]	N [b]	Difference, %		Outliers, %		HORRAT		
				CONS	IUPAC	CONS	IUPAC	NIST	CONS	IUPAC
Copper	Tomato	11	M	58	0.0	0.9	12	2	0.8	1.6	2.3
	Pine Nd	3.0	M	49	0.0	17	18	4	0.7	1.0	3.3
	Bovn Li	193	M	183	1.6	2.1	10	3	0.7	0.6	1.0
	Bovn LiA	158	M	34	5.7	10	12	1	0.6	1.3	2.4
Iron	Oyster	195	M	26	0.0	0.5	15	4	2.3	0.8	1.0
	Wheat F	18.3	M	21	2.7	3.8	17	10	0.5	0.6	0.7
	Spinach	550	M	43	1.8	1.5	16	9	0.6	0.9	1.1
	Orchard	300	M	165	4.7	5.7	11	4	1.0	1.4	1.8
	Tomato	690	M	53	16	19	19	0	0.6	3.2	5.7
	Pine Nd	200	M	44	7.5	9.0	16	5	0.7	2.0	3.2
	Bovn Li	268	M	152	1.1	0.4	11	5	0.4	1.0	1.4
	Rice Fl	8.7	M	15	8.0	5.7	7	0	0.6	1.3	1.5
	Bovn LiA	194	M	31	20	23	19	0	1.4	1.5	2.9
Lead	Oyster	480	B	21	0.0	1.1	10	5	0.5	0.4	0.4
	Spinach	1.2	M	33	0.8	16	15	3	1.1	1.4	2.8
	Orchard	45	M	144	2.2	2.2	12	5	0.7	0.8	1.2
	Tomato	6.3	M	46	6.3	4.8	11	2	0.4	1.1	1.4
	Pine Nd	10.8	M	35	0.9	0.0	14	3	0.4	0.4	1.0
	Bovn Li	340	B	79	2.9	6.2	11	1	1.2	0.8	1.0
Magnesium	Oyster	1280	M	13	3.9	2.3	8	0	1.3	1.4	1.9
	Orchard	6200	M	80	2.4	2.2	11	2	0.7	1.5	1.95
	Bovn Li	604	M	58	0.7	1.0	14	9	0.2	1.2	1.3
Manganese	Oyster	17.5	M	26	2.9	2.3	15	8	0.7	1.0	1.4
	Wheat F	8.5	M	23	1.2	1.2	9	0	0.5	0.6	1.0
	Rice Fl	20.1	M	17	2.0	2.0	6	6	0.2	0.5	0.5
	Spinach	165	M	50	0.6	0.0	22	12	0.5	0.5	0.7
	Orchard	91	M	162	2.2	2.2	14	4	0.5	0.7	1.0
	Tomato	238	M	50	5.9	5.5	12	4	0.4	0.8	1.1
	Pine Nd	675	M	39	3.7	1.8	8	13	0.4	1.5	1.0
	Bovn Li	10.3	M	150	1.0	1.0	8	4	0.9	0.6	1.2
	Bovn LiA	9.9	M	35	0.0	1.0	3	0	0.7	0.4	0.5
Mercury	Orchard	155	B	100	0.0	0.6	13	2	0.5	0.4	0.6
	Bovn Li	16	B	50	2.5	5.0	14	8	0.4	0.3	0.4
Molybdenum	Orchard	300	B	34	3.3	156	26	9	1.7	1.3	8
Nickel	Oyster	1.03	M	12	1.9	9.7	25	0	1.2	0.6	1.4
	Orchard	1.3	M	70	0.0	7.7	14	9	1.0	1.0	1.2
Nitrogen	Orchard	2.76H		17	1.4	1.8	6	0	0.5	0.4	0.6
Phosphorus	Spinach	0.550H		27	6.2	8.7	11	4	0.8	1.4	1.8
	Orchard	0.210H		63	4.8	4.2	11	3	1.0	1.8	2.5
	Tomato	0.340H		31	0.9	2.5	10	3	1.2	1.4	1.9
	Pine Nd	0.120H		29	0.8	1.7	14	3	3.0	1.7	2.7

Continued on next page

Table III. Continued

Analyte	SRM	NIST concn[a]	N[b]	Difference, %		Outliers, %		HORRAT		
				CONS	IUPAC	CONS	IUPAC	NIST	CONS	IUPAC
Potassium	Oyster	0.969H	14	4.1	4.0	14	14	0.1	1.9	1.9
	Wheat F	0.136H	15	4.4	3.7	20	7	0.5	1.3	1.9
	Rice Fl	0.112H	11	6.2	2.4	18	0	0.3	1.6	2.2
	Spinach	3.56 H	31	0.0	0.6	19	10	0.8	1.3	1.8
	Orchard	1.47 H	97	2.0	2.0	14	1	0.5	1.3	2.2
	Tomato	4.46 H	36	0.4	2.9	22	3	0.2	1.7	4.2
	Pine Nd	0.370H	26	0.8	6.7	12	4	1.2	1.8	4.6
	Bovn Li	0.97 H	75	1.0	12	17	0	0.0	1.5	3.3
Rubidium	Oyster	4.45 M	8	1.1	1.1	12	12	0.2	0.9	0.9
	Spinach	12.1 M	8	5.2	1.6	25	12	0.2	0.7	1.7
	Orchard	12 M	74	5.3	4.3	8	7	0.8	1.0	1.0
	Tomato	16.5 M	8	4.6	4.6	12	12	0.1	1.4	1.4
	Bovn Li	18.3 M	67	0.5	0.5	13	6	0.5	0.6	0.9
Selenium	Oyster	2.1 M	21	1.0	1.0	10	0	1.7	0.7	0.8
	Wheat F	1.1 M	38	6.4	8.2	8	0	1.2	0.5	0.7
	Rice Fl	400 B	34	5.0	4.3	0	0	1.4	0.7	0.7
	Orchard	80 B	107	1.2	2.5	10	3	0.5	0.5	0.8
	Bovn Li	1.1 M	187	0.9	0'.0	8	5	0.6	0.5	0.6
Sodium	Oyster	5100 M	12	2.9	4.3	17	8	1.3	1.0	1.4
	Orchard	82 M	65	8.5	17	25	5	0.9	2.0	4.1
	Bovn Li	2430 M	64	1.4	1.2	8	3	1.1	1.7	2.1
Strontium	Oyster	10.36M	9	2.5	2.5	11	11	0.5	0.6	0.7
	Spinach	87 .M	8	8.0	7.6	12	12	0.3	0.8	0.8
	Orchard	37 M	65	2.7	2.7	18	6	0.3	0.9	1.5
	Tomato	44.9 M	14	6.5	1.6	12	12	0.1	1.3	2.0
	Pine Nd	4.8 M	9	4.2	17	22	11	0.3	0.6	2.6
Sulfur	Citrus	4070 M	11	0.2	0.7	36	0	0.5	1.0	1.4
Thorium	Orchard	64 B	16	9.4	16	19	0	0.4	0.9	1.8
Uranium	Orchard	29 B	23	0.0	0.7	9	4	0.6	0.4	0.5
Vanadium	Oyster	2.3 M	9	17	13	11	0	0.3	0.5	1.2
Zinc	Oyster	852 M	26	0.2	2.1	19	4	0.3	0.5	1.0
	Wheat F	10.6 M	21	0.0	3.8	19	0	0.8	0.3	1.0
	Rice Fl	19.4 M	19	1.5	1.5	16	5	0.5	0.3	0.5
	Spinach	50 M	49	0.0	2.6	9	2	0.4	0.9	1.4
	Orchard	25 M	212	0.0	6.4	11	4	1.2	0.8	1.3
	Tomato	62 M	56	1.6	0.0	20	5	1.1	0.8	1.3
	Bovn Li	130 M	210	0.0	0.3	9	2	1.3	0.7	1.0
	Bovn LiA	123 M	29	0.8	0.8	7	0	0.8	0.4	0.6

[a]Concentration units: parts per hundred (10^{-2}), H; parts per million (10^{-6}), M; and parts per billion (10^{-9}), B.

[b]Number of quantitative values listed in compilation before removal of extreme values.

value for the consensus values is 0.8, slightly better than expected from the Horwitz curve (1.0), ranging from 0.3 for strontium and rubidium to 1.5 for phosphorus. The number of extreme values removed by the compilers to achieve the consensus value (from Table IV) averaged 14%, ranging from 7% for selenium to 25% for chromium. The scatter of values within a specific analyte of interest can be determined from Table III.

Discussion

Our interest in the NIST certified values lies not so much in their absolute accuracy but in the clues that the use of the certified value statement can supply as to the practical reliability of analytical chemistry as a measurement science.

NIST is charged by law (11) to prepare and distribute standard samples, which are now called Standard Reference Materials (SRMs). Consequently, the certified concentration value assigned by NIST is the "true value" used in the definition of [in]accuracy -- the difference from the true value. Descriptions published by NIST of the work that has gone into the choice and validation of the certified value, the infrequent appearance of papers disputing an assigned value, and the usual closeness to the certified values and the single-peak distribution of subsequently published literature values attest to the scientific soundness of the NIST assignments.

There is less assurance with respect to the meaning of the uncertainty statement attached to the NIST certified values. Taylor (12) indicates that the uncertainty limits accompanying certified analytes are for the "95% level of confidence and include allowances for the uncertainties of known sources of systematic error as well as the random error of measurement." Unfortunately this definition does not permit a practical interpretation of the stated uncertainty since the quoted statement raises more statistical questions than it answers. For example, if "95% limits of confidence" is meant to apply to "95% confidence limits," these depend on the number of observations entering into the mean, which is not given for NIST values. Furthermore, uncertainties presented as (± values) cannot properly depict unsymmetrical distributions, which would be formed if systematic error, which is always one-sided, is included. NIST also has difficulties in formulating a definition since different certificates provide different definitions. Nevertheless, in order to avoid extensive exploration of the meaning of uncertainty as applied to certified values, the uncertainty value given in the compilation (5) for the NIST certified value is treated as a standard deviation so that other users may reinterpret the value according to their own understanding.

That these uncertainty values attached to the certified values approximate standard deviations is suggested by Figure 1, where the uncertainties, considered as standard deviations, are transformed to relative standard deviations (by dividing by the mean and multiplying by 100) and plotted against (-log concentration). The similarity of Figure 1 to Figures 2 and 3 of the paper by Boyer et al. (10) is striking. All the figures use the same variables and range of relative standard deviations between laboratories (RSD_R) plotted against (-log concentration) for many of the same analytes

Table IV. Averages of Absolute Differences Between NIST
Certified Values and Consensus Values as Developed by Compilers (5),
Expressed as Percentages of NIST Values (Deltas); Average
Horwitz Ratios (HORRATs) of Consensus Values; and Average
Percent Outliers Removed to Achieve Consensus Values[a]

Element	Concentration Range			No. of Data Sets	Av. Delta	Av. HORRAT	Av. % Outliers
Arsenic	10	-	300	M	7	2.9	1.1	11
Cadmium	0.03-		4	M	5	5.5	0.6	11
Calcium	0.01-		3	H	9	3.0	1.0	14
Chromium	0.09-		2	M	7	8.7	0.8	25
Copper	2	-	200	M	9	1.8	0.9	12
Iron	9	-	700	M	9	6.9	0.9	15
Lead	0.3	-	45	M	6	2.2	0.7	12
Magnesium	0.06-		0.6	H	3	2.3	0.8	11
Manganese	10	-	700	M	9	2.2	0.5	11
Phosphorus	0.1	-	0.6	H	4	3.2	1.5	12
Potassium	4	-	0.1	H	8	2.4	0.5	17
Rubidium	4	-	18	M	5	2.4	0.3	14
Selenium	0.01-		2	M	5	2.9	1.1	7
Sodium	0.01-		0.5	H	3	4.3	1.1	17
Strontium	5	-	90	M	5	4.8	0.3	15
Zinc	10	-	900	M	8	0.5	0.8	14
Av.						3.5	0.8	14

[a]Only elements with at least 3 certified values in Table III were
used for the calculations.

in comparable biological matrices. The 2 sets of figures
qualitatively show the same expansion of RSD_R with decreasing
concentration (increasing -log C) and the same infrequent scatter of
values above the upper limit of twice the Horwitz curve.

Comparison of Consensus and IUPAC Values with NIST Values. The
relationship between the NIST values (both certified and
informational) and the consensus values is given in Figure 2. The
negative logarithm of the consensus values is plotted against the
negative logarithm of the NIST values in order to present in a
single plot the concentration range covered by the SRM analyte
values over 7 decades. The absence of substantial deviations from
linearity in the 117 pairs of values from the database supports the
assumption of the soundness of the NIST values.

Figure 3 is a similar log-log plot of the analyte values
calculated by the extended IUPAC-1987 protocol against the NIST
values. At least 3 substantial deviations from the straight line
relationship of the consensus values are apparent.

Figures 2 and 3, covering 7 decades of concentration values on
a logarithmic scale, are not sensitive to the numerical differences
between the NIST values and the consensus values established by the
compilers, or to the differences between the NIST values and the
values calculated by IUPAC-1987. The absolute numerical differences
between the NIST certified values and the consensus values (or
between the NIST certified values and the IUPAC-1987 values) as
percentages of the NIST values are shown in Figure 4 for the 117
certified analyte values. The two histograms are roughly comparable
except for a long tail in the IUPAC set, for which the 4 values
given numerically are off the scale. Approximately 2/3 of both sets
of values (1 standard deviation) show a difference of less than 4.0%
from the NIST value; approximately 95% of the consensus differences
and approximately 90% of the IUPAC differences are less than 10%.
All of the highest consensus differences (maximum 31%) can be
considered analytically reasonable; the 3 highest differences
calculated by the IUPAC protocol are analytically unreasonable (64%,
Cr in Spinach; 156%, Mo in Orchard Leaves; and 364%, Cr in Bovine
Liver). This confirms that even the extended IUPAC-1987 outlier
removal procedure is inadequate to trim the values to approach the
NIST certified values, when compared with the analytical judgment
exercised by the compilers. However, the IUPAC-1987 outlier removal
protocol was not intended to apply to data grossly contaminated with
outliers. The protocol was intended to apply to "valid data"
defined (8; §3.1) as "those data that would be reported as resulting
from the normal performance of laboratory analyses; they are not
marred by method deviations, instrument malfunctions, unexpected
occurrences during performance, or by clerical or typographical
errors."

Outliers. The mean (consensus) values obtained from the examination
of the data reported in the literature for the certified analytes
can depend greatly on the procedure used to remove extreme values,
although it becomes less influential as the number of values in the
set examined increases, particularly when the number of values
exceeds 100. Although the compilers published all the data without

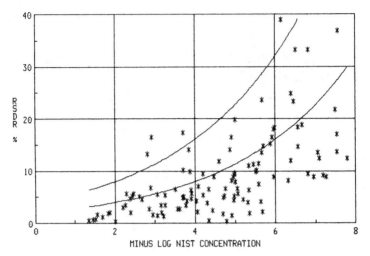

Figure 1. The uncertainties of the certified analytes of the
NIST biological SRMs transformed to relative standard deviations
(RSDR) and plotted against (-log certified concentration). The
lower line is the Horwitz curve and the upper line is twice that
curve.

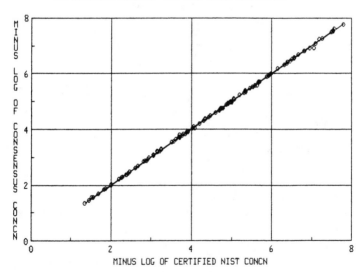

Figure 2. The negative logarithms of the consensus mean values
of certified analytes plotted against the negative logarithms of
the NIST values.

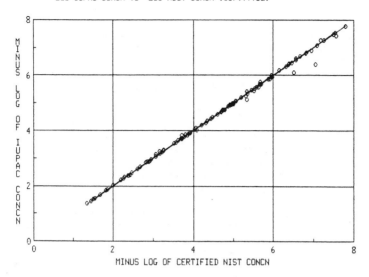

Figure 3. The negative logarithms of the analyte mean values calculated by the extended IUPAC-1987 protocol plotted against the negative logarithms of the NIST values.

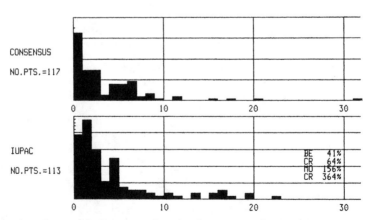

Figure 4. Upper histogram: The absolute numerical differences between the NIST certified values and the consensus mean values as percentages of the NIST values for the 117 certified analyte values. Lower histogram: Same as the upper histogram, but using mean values generated by the IUPAC-1987 protocol in place of consensus mean values. The vertical scales are different in the 2 histograms, but the horizontal % scales are the same.

censoring, they used a rather liberal approach to remove values,
i.e., they removed clearly unacceptable values on analytical grounds
as well as values outside 2 standard deviations from the mean of the
trimmed data; then the mean and standard deviation of the remaining
values were calculated and presented as the consensus mean and
consensus standard deviation. This procedure resulted in the
removal of some values by the compilers from almost every data set
as shown in the lower histogram of Figure 5 -- frequency of outlier
removal from an analyte-SRM data set against percent outliers
removed by the compilers. From the 117 certified values, 16 (13.7%)
data sets had more than 22.2% (maximum 39%) of the values removed by
the compilers. The significance of the 22.2% cutoff point is that
the IUPAC-1987 protocol abandons outlier removal if more than 2/9
(22.2%) of the values must be removed from method-performance
(collaborative) studies. In such cases, either the estimate of the
precision parameters must be considered acceptable with fewer
outliers removed, as providing a "ceiling" parameter, or the method
must be improved.

The distribution of the percent outliers removed by the
IUPAC-1987 protocol is shown in the upper histogram of Figure 5.
The pattern is completely different from the outlier removal pattern
of the consensus values, as expected.

Although different reviewers probably would not select the same
values from all analyte-SRM combinations for removal, the choices of
the compilers appear to be reasonable and fairly consistent.
Although we reviewed only superficially the pattern of outlier
removal by the compilers, certain values, from the same or
similar studies by the same type of method, for example, X-ray
fluorescence and optical emission spectroscopy, applied to a number
of different elements in the same SRM and to the same element in
different SRMs always seemed to appear near the extremes of the
distribution. For the present purpose, it was not necessary to
review the original documents to determine possible sources of the
systematic error or to determine if the excursions are inherent in
the methods themselves. Undoubtedly some explanations may be
discovered for systematic errors of a single analyte in an SRM --
for example, if different methods of dissolution of an SRM were
considered as independent observations when actually the same
calibration curve may have been used to obtain all of the values in
a study.

In any case, as shown in the lower histogram of Figure 5, only
1 data set from the 117 certified analyte-matrix combinations
examined (of those having at least 8 values) had no outliers removed
by the compilers; the average proportion of outliers removed was 14%
(mode, 11%), with a maximum of 38% in the case of chromium in SRM
Spinach. The existence of such a large proportion of outliers in
analytical work published in peer-reviewed journals was not thought
possible (13).

The outlier removal procedures of ISO (2), the IUPAC-1987
protocol (8), and the AOAC Guidelines (14) remove extreme values
only at the alpha (probability of falsely removing a good value)
= 0.01 level. Youden (15), however, had no qualms regarding removal
of values that would distort "the picture that would occur from
keeping a result so far out of line that the estimate of error does

not mirror the real merit of the analytical method." Chemists have
not been willing to admit to the frequent appearance of extreme
values in their output despite the existence, in the clinical and
drug testing areas, of errors of such a magnitude that they have
attracted the attention of the public media.

One school of thought with respect to outlier removal holds
that outliers should not be removed unless an explanation is
available for the extreme values. Such a concept is untenable for
the SRM data sets because it would leave, as acceptable, values that
would result in unrealistic precision parameters. After-the-fact
examination of the validity of values from defunct laboratories and
unavailable analysts is not practicable. The only evidence often
available must be inherent in the data themselves or in the
accompanying documentation. Many outlier-removal decisions must be
arbitrary or based on the recognizable distribution pattern of the
remaining population, or on other available information or
inferences.

A second school of thought permits retention of outliers in the
data set as long as they do not affect the precision parameters
significantly (10). "Significantly" was found by experience to be a
reduction in precision parameters by at least 40%. This procedure
is most effective with small numbers of values (fewer than 20).
This "delta 40%" rule was found to be related to the Grubbs test and
could be abandoned when the IUPAC-1987 protocol advocated replacing
the previously used Dixon outlier removal test with the Grubbs
tests, particularly if the Grubbs tests are expressed in terms of
percentage reduction in standard deviation caused by the extreme
value removal.

The IUPAC-1987 protocol exemplifies a third approach in which a
reasonable, scientifically based set of rules is utilized in the
interest of harmonization. All practitioners applying the same
protocol should obtain the same answers. The protocol recommends
calculation of the statistical parameters of mean and relative
standard deviation both (1) without outlier removal and (2) with
outlier removal by the Cochran test, when replication is present,
and then by the Grubbs tests, single and paired, at a critical
2-tail probability level of 1%. It also permits, prior to
application of the prescribed outlier tests, prescreening for
invalid data. This procedure has worked fairly well with the small
number of laboratories (usually 8-12) that are involved in the very
well-controlled method-performance (collaborative) studies. But it
is obvious that this procedure is not satisfactory if applied
blindly to material-certification studies, where a "true value" is
available against which any given value can be judged. Knowledge of
the true value serves as a useful tool to screen out invalid data.

A fourth approach to the removal of extreme values is the
combination used by the compilers (3) -- using analytical judgment
to eliminate obviously invalid results, removing values outside of 2
standard deviations from the mean of the trimmed values, and using
the remaining values for the calculation of the consensus mean and
standard deviation. The extreme values thus removed are usually
values that a casual review of the data would have eliminated
anyway. The procedure is used to provide a benchmark for judging
the acceptability of method performance and laboratory performance.

As such, the procedure can also provide the reference standard for
the acceptability or rejection of extreme values.

Unfortunately, in any discussion of reliability, outliers must
be given far greater prominence than they deserve. They are very
conspicuous in any set of data and automatically attract attention.
But this very visibility should cause any analyst to pause and
reexamine the support for aberrant values before they are reported.
Nevertheless, the persistent reporting of extreme values in the face
of the availability of a true value is disturbing.

Distribution of Estimates of Precision of the Consensus Values of
the Certified Analytes. The distribution of the estimates of the
variability of the 117 values accompanying the certified values for
the 28 elements from 11 biological SRMs where at least 8
quantitative values are available cannot be examined directly
because of the wide range of concentrations displayed by the data
(from 5 ng/g to 4%). Removing the effect of concentration by
normalizing the relative standard deviations against the Horwitz
curve permits an examination of the spread of the consensus data.

The lower histogram of Figure 6 shows the HORRAT values of the
117 consensus analyte-SRM combinations. Only 2 HORRAT values are
above 2.0 -- calcium in Oyster Tissue and iron in Tomato Leaves.
This suggests that the compilers' choice of protocol for removing
extreme values in an uncontrolled materials-performance study
results in a distribution of concentration-normalized precision
values that closely matches the precision distribution of closely
controlled methods-performance studies. If the same type of
histogram (top portion of Figure 6) is prepared for the HORRAT
values calculated from the application of the IUPAC-1987 protocol to
the unscreened data, a flatter distribution with a very long
positive tail results, extending up to a HORRAT of 8.1 for
molybdenum in Orchard Leaves. If the IUPAC protocol is applied to
the data prescreened by the consensus protocol, only 3 analyte-
material combinations are affected. However, the effect is
inconsequential since the precisions of the affected analytes were
already acceptable by the consensus protocol: cadmium in Wheat
Flour, HORRAT = 0.62; chromium in Brewer's Yeast, HORRAT = 0.90;
and lead in Spinach, HORRAT = 1.35.

For these calculations, the "uncertain" uncertainty values
associated with the NIST certified concentrations are not used.
Therefore, the assumption that the values are standard deviations is
immaterial for the discussion of the precison parameters of the
consensus values, which we suggest reflect the best current
analytical performance.

Conclusion

The 206 analyte-matrix combinations produce a remarkable array of
precision variability, even when normalized for concentration by use
of the HORRAT ratio. Direct application of the IUPAC rules to the
raw data does not characterize many of the extreme values as
analytically unlikely. Therefore, it is important that analytical
judgment be applied to the reported values to remove invalid data
before an outlier procedure is used.

HISTOGRAMS OF PCT. OF LABS DROPPED (certified NIST and consensus)

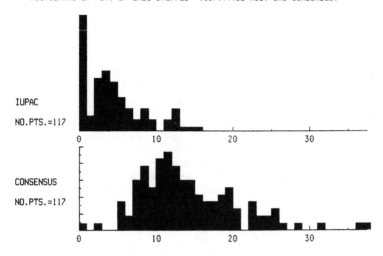

Figure 5. Upper histogram: The frequency of outlier removal from the 117 data sets of the certified analytes against percent outliers removed by the IUPAC-1987 protocol (8) using the Grubbs tables extended to 325 values. Lower histogram: Same as the upper histogram, but using the frequency of outliers removed by the compilers, Gladney et al. (5). The vertical scales are different in the 2 histograms, but the horizontal % scales are the same.

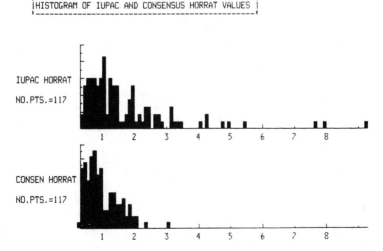

Figure 6. Upper histogram: HORRAT values calculated after removal of outliers by the IUPAC-1987 protocol (8). Lower histogram: Same as the upper histogram, but using the consensus mean and RSD_R values generated by the compilers, Gladney et al. (5).

Statisticians balk at consecutive application of outlier tests
to data because once an outlier removal procedure has been applied
on the basis of probability, the probabilities for the occurrence of
additional outliers change in a complex way. Moreover, the blind
application of statistical rules for removal of extreme values is
often inappropriate. Application of a rule of "inconsistency with
the bulk of a population" composed of independent values obtained in
numerous laboratories at different times, by different techniques,
reasonably scattered about a certified value, or even an
informational value, is sufficient reason for removal of reported
results. This is particularly true when a pattern of consistent
deviations appears for a given analyte in different SRMs.

The concentration estimates of the constituents of NIST SRMs as
reported in the literature have a large variability. Part of this
variability is clearly due to "sore-thumb" outliers, grotesquely
deviant values, obtained when the analyst unknowingly was out of
control. Since no a priori distribution is conceivable for such
outliers, i.e., since the appearance of such outliers is not based
on any describable process, no conventional statistical tests are
possible for flagging and removing them. The data compilers (5)
used their own common sense to screen out the truly obvious
outliers. Then they refined the data further by removing from the
reduced data all values more than 2 standard deviations from the new
average. This approach is quite close in spirit and gives results
comparable to those obtained by the IUPAC protocol which applies a
Grubbs test (which, in effect, removes as outlying values results
that are more than a certain number of standard deviations from the
average) to data deemed to be valid (8, §3.1; 14, §6.1) -- i.e., to
data purged of blatant outliers for reasons apparent in the
supporting documentation or for gross inconsistency with the true or
assigned value.

Even after outlier removal and the consequent narrowing of the
variability, no guarantee as to the accuracy of the concentration
estimates exists. For results to be taken seriously as valid
additions to the database estimates of concentrations, authors must
report their quality control procedures and their efforts to track
down sources of extreme values that do not agree with the certified
value. Built-in quality control with frequent recourse to known or
formulated reference materials must become the standard operating
procedure for all analytical laboratories. An example of a system
to evaluate analytical results is given by Holden et al. (16).

Literature Cited

1. Currie, L. A. In Treatise on Analytical Chemistry Part I, 2nd
    Ed.; Kolthoff, I. M.; Elving, P. M., Eds.; Wiley: New York,
    1978; Vol. 1, Chapter 4.
2. ISO-5725-1986. Circulated proposed revision in 6 parts,
    International Organization for Standardization, Geneva,
    Switzerland. Available from national standards organizations.
3. Iyengar, G. V. Elemental Analysis of Biological Systems; CRC
    Press: Boca Raton, FL, 1989; Vol. I, pp 200-29.
4. Horwitz, W.; Kamps, L. R.; Boyer K. W. J. Assoc. Off. Anal.
    Chem. 1980, 63, 1344-54.

5. Gladney, E. S.; O'Malley, B. T.; Roelandts, I.; Gills, T. E. Standard Reference Materials: Compilation of Elemental Concentration Data for NBS Clinical, Biological, Geological, and Environmental Standard Reference Materials, National Bureau of Standards Special Publication 260-111, Superintendent of Documents: Washington, DC 20402-9325, 1987.
6. Byrne, F. P. In Accuracy in Trace Analysis, NBS Special Publication 422, Superintendent of Documents, Washington, DC 20402-9325, 1976.
7. Thompson, M. In Proceedings Third International Symposium on the Harmonization of Quality Assurance Systems in Chemical Analysis, 19-20 April 1989, Washington, DC, Prepared by International Organization for Standardization as ISO/REMCO 184, 1989, pp 183-189.
8. Horwitz, W. Pure Appl. Chem. 1988, 60, 855-64.
9. Kelly, P. J. Assoc. Off. Anal. Chem. 1990, 62, 58-64.
10. Boyer, K. W.; Horwitz, W.; Albert, R. Anal. Chem. 1985, 57, 454-9.
11. Act of 22 July 1950; 64 Stat. 371 (Public Law 619, 81 Congress) Amending Sec. 2 of the Act of Mar. 3, 1901, establishing the National Bureau of Standards.
12. Taylor, J. K. Standard Reference Materials: Handbook for SRM Users. US Government Printing Office: Washington DC 20402-9325, 1985; p 27.
13. Unrecorded discussion at Third International Symposium on the Harmonization of Quality Assurance Systems in Chemical Analysis, 19-20 April 1989, Washington, DC; International Organization for Standardization, Geneva, Switzerland.
14. Guidelines for Collaborative Study Procedure to Validate Characteristics of a Method of Analysis. J. Assoc. Off. Anal. Chem. 1989, 72, 694-704.
15. Youden, W. J. Statistical Manual of the AOAC; AOAC: Arlington, VA 22201, 1975; p 30.
16. Holden, J. M.; Schubert, A.; Wolf, W. R.; Beecher, G. R. In Food and Nutrition Bull. Supplement 12, Food Composition Data: A Users Perspective; W. M. Rand; C. T. Windham; R. W. Wyse; V. R. Young, Eds.; United Nations University: Tokyo, 1987; pp 177-93.

RECEIVED July 16, 1990

Chapter 6

The Importance of Chemometrics in Biomedical Measurements

Lloyd A. Currie

Center for Analytical Chemistry, National Institute of Standards and
Technology, Gaithersburg, MD 20899

Chemometrics as a discipline blends modern mathematical and
statistical techniques with chemical knowledge for the design,
control, and evaluation of chemical measurements. For complex
systems, such as those involving biomedical trace element research,
the multidisciplinary efforts toward problem formulation and
measurement process design and evaluation can be substantially
aided by exploratory chemometric approaches. Following a brief
overview of potential chemometrics contributions, primary
attention is given to exploratory multivariate data analysis
techniques, which can capture the essence of a complex data set in a
few, visualizable dimensions. Such techniques are appropriate
because nuclear-related measurements, quality control samples and
data, global dietary intakes, and biological compositions *all* comprise
multiple chemical species, frequently exhibiting correlated behavior.
Applications of some of the more powerful techniques, such as
principal component factor analysis, are illustrated by multivariable
interlaboratory quality control, assessment of pollutant origins, and
exploration of daily dietary intake data.

Multidisciplinary and multinational scientific studies are becoming
increasingly important as we face issues involving the global environment and
global nutrition. Such studies frequently have a central chemical component
which, in turn, is multivariable. Chemometrics, itself multidisciplinary,
provides a number of advanced computational and statistical tools that can
greatly benefit these studies, from planning to interpretation. An excellent
introduction to the methods and multivariate research applications of
chemometrics will be found in reference 1. The most recent textbook on the
subject is reference 2, and reference 3 gives a brief overview of its history,
content and relationship to chemical standards.
    The multivariable aspects of modern biomedical and bioenvironmental
research are manifest in: (a) the several biological and environmental variables
that can influence the sampling and/or chemical measurement processes, and

(b) the chemical elements or compositional variables that characterize samples and reference materials. Central to the realization of each of these aspects is a matrix, where rows denote samples and columns denote variables. In the first instance, the matrix is labeled the *design matrix* of independent variables or factors; in the second, the *data matrix* of measured (response) variables. The importance of these matrix representations lies in the fact that actual measurement systems and actual bioenvironmental systems rarely exhibit "one-at-a-time" behavior. That is, complex interactions among variables generally occur. If we restrict our view to the effects of one variable or factor at a time, we may come to erroneous conclusions and fail to understand the nature of the overall system. The multivariate perspective is therefore vital for augmenting our knowledge of the univariate structure. (Note that the term "variate" denotes a variable which exhibits random character.) In the following sections, we review chemometric approaches to multivariable *design,* multivariable *control* of measurement accuracy, and *evaluation* of the resulting multivariate chemical data -- specifically, data from the International Atomic Energy Agency (IAEA) Coordinated Research Programme on Human Dietary Intakes, known as the Daily Dietary Study (DDS) (4).

Multivariable Experimental Design

Identification of the critical variables and their important levels, based on scientific knowledge of the problem, constitutes the first phase of the design process. Inattention to this matter may lead to experimental results of inadequate precision, or worse, to serious bias and/or lack of control. A biomedical case in point is the sampling of human serum, without giving adequate attention to factors such as stress, diet (and fasting), diurnal effects, and body position (5). This illustrates one of the objectives of designed experiments: identification of influential variables or "ruggedness testing." (6). Factorial design (described below) accomplishes this end, by indicating variations that can be tolerated without degrading statistical control. In addition to the purpose of screening multiple variables and their interactions, designed experiments are important for estimating shapes and finding optima of response surfaces, and for chemical model-building and parameter estimation. In selecting ranges of the experimental variables to be ultimately explored, it very important to "span the factor space" so that an underlying physical or biological process (model) may be adequately assessed. The latter objective is the basis for the design of a current study, from the author's laboratory, of urban sources (fossil fuel, wood burning) of carbon monoxide using isotopic measurements (G.A. Klouda, personal communication, 1990).

Factorial designs, which define experiments at each of the possible combinations of factors and levels, have had a major impact on efficiency and accuracy in experimentation -- such as testing for (screening) or estimating main effects and interactions among factors, and exploring response surfaces (2, 7). Special characteristics of such designs are that they provide independent and simultaneous estimates of all factor effects, with the same precision that would have been obtained had all the experiments been performed to estimate the effect of just a single factor. As an illustration of a 2^k design (k factors at 2 levels each), a slightly (didactically) modified version of the forementioned carbon monoxide experimental design is given in Table I. Variable "B" in the table can be viewed in two different ways: (a) as a blocking variable for a 2^3 factorial, specifying two sets of optimal (orthogonal, "space spanning") fractional factorial designs, to be performed separately, if external variables cannot be adequately controlled over the period required for a single, complete set of eight experiments; (b) as a level index (1 = -, 2 = +) for a fourth variable (X_4) for a 2^{4-1} half factorial. The implication in the latter case is that for the same amount of work, 8 experiments, one can estimate either the (main) effects for four factors, or the main effects and interactions for three factors. Categorical designs of this sort are interesting not only for planning univariate (single response variable) investigations, but also for experiments which generate multivariate data matrices.

Table I. Factorial Design (2^3) -- Source Apportionment of Urban CO
(Winter, 1989-90, Albuquerque)

Factors (Variables) and Sampling Conditions[*]

X_1:	Sampling Period: day (-) vs night (+)
X_2:	Forecast/Meteorology: dynamic (-) vs stagnant (+)
X_3:	Sampling Location: residential (-) vs highway (+)
B:	Blocking variable; used to define two fractional factorials, to reduce the potential impact of an external variable, such as analytical laboratory, month of collection, etc.

Sampling order	X_1	X_2	X_3	B
8	-	-	-	1
1	+	-	-	2
4	-	+	-	2
3	+	+	-	1
2	-	-	+	2
5	+	-	+	1
7	-	+	+	1
6	+	+	+	2
a	+	+	-	
b	+	+	+	

[*]The response variable is the isotope ratio, $^{14}C/^{12}C$. (In a multivariate version of the experiment the response variables could be concentrations of a set of selected elements or organic compounds.) Sampling order should be randomized; one such randomization is illustrated. If blocking is used, randomization should take place within each block. Samples a and b might be added to replicate pollutant conditions of major interest.

When starting a new investigation, it can be very beneficial to employ *fractional* factorial designs to screen a potentially large number of variables for those which may be of consequence. The efficiency of fractional factorial designs for screening becomes impressive as the number of factors grows: for example, with a 2^7 factorial -- 2 levels each, of 7 factors -- the main effects, free from any 2-factor confounding, may be estimated from a fraction (2^{7-3}) comprising just 16 of the possible 128 experiments. Guidance on the structure and use of fractionals, including the very efficient "saturated" fractional designs, may be found in chapter 12 of Reference 7.

When the factors are quantitative, full 2^k factorials are effective for empirical (linear) modeling and indicating a direction (steepest ascent) for subsequent experiments. Central composite designs are appropriate for non-linear response surfaces (locally quadratic), and in the region of an optimum. The process of repeated experimentation to move up a response surface is known as sequential experimental design. A common approach involves the use of sequential factorials and regression methods for response surface fitting and optimization. An alternative, popular in chemical applications, employs the rather rudimentary but efficient non-regression technique of sequential simplex design. Here the simplex is a geometric figure in k-factor space with k + 1 vertices. The simplex search moves one experiment at a time in a preferred direction based on the prior simplex and a specific set of rules. At each step, the vertex with the poorest response is replaced by a new one -- generally the mirror image of the one dropped. Figure 1 depicts a multi-step simplex search for the best experimental conditions for the automated determination of formaldehyde. Interestingly, this same search technique proves useful in the data analysis context, e.g., in finding the optimum solution for complicated non-linear least squares spectrum fitting (8).

Another approach to design optimization employs a theoretical model-based objective function -- i.e., an expression, such as relative standard error, detection limit, etc. -- which one wishes to optimize. Figure 2 illustrates this technique, which we applied to the selection of an optimal subset of atmospheric particulate samples for source apportionment model validation via (expensive) measurements of ^{14}C by Accelerator Mass Spectrometry (9). The objective function chosen was the "Fisher information" (determinant of the least squares normal equations matrix). The optimizing algorithm had a large task; the "best" subset of 10 samples out of 20, for example, must be selected from nearly 185 *thousand* possible combinations.

The special relevance of multivariable experimental design to biomedical trace analysis arises from the complexity of biological systems, which *ensures* the existence of multivariable interactions. An example comes from the work of Gordon (10) in which the results of a 3-level factorial experimental design suggested important interactions among iron, zinc and copper affecting their bioavailability.

Exploratory Data Analysis (EDA)

Before discussing methods of Multivariate Control and Multivariate Evaluation, it is useful to introduce some basic rules for Exploratory Data Analysis. These rules, which are important for univariate data analysis, become critical for multivariate analysis; their purpose is to link the enormous pattern recognition power of the (trained) human observer with patterns inherent in the data. An implicit, zeroth rule is that EDA should precede attempts at statistical modeling and analysis. The first four rules (Figure 3) are self explanatory. (ANOB means Analysis of Blunders.) The fifth refers to the

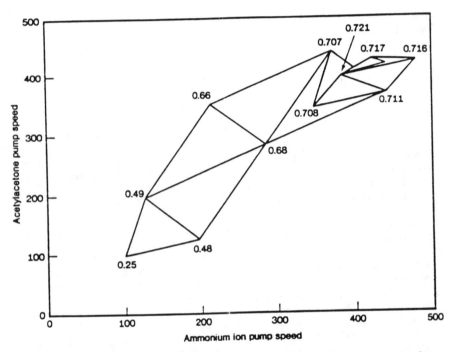

The Problem: Selection of the best subsets (10 out of 20) of air particulate samples from each of four urban sites for C-14 measurement.

Goal: To validate and calibrate an inexpensive dual-tracer method (K, Pb) for pollutant carbon source apportionment, using an expensive, accurate, and absolute tracer (C-14) technique.

Model: $C_i = b_0 + b_1 \cdot K_i + b_2 \cdot Pb_i$ or $C = b X$ (matrix equation)

where: C_i, K_i, and Pb_i represent the carbon, potassium, and lead concentrations in sample-i. (C-14 gives *independent* parameter estimates, of demonstrated validity.)

Approach: Select the "best" subset of samples, based on already completed measurements of Pb and K, to estimate the regression coefficients (b_0, b_1, b_2) and test the above model using (expensive) carbon and C-14 measurements.

Statistical Solution (D-optimal Design):

For each of the four sites, the best 10 sample subset is given by the design matrix X that maximizes the "Fisher Information." (This is equivalent to seeking the subset that yields the best overall precision for the estimated regression coefficients, based on least squares.)

The Fisher Information is defined by Det ($X'X$),

where $X' = (1\,1\,1\,1\,....)$
$(K_i ....)$
$(Pb_i ....)$

Fig. 2.   Design: Optimal sample subset selection using a model-based objective function (9).

1. Plot the data (Filliben)

2. Remove what you know (Tukey)

3. Examine what remains from every possible perspective (Filliben, Shaler, ...)

4. Outliers are generally present: ANOB (Currie)

5. df = observations - parameters < 0, *always;*

   therefore, Scientific Intuition is essential (Currie)

6. Perform subset analysis (Filliben)

   If a model is to be valid for *all* the data, it must be valid for interesting *subsets* of the data

7. Univariate - multivariate exploration (Filliben)

   a) Understand low-dimensional structure before attempting to grasp multivariate structure

   b) Realize that *some* aspects of multivariate data can *only* be perceived through multivariate techniques

   Fig. 3.   The 7 Rules of Exploratory Data Analysis.

fact that complex physicochemical processes *always* exhaust our ability to describe them by simple mathematical models. We *must* rely on "scientific intuition" (or scientific insight) or assumptions to derive unique solutions from experimental data sets. The last two rules are specially pertinent to data characterized by multiple subsets (including clusters) and multiple variates. They will be illustrated in the following sections of this text. The first five EDA rules are illustrated by a univariate example: a calibration curve for the X-ray fluorescence analysis of calcium (11). We "plot the data" in Figure 4, and "subtract what we know" -- i.e., the fitted regression line -- leaving residuals (Figure 5). The three sets of residuals in Figure 5 have arisen from three perspectives: ordinary least squares fitting of the data; weighted least squares (WLS) fitting using Poisson counting statistics for weights, plotted against CaO concentration; and WLS after deletion of a major outlier (exposed in the second plot), but plotted vs sample order. These alternative ways of viewing the data revealed the major outlier (perspective-B), as well as a quadratic systematic error (perspective-C), which later was learned to be due to a distorted sample holder wheel. This example illustrates patterns that the human observer can often find through graphical EDA, but which may not always be discerned by preset, classical statistical tests. Enormous progress in interactive statistical graphics, illustrated by Figures 6 and 7, now makes possible *very efficient and convenient* application of the remarkable human pattern recognition abilities with corresponding multifold increases in data analyst insight (12).

Multivariate Quality Control

Multivariate approaches to quality control may be introduced through a bivariate example (13). Figure 8 shows a Youden-type plot for testing the accuracy and precision of vanadium analyses by collaborating laboratories. The dashed region represents the "truth"; the 45 degree line is the locus for laboratory systematic error ("between-lab" variation); the scatter about the line is a measure of precision ("within-lab" variation); and the lab-x result displaced from the line implies a within-lab control problem. This very simple tool, where a group of laboratories measure the same pairs of samples, is one of the most powerful graphical means for assessing laboratory performance "at a glance."

We wish to extend the bivariate graphical assessment of quality control to the multivariate case. To proceed, we must first introduce the concept of reduced-dimensional projections, as accomplished through *Principal Component Analysis* (PCA) (2). A key objective of PCA is to create the best 2- or 3-dimensional projections (plots) of a multivariate data set, so that the patterns in the data can be visually grasped. The first step is to form the data into a "data matrix," where each row represents a vector or sample, and each column, a variable. An example that we shall refer to later is given in Table II, where the rows represent individual composite food samples labeled by sample numbers and (coded) Population Groups (A - E), and the twelve columns represent ten elemental variables plus fiber (FB) and phytate (PA). (The data were drawn from a multinational study of daily dietary intakes coordinated by the International Atomic Energy Agency (4).) The data matrix in Table II represents the largest array available having no missing values, as of May 1989. As the DDS database is not yet complete, this portion *should not* be viewed as

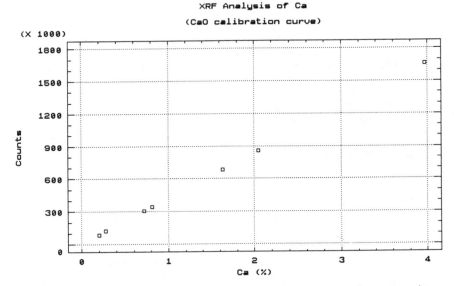

Fig. 4.   X-Ray Fluorescence calibration curve: counts observed vs Ca concentration (11).

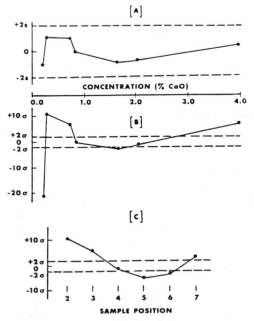

Fig. 5.   Residual plots for the XRF calibration curve: (A) residual pattern from ordinary (unweighted) least squares fitting; (B) residuals from weighted least squares (WLS) fit using Poisson weights; (C) residuals from Poisson-weighted WLS after outlier deletion, plotted vs position in sample holder. (Reproduced from Ref. 11; 1977, American Chemical Society.)

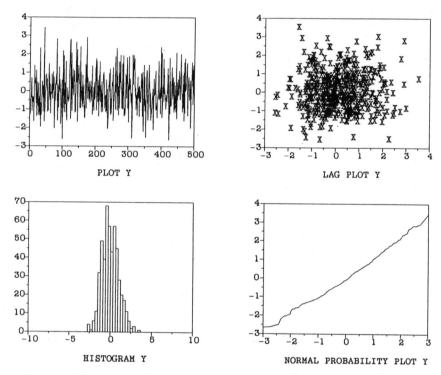

Fig. 6.   4-Plot DATAPLOT command applied to random numbers (12). This command, "4-Plot," recommended as an initial step in univariate data analysis, provides an immediate graphical test of assumptions concerning univariate patterns, independence, and frequency distribution (histogram, normal probability plot).

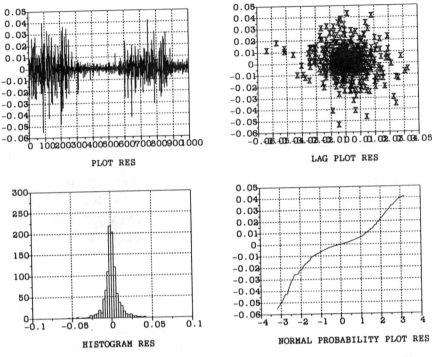

PLOT RES

LAG PLOT RES

HISTOGRAM RES

NORMAL PROBABILITY PLOT RES

Fig. 7.   4-Plot DATAPLOT command applied to standard resistor measurements (12).
Residuals from an 11-point moving average fit to a 5 year resistor time series
are examined for homogeneity of variance, randomness, and distributional
form.

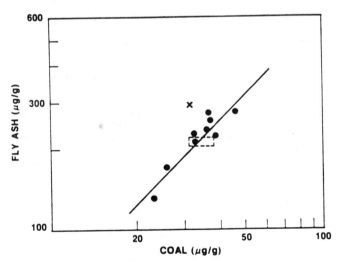

Fig. 8.   Two dimensional Youden-type correlation plot of interlaboratory
measurements of vanadium in Fly Ash and Coal SRMs. Dashed region
encloses the certified values. (Reprinted from Ref. 13; 1986, National Bureau
of Standards.)

representative. It is introduced here strictly to illustrate some of the relevant chemometric approaches. The form of the data matrix in Table II represents one of several that may be extracted from the DDS database by the program *IAEADM*, written in our laboratory. The program permits the operator to select interactively: (a) a list of nations and population subgroups that comprise the rows of the data matrix, and (b) a list of chemical variables for the columns. Two types of output can be chosen, either a triplet of data matrices representing laboratory classes (reference, backup, and information-only), or a sextuplet of data matrices representing alternative analytical measurement techniques.

Table II.   Data Matrix for 12 Chemical Species and 5 Population Groups

(extracted from the DDS database; units: g/kg [Ca, FB, PA],
mg/kg [Ni, Zn], µg/kg [all others])

Sample (PopGp)	As	Ca	Cd	Cr	FB	Hg	I	Ni	PA	Pb	Se	Zn
49(A)	16.2	1.65	72.0	148.0	27	6.	121.0	0.159	1.530	1300.	70.0	17.60
50(A)	75.4	3.95	50.0	251.0	29	3.	234.0	0.468	1.240	375.	122.0	29.90
51(A)	16.5	1.99	50.0	271.0	34	4.	67.4	0.235	1.080	323.	92.1	18.80
52(A)	14.6	1.45	24.0	152.0	26	7.	191.0	0.200	1.140	200.	75.5	21.00
53(B)	442.0	1.40	44.7	110.0	24	165.	255.0	0.158	1.080	930.	280.0	11.80
54(B)	395.0	2.73	39.7	168.0	21	11.	294.0	0.182	1.090	746.	138.0	22.70
55(B)	294.0	1.75	55.9	154.0	20	32.	314.0	0.282	0.730	98300.	198.0	16.40
56(B)	16.3	1.69	37.8	86.1	24	22.	194.0	0.136	0.960	275.	83.0	25.10
57(C)	25.2	3.21	13.5	272.0	32	7.	325.0	0.228	0.793	156.	45.0	20.30
58(C)	252.0	1.80	54.5	85.5	36	14.	150.0	0.326	1.060	583.	188.0	22.80
59(C)	68.7	0.83	16.1	145.0	23	13.	1960.0	0.204	0.934	442.	63.0	29.20
60(C)	2060.0	4.00	36.0	141.0	31	120.	344.0	0.210	0.798	165.	113.0	18.90
61(C)	402.0	1.62	42.0	69.5	28	25.	193.0	0.176	0.683	536.	195.0	16.00
62(C)	445.0	1.62	30.0	119.0	33	8.	216.0	0.361	2.380	173.	116.0	33.30
63(D)	409.0	2.34	26.9	176.0	28	45.	427.0	0.237	0.905	303.	205.0	15.40
64(D)	488.0	3.14	20.9	193.0	44	29.	274.0	0.450	1.530	219.	225.0	17.50
65(D)	30.5	1.98	15.3	80.8	19	1.	275.0	0.321	0.870	142.	102.0	42.40
66(D)	1390.0	2.36	18.3	151.0	29	23.	929.0	0.319	1.720	129.	115.0	16.60
67(D)	23.4	1.10	22.0	63.3	28	17.	495.0	0.257	1.010	156.	147.0	31.20
68(D)	26.2	1.61	28.5	300.0	16	13.	185.0	0.229	0.612	387.	73.0	13.10
69(E)	277.0	1.63	25.0	124.0	47	168.	193.0	0.684	0.668	102.	121.0	1.85
70(E)	37.7	1.36	14.0	264.0	70	7.	65.1	0.698	1.760	94.	27.7	25.80
71(E)	72.8	1.46	16.0	177.0	37	80.	178.0	0.880	0.634	172.	50.4	14.00
72(E)	43.3	1.52	20.0	185.0	58	46.	79.9	0.970	0.543	301.	71.3	26.60
73(E)	53.6	1.60	13.0	284.0	70	29.	148.0	1.130	2.760	406.	45.7	20.10
74(E)	73.0	2.66	21.0	221.0	49	54.	337.0	0.930	1.400	172.	68.2	31.60

The mechanics of PCA consists of two parts: representation of each data vector (or spectrum) as a point in "parameter space," where the coordinates denote the several observed species or variables (elements, masses, ...); and linear transformation in which the coordinate system may be translated, scaled or standardized, and rotated so that most of the "structure" in the data can be displayed in a few orthogonal dimensions. The dimension containing the largest dispersion (variation) in the data is the first principal component; the orthogonal dimension having the next largest dispersion, is the second principal component, etc. A trivial example would be a set of data in 3 dimensions (e.g., concentrations of three elements measured in several samples) which, when plotted in 3 dimensions, formed a straight line, like a pencil. By moving to the center of the "pencil," and rotating it (to form a new "x"-axis), one would display the most prominent information in the data along this new "x"-axis, which is the first principal component. The data projections on this axis are called the "scores" on the first principal component, and the first "eigenvalue" is a measure of their dispersion (variance) in this direction. (PCA is sometimes called "eigenanalysis," and the principal components, "eigenvectors.") Another central point in PCA is the chemical meaning of the components. The principal components are defined by "loadings" (or coefficients) related to the underlying chemical variables. By plotting loadings and scores on the same diagram, one can infer associations among variables, and connect variables with patterns in the data.

PCA is generally recommended as a *first step* in *all* investigations of multivariate structure, so those who wish to apply visual pattern recognition to multivariate data are urged to gain a solid understanding of PCA concepts from an elementary text, such as reference 2. The linear algebra of PCA, also known as "eigenvector analysis" and "singular value decomposition" is essential for computation, but graphical EDA can be very well accomplished if the geometric formulation is grasped.

At this point it is appropriate to consider a multivariate extension of the Youden plot, as applied to the DDS data. The principal component plot in Figure 9 represents the most efficient 2-dimensional projection of the 3-dimensional (3-laboratory) Zn data obtained from measurements of 13 DDS samples. Had all measurements been under control, we would have expected to see the 1-component "pencil" projection referred to above -- the principal variation arising from the differing zinc concentrations. Figure 9 shows that this is not the case. Clearly one laboratory [R] and one sample [#69] depart from the rest; and the PCA projection showed immediately which sample and laboratory differed. This multivariate outlier detection scheme becomes clearer when we examine the more familiar 3-dimensional perspective of the same data (Figure 10). Detection of an apparent outlier, of course, is only the first step. When we examined the data for Lab-R/sample #69, we found that it could be brought into consonance by shifting the decimal point! The power of the PC-Youden plot can be appreciated if the extension to 4, 5, 6 or more laboratories is envisaged. We need not look all twenty 3-dimensional projections of 6-space to spot the outliers; the single 2-dimensional PC plot will suffice. Information on more general techniques of multivariate control based on estimates for the multivariable mean and variance-covariance matrix may be found in Alt (14).

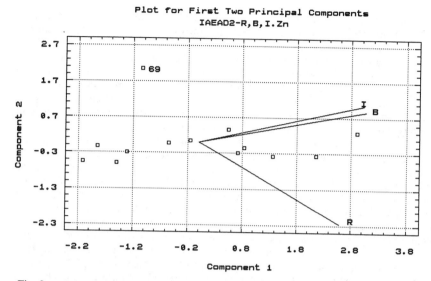

Plot for First Two Principal Components
IAEAD2-R,B,I.Zn

Fig. 9.   Principal component projection of three dimensional interlaboratory measurements of zinc (standardized). I, B, and R are the three laboratory codes; the result for sample-#69 from laboratory-R is outlying [IAEA - DDS data].

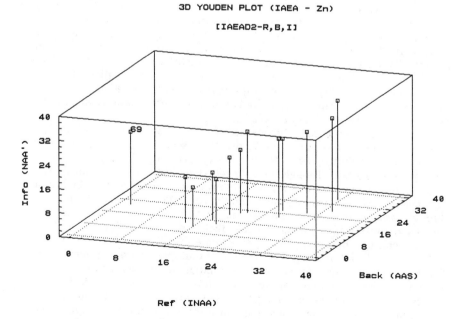

3D YOUDEN PLOT (IAEA - Zn)

[IAEAD2-R,B,I]

Fig. 10.   "Real" (chemical) variable three dimensional plot of the zinc interlaboratory data. Concentrations are given as mg/kg for each of the three laboratories [IAEA - DDS data].

Multivariable (multielement) natural matrix certified reference materials (CRMs) are also relevant. Such materials afford the opportunity to control measurement accuracy, not just from the perspective of "multi-univariate" analysis, but also from the true multivariate chemical analysis perspective, in that interactions among elements and interactions between elements and the chemical matrix become part of the test of analytical control. Interestingly, the techniques of multivariate analysis (principal component and cluster analysis) have been applied to assess the multivariate (chemical) adequacy of CRMs in serving as surrogates for actual food and biological samples (15).

An issue of increasing importance in the area of multivariate quality control is the matter of accuracy and precision of computations or data reduction involving multivariate observations. This is a very serious issue, because the "chemical" models employed for multivariate analysis (MVA) are often very complex, incomplete, and assumption-ridden. With the growth in the use of "canned programs," and with automatic MVA software being built into instrumentation systems, erroneous results and erroneous uncertainty estimates (precision) may abound without the operator's knowledge. One quality control solution, which has well-served many laboratories engaged in multi-nuclide gamma ray spectroscopy and multi-element atmospheric particle source apportionment, is to use computer simulated "Standard Test Data" to assure MVA accuracy, much as Standard Reference Materials are used to assure measurement accuracy (13).

Multivariate Data Evaluation

Multi-univariate techniques. The examination of multiple one and two dimensional representations of the data in "real" variable space -- in contrast to the seemingly arcane principal component space -- is an effective method for understanding *chemical* relations. Iteratively viewing the data in both real space and PC space, however, is an even more effective means of gaining insight. To illustrate, multiple, single-element distribution plots from the DDS are given, in the form of histograms, in Figure 11. Such plots are very informative, especially when comparisons are made among various population groups; but 2, 3, or higher dimensional relations among the elements are, of course, totally concealed.

To illustrate what may be learned from various forms of multiple distribution plots, we see in Figure 11 that the essential element Zn appears to be normally distributed, whereas the toxic element Hg is clearly asymmetric with quite a long tail. Though the data reflect typical dietary intakes derived from plant and animal matter, it is interesting that in normal biological systems essential elements exhibit tightly controlled, normal distributions through homeostasis, while non-essential and toxic elements may vary in a somewhat uncontrolled manner in that their "true normal levels should be close to zero" (5). In contrast to the other two elements, Se exhibits multimodal structure, (correctly) implying the presence of multiple underlying populations. This type of univariate information is very important in itself, and it should be examined *before* any multivariate exploration takes place. In fact, lower dimensional (univariate, bivariate, ...) EDA can give considerable insight concerning the validity and interpretation of certain multivariate approaches

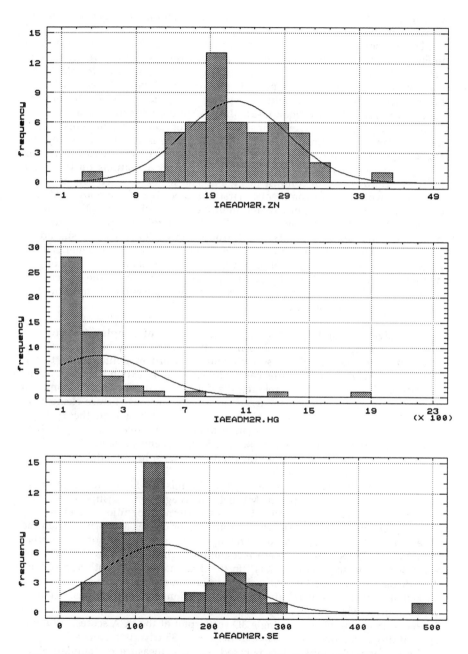

Fig. 11. Frequency histograms for zinc, mercury, and selenium; laboratory-R, units are mg/kg for zinc, and μg/kg for mercury and selenium [IAEA - DDS data].

which depend on assumptions involving population homogeneity, model linearity, measurement error covariance, etc. (16).

Three modes of examining univariate element distributions are illustrated in Figure 12, using the Se data. The popular, concise box plot is given at the top; the Se histogram from Figure 11 appears next; and a "point density plot" is at the bottom. The three plots exhibit a range of "data compression." For homogeneous data, they would be equally satisfactory, but in exploratory studies it is very useful to "look inside the box" to discern subpopulations. The point density plot looks even inside the histogram, for equal histogram intervals can obscure fine structure. (Fine structure is preserved also by the closely-related, empirical cumulative distribution function.) In Figure 12, for example, the point density plot, which involves no operations on the data, hints at a third mode, in contrast to the bimodal histogram. To conclude this discussion we offer two comments. First, the objective of the above reduced data compression is to permit visual, intuitive data exploration rather than to generate proof. Second, the actual DDS data involved known (human) population subgroups. If reliable subgroup knowledge exists, it should of course be utilized; if not, the above techniques may prove beneficial in revealing such substructure. (Future work will: (a) evaluate the utility of gap distributions in the point density plot for graphically testing for the presence of multimodal structure, and (b) further explore the application of multi-bi- and tri-variate plots to complement multi-univariate EDA.)

True multivariate methods (overview). Collections of samples consisting of several measured variables, comprising a data matrix, may be evaluated by means of "hard modeling" or "soft modeling." Hard modeling refers to the case where a mathematical model can be constructed, based on sound physical-chemical-biological principles, relating the data to the underlying phenomenon of interest -- e.g., the mechanistic model for the interactions of trace elements in the biological system. Models of this sort may be based on theory (e.g., exponential radioactive decay), or more commonly extensive laboratory or field experiments, to characterize the respective physicochemical interactions. For a fundamental understanding of multivariable relations, hard models are always the models of choice, if available.

"Soft" models, which are really the models of EDA, are commonly employed in the two varieties of MV-pattern recognition: "unsupervised" and "supervised." The first category includes pattern recognition techniques -- primarily cluster analysis -- in which groupings or classes are formed from the observed multivariate associations among the data, with little or no external information. "Supervised" pattern recognition begins with known or assumed substructure (classes, "training sets") in the data matrix. For both approaches, PCA display is an important early step, as one can quickly visualize the main (systematic) relations in relatively low-dimensional PC-projections. (For 7-variable space, for example, one can examine the "best" (principal component) 3-dimensional projection in a single 3D scatterplot of the PCA scores. Without optimal projection, one would have to inspect all 35 trivariate plots.)

Cluster analysis, unlike PCA, is based on some measure of "similarity" or "distance" in the full dimensional parameter space (2, 20). A popular measure is the Euclidian Distance, which is simply the root mean square distance taken over all dimensions (variables). There are two primary approaches to cluster

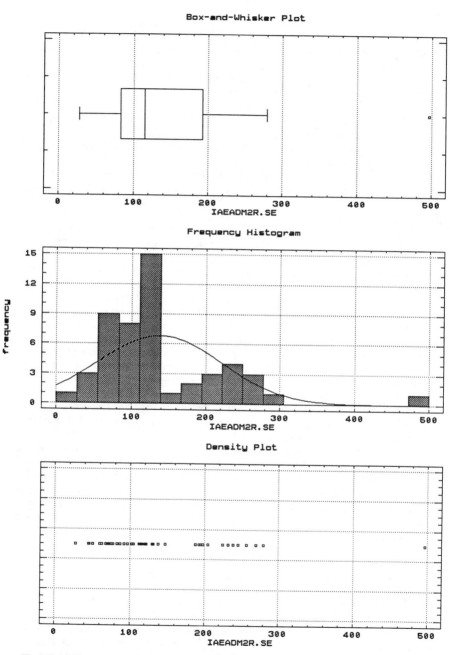

Fig. 12.   Three representations of the selenium distribution: box plot, frequency histogram, and point density plot; units μg/kg [IAEA - DDS data].

analysis: hierarchical, commonly represented as a dendrogram (or tree) as in Figure 13; and non-hierarchical or partitioning methods. Within these two primary schemes, one finds many others. Unfortunately there is no single method of choice; furthermore, different methods rest on different fundamental assumptions and lead to different results. Cluster analysis, therefore, is truly a (very powerful) *exploratory* technique, best used for EDA, and in conjunction with a skilled human observer.

Supervised pattern recognition begins with predefined classes. Its objective is to discriminate and/or model the classes, and to define a strategy for assigning new objects (samples, data vectors) to the extant classes. This part of EDA may be considered at three levels: (1) the usual PCA display of the overall data structure; (2) discrimination or modeling of classes; and (3) mixture analysis. Level-2 analysis commonly employs Linear Discriminant Analysis (LDA), but this technique should be applied with prudence, especially if the system is not linearly separable, if class "shapes" differ greatly, or if outliers are likely to be present. Preferred modeling techniques use multivariate normal models and *local* PCA models. A word of caution regarding the former: models and statistical tests based on the assumption of (multivariate) normality -- whether for supervised pattern recognition or for cluster analysis -- can be *very* misleading unless each class so modeled has a large number of members. Otherwise, the estimated covariance matrix may be ill-defined. Local, or class-specific PCA models are more robust, since the PC's are orthogonal, and classes may often be adequately modeled by just the first few eigenvectors. These so-called SIMCA models have proven very effective for class modeling and discrimination for modest data sets (17). Figures 14 and 15 illustrate the construction of multivariate normal and SIMCA class models, respectively.

Level-3 of supervised pattern recognition concerns itself with intraclass mixture analysis. Qualitatively, the variable loadings on the first few principal components for each class model carry important clues as to the chemical meaning of those components. The *number* of significant principal components (beyond the noise), in turn, indicates the number of underlying constituents in a regular mixture. These observations serve as the basis of intraclass multivariate analysis of mixtures, using techniques such as "target factor analysis" and "self-modeling" (18,19). Further discussion is beyond our scope, but two caveats should be mentioned. First, mixture analysis generally assumes linear systems; one must be alert to non-linearities masquerading as extra constituents. Second, widely used ad hoc rules for rotating principal components "toward more meaningful solutions" (e.g., varimax rotation) often provide pitfalls. An infinite set of possible rotations always exists, and meaningful selection follows only from *chemical* (or physical) knowledge, not from convenient "rules of thumb."

Thus, supervised pattern recognition stands intermediate between unconstrained exploratory techniques, such as cluster analysis, and the more rigorous hard, physicochemical modeling. Each multivariate data evaluation approach must be considered in the light of available knowledge; and interactive human experts in discipline-oriented pattern recognition are essential for reliable conclusions. EDA approaches may, however, be charted, just like computer programs. Figure 16 is presented as a general guide through the paths of "soft" modeling (20).

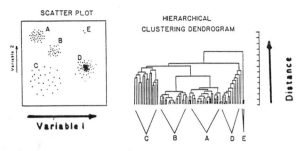

Fig. 13.   Cluster analysis representations in exploratory data analysis. Note that cluster analysis links objects according to minimum MV distance, while PCA rotates (projects) the MV coordinates according to maximum variance. (Reproduced with permission from Ref. 20. Copyright 1988 Elsevier Science Publishers.)

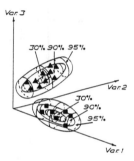

Fig. 14.   Multivariate class models based on multivariate normal density functions. Objects are classified to the class with the highest probability density at the object point. (Reproduced from S. Wold, et al., "Multivariate Data Analysis in Chemistry," ch. 2 in Ref. 1, with permission. Copyright 1984 Kluwer Academic Publishers.)

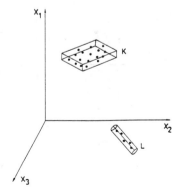

Fig. 15.   Multivariate class models based on principal components. SIMCA: a two-class classification. K is described by a 2-component model, and L, by a 1-component model. (Reproduced with permission from Ref. 2. Copyright 1988 Elsevier Science Publishers.)

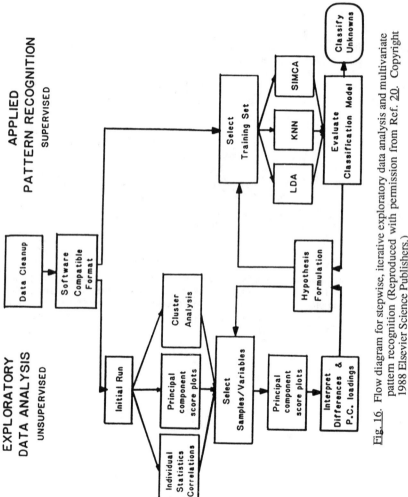

Fig. 16. Flow diagram for stepwise, iterative exploratory data analysis and multivariate pattern recognition (Reproduced with permission from Ref. 20. Copyright 1988 Elsevier Science Publishers.)

MV exploration of the DDS data. As an initial step toward understanding the DDS data, we performed principal component analysis on the eight non-toxic elements in the data matrix of Table II. (The non-toxic portion was selected to limit the present tutorial discussion to data having fewer large excursions.) Mean-centered and standardized ("autoscaled") data were used in the computation to aid in interpretation and to make the results independent of the units of measurement. The first conclusion from this analysis, which comes from examining the relative magnitudes of the ordered eigenvalues, is that the data are indeed heterogeneous, and not clearly "factor analyzable" (18). This supports the inference from the earlier one-dimensional analyses, together with what is known about the actual sampling groups. The conclusion revolves about the question of dimensionality -- i.e., the division into systematic "factor" dimensions and dimensions reflecting primarily the noise. Many popular rules have been offered for dimensionality decisions, one of the most reliable (if certain assumptions are fulfilled) being the F-test (21). When the break between signal and noise eigenvalues is sharp, however, the visual test and most of the alternative rules lead to the same decision as the F-test. Such a break is seen in the MV-Youden test eigenvalues. The three ordered eigenvalues before and after removal of the erroneous point [#69] are:

before: 2.48 0.47 >> 0.05   after: 2.83 >> 0.12 0.05

The drop of a factor of ten or more, accompanied by eigenvalues significantly less than unity are strong indicators of the onset of noise dimensions. Thus, before removal of point #69, there are two significant dimensions; afterwards, there is but one. Physically, the first (significant) dimension corresponds to the variation of Zn concentrations in the samples. The noise dimensions correspond to random measurement error.

When the 8 dimensional non-toxic element data matrix (Table II, excluding As, Cd, Hg, Pb, and sample #69) is examined, we find the following series of eigenvalues:

2.81 1.46 1.06 0.96 0.79 0.55 0.23 0.14

Absence of a distinctive drop (at least a factor of three) indicates that these data lack a clear division between signal and noise. Intrinsic heterogeneity prevents the development of a low-dimensional factor model that might be used for classification or mixture analysis. Nevertheless, PC analysis *is* useful for *assumption-free* MV exploration, since the ordered eigenvalues guarantee the most efficient data projections, using the corresponding eigenvectors. Scores (projections) of the data on the first two eigenvectors (Figure 17), representing 53% of the variance from the eight dimensions, will be used for this purpose. Loadings (direction cosines) of the original chemical variables on the two eigenvectors add chemical meaning to the scores. The major loadings (absolute value > 0.5) are as follows:

	Ca	Cr	FB	I	Ni	PA	Se	Zn
PC1:	---	0.7	0.9	----	0.9	0.6	-0.6	---
PC2:	-0.6	---	----	0.6	---	----	----	0..7

Exploratory work can begin fruitfully by scrutinizing the best low-dimensional projections -- i.e., PC projections -- for isolated points ("outliers"), clusters of points, or hints of mixing lines. The first two PC's in Figure 17 contain such hints. Looking for signs of structure along the largest variance axis (PC1), we see the suggestion of some linear behavior extending parallel to the PC1 axis. This implies a systematic variation among the samples involved (#70 - #74) related to the major PC1 variables (+[Cr, FB, Ni, PA], -[Se]). Curiously, these samples are the very ones that comprise PopGp (Population Group)-E. Similarly, the most pronounced feature of the second dimension (PC2) is the isolated point #59. It turns out that this point is almost exactly in the direction of the loading vector for iodine, implying that an iodine outlier may be indicated. (Excess Zn and/or deficient Ca cannot be ruled out at this stage, however; nor can we, on the basis of these data, distinguish between an unusual sample and an experimental artifact.)

Cluster analysis serves as a useful ancillary to PCA. The drawback of cluster analysis is that a priori knowledge of the correct number of clusters is generally lacking, yet necessary for the analysis; and alternative cluster algorithms yield different results. The great advantage of cluster analysis is that it treats the entire 8-dimensional variable space. We are not constrained to a 2- or 3-dimensional projection to assess its results. Application of cluster analysis to the non-toxic DM yields: at the 2-cluster level the clusters {#70, #73} and {all else}; at the 3-cluster level {#59}, {#70, #73}, and {all else}. (Braces are used to denote individual clusters; sample numbers, to denote cluster membership.) Through several higher levels of clustering, {#59} remains the only singleton, cluster {#70, #73} remains intact, and other pairs appear ({#62, #74}, {#71, #72}). Cluster {#59} is clearly the most isolated sample, with a minimum Euclidean distance of 4.1, and a median of 5.6. These distances may be interpreted as the root of chi square. If the data matrix were multivariate normal, the expected value for the 8-dimensional standardized distance would be $\sqrt{8}$ or 2.83. The median distance for a "typical" sample (#55) was 2.92.

Since cluster analysis and PCA are mutually supportive in this example, it makes sense to return to the real (chemical) dimensions for further possible interpretation. The PC2 projection suggested that we look to iodine for the explanation of sample #59. In fact, examination of the distribution of iodine concentrations (which can be done by inspection of Table II) shows that sample #59 is the major outlier; its concentration exceeds that of the nearest point by more than a factor of two, and the median by a factor of eight. It was not a complete blunder, at least in the analytical sense, based on a parallel information value from one of the quality control laboratories. The meaning and/or source of such apparently real excursions as this of course lie at the heart of the DDS program.

The multiply-correlated chemical variables and the associated PopGp-E samples would not have been so readily spotted without the help of the PC projection. Following that lead, however, it is interesting to examine the data from the perspective of three of the primary chemical dimensions: FB, Ni, Se. (The first two showed strong positive correlation with one another and with the "E" samples; the third, showed strong negative correlation.) This selected 3-dimensional view from 8-space is given in Figure 18. Both the 3-variable correlation and the special position of PopGp-E samples (#69 - 74) are evident. (Sample #69 is included here because its erroneous Zn-value does not affect

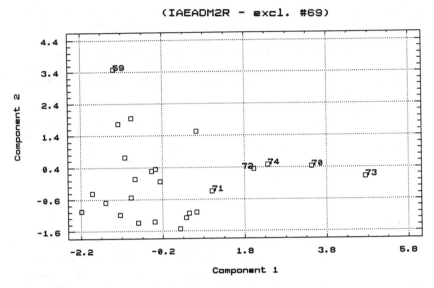

Fig. 17.   Principal component projection of the eight variable "non-toxic" segment of the data matrix in Table II; data were first standardized [IAEA - DDS data].

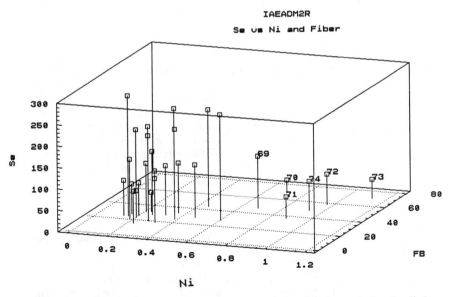

Fig. 18.   "Real" (chemical) variable three dimensional plot of the inter-relations implied by the PC-projection of Fig. 17; units are g/kg (FB), mg/kg (Ni), and μg/kg (Se) [IAEA - DDS data].

this plot.) At this point it becomes very interesting, indeed, to ask our multidisciplinary colleagues what might be the underlying bioenvironmental meaning to such apparent multivariable relationships. Is it possible, for example, that local soil composition may be a factor, or that limited plant availability of Se (considering the vegetarian diet of PopGp-E) may account for the inverse correlation with regional dietary fiber levels? (22; G. V. Iyengar, personal communication, 1989).

A final note: with heterogeneous data one should be quite cautious of going much beyond the tentative type of exploration indicated above. The next steps involving *quantitative* multivariate work, such as class modeling and discrimination, or mixture analysis, or bioenvironmental modeling, should be taken only after homogeneous subgroups have been extracted for individual study.

Conclusion

It was expected and it is evident that important multivariable relations exist in complex environmental and biological systems, as represented by the IAEA's DDS project. Interactive, exploratory univariate and multivariate methods of searching for patterns -- clusters, mixing curves, extrema -- in such data can serve two vital purposes: multivariate control of analytical quality, and generation of questions or even hypotheses concerning the underlying structure of the data. Multivariate data evaluation, though perhaps the most intellectually stimulating facet of the multivariate activities, *must* take its place after the essential prerequisites of multivariate design and multivariate control of accuracy. Beyond that, the exercise can be totally misleading without the benefit of expert, multidisciplinary colleagues.

Acknowledgment

This work would not have been undertaken without the multidisciplinary stimulation of Venkatesh Iyengar, Robert Parr, and Wayne Wolf. I am grateful to Dr. Iyengar for guiding me to some of the relevant literature, and for introducing me to some of the subtleties of bioavailability. Dr. Parr provided encouragement, plus an essential ingredient for this investigation: an early version of the DDS database. Dr. Wolf shared his enthusiasm for biomedical multivariate analysis, and pointed me toward critical literature in this field. Special acknowledgment is due also to Jim Filliben for Figures 6 and 7, and for important guidance in experimental design and the art of univariate, bivariate, and multivariate exploratory data analysis.

Literature Cited

1.   Kowalski, B. R., Ed. Chemometrics: Mathematics and Statistics (Reidel Publishing Co.) 1984.
2.   Massart, D. L., Vandeginste, B. G. M., Deming, S. N., Michotte, Y., and Kaufman, L. Chemometrics: a textbook (1988) Elsevier, Amsterdam.
3.   Currie, L. A. "Chemometrics and Standards," J. Res. Natl. Bur. Stand. (USA), (May/June 1988) 93, 193.

4. International Atomic Energy Agency (1988), R M Parr, Coordinated Research Programme on Human Dietary Intakes.

5. Iyengar, V. and Woittiez, J. "Trace Elements in Human Clinical Specimens: Evaluation of Literature Data to Identify Reference Values," Clinical Chemistry (1988) 34, 474-481; Iyengar, G. V. "Normal Values for the Elemental Composition of Human Tissues in Body Fluids: A New Look at an Old Problem," in D. G. Hemphill, Ed., Trace Substances in Environmental Health - XIX, (Environmental Trace Substances Research Center, Univ. Missouri, 1985) 277-295.

6. Youden, W. J. "Critical Evaluation (Ruggedness) of an Analytical Procedure," Encycl. of Industrial Chemical Analysis (Wiley-Interscience, New York, 1966) 755-788.

7. Box, G. E. P., Hunter, W. G., and Hunter, J. S. Statistics for Experimenters - An Introduction to Design, Data Analysis, and Model Building (John Wiley & Sons, New York) 1978.

8. Ritter, G. L. and Currie, L. A. "Resolution of Spectral Peaks: Use of Empirical Peak Shape," Proc. of the Topical Conference on Computers in Activation Analysis of the American Nuclear Society, (B. S. Carpenter, M. D. D'Agostino, and H.P. Yule, Eds.), DOE Sympos. Series 49, 39 (1979).

9. Currie, L. A., Beebe, K. R., and Klouda, G. A. "What Should We Measure? Aerosol Data: Past and Future," Proc. of the 1988 EPA/APCA International Sympos. on Measurement of Toxic and Related Air Pollutants, (Air Pollution Control Assoc., 1988) 853-863.

10. Gordon, D., "Interactions Among Fe, Zn and Cu Affecting Their Bioavailability," Abstract, This Symposium, ACS, 1989.

11. Currie, L. A. and DeVoe, J. R. "Systematic Error in Chemical Analysis," Chapter 3, Validation of the Measurement Process, p. 114-139, J. R. DeVoe, Ed., (Am. Chem. Soc. Sympos. Series 63; ACS, Washington, DC, 1977).

12. Filliben, J. J. DATAPLOT: Introduction and Overview (1984) NBS Spec. Publ. 667. See also "Dataplot: new features, Version 87.1."

13. Currie, L.A. "The Limitations of Models and Measurements as Revealed through Chemometric Intercomparison, J. Res. Natl. Bur. Stand. (USA) (1986) 90, 409.

14. Alt, F. B. and Smith, N. D. "Multivariate Process Control," Ch. 17 in P. R. Krishnaiah and C. R. Rao, Eds. Quality Control and Reliability (North-Holland Press, Amsterdam) 1988.

15. Wolf, W. R. and Inhat, M. "Evaluation of Certified Biological Reference Materials for Inorganic Nutrient Analysis," Ch. 6 in W. R. Wolf, Ed., Biological Reference Materials (John Wiley & Sons, New York) 1985.

16. Currie, L. A., "Metrological Accuracy: Discussion of 'Measurement Error Models' by L. J. Gleser," Chemometrics and Intelligent Laboratory Systems (1990) in press.

17. Wold, S. and Sjöström, M. "SIMCA: A Method for Analyzing Chemical Data in Terms of Similarity and Analogy," Ch. 12 in Kowalski, B., Ed, Chemometrics: Theory and Application, ACS Sym.Ser. 52 (Amer Chem Soc, Washington, DC 1977).

18. Malinowski, E. R. and Howery, D. G. Factor Analysis in Chemistry, (1980) Wiley, New York.

2

- 99.

0BIOLOGICAL TRACE ELEMENT RESEARCH

19. Hamilton, J. C. and Gemperline, P. J. Mixture Analysis Using Factor Analysis II: Self-Modeling Curve Resolution, J. of Chemometrics, 4 (1990) 1-13.
20. Meglen, R. R. "Chemometrics: Its Role in Chemical and Measurement Sciences," Chemometrics and Intelligent Laboratory Systems, 3 (1988) 17.
21. Malinowski, E. R. Statistical F-Tests for Abstract Factor Analysis and Target Testing, J. of Chemometrics, 3 (1988) 49-60.
22. Allaway, W. H. "Soil-Plant-Animal and Human Interrelationships in Trace Element Nutrition," Ch. 11, V.2 in W. Mertz, Ed., Trace Elements in Human and Animal Nutrition - Fifth Edition (Academic Press, San Diego) 1989.

RECEIVED August 20, 1990

Chapter 7

Establishing Quality Measurements for Inorganic Analysis of Biomaterials

James R. DeVoe

Inorganic Analytical Research Division, Center for Analytical Chemistry, National Institute for Standards and Technology, Gaithersburg, MD 20899

While there have been a number of excellent studies correlating the effects of trace element concentrations in biomaterials with a number of diseases and other health related processes, the inconsistencies of quantitative analytical data still limit progress. It is suggested that a well integrated system of laboratories that is dedicated to analytical chemistry quality assurance can provide an interlaboratory structure that will improve the quality of the measurements. Quantitative measurement of the elemental-bound compound appears to be greatly needed, and significant effort needs to be expended in this most difficult area. Progress can be greatly enhanced by including the analytical chemist in the biomedical research team.

A large number of the inorganic elements play an important role in the functioning of biological systems (1-3). Many of the elements in the middle of the periodic table have beneficial effects at very low concentration levels; often at the parts per billion level by weight. Furthermore, these same elements at higher levels have been shown to be toxic to the same biological systems (4,5). While, there have been a number of studies correlating trace and ultra-trace levels of the elements to disease, inconsistencies in the data still limit progress in conclusively understanding the role that these trace elements play (6-8). There are several reasons for this. The systems in which these elements operate are extremely complex and rarely can be studied without a detailed understanding of interactions with a myriad of other biological systems. In addition, the quantitative measurement of these elements in biological materials, at these ultra-trace levels, is very difficult. In many cases it is known that only certain forms of a given trace

element exhibit the sought-after biological activity, and if total
elemental concentration is to be correlated to these biological
effects, extremely good measurement reproducibility is required
within a given laboratory study. In addition, it is very important
to be able to measure the absolute quantity (e.g., as close to the
true quantity as possible) because correlation between laboratory
studies will help to clarify the complexity of the biological
interactions. Unfortunately, the accuracy of these measurements is
insufficient to result in the determination of meaningful mechanisms
for many of the ultra-trace elemental biological processes.

The purpose of this work is to briefly describe the author's
suggestion for establishing a network of laboratories which have the
capability of making accurate trace element measurements in biomaterials and to indicate how the Inorganic Analytical Research Division
(IARD), Center for Analytical Chemistry (CAC), National Institute of
Standards and Technology (NIST), can assist in establishing good
measurement methods. Finally, once reliable measurements have been
established, a few recommendations for future studies are described.

Establishing Quality Measurements. We have identified a specific
set of operational procedures which will help to establish measurement quality assurance in analytical laboratories. If such procedures are adopted, it should solve long-standing problems that exist
involving a complex array of analytical procedural methodologies
which extend across biomedical research laboratories, industrial
laboratories, and regulatory government agencies. The following
paragraphs describe a means for measuring, continually evaluating
and improving Analytical Chemistry Quality Assurance (ACQA).

Defining the requirements for the generation of ACQA is
straightforward. The establishment of statistical control of the
analytical method is the first requirement, such that reproducibility of the measured concentrations of the elements in biomaterials
can be determined. It is important to note that the reproducibility
is affected by the heterogeneity of the aliquots of the material
being analyzed. It is mandatory that the aliquots of the real
sample have a known compositional uniformity, and this requires
knowledge of the reproducibility of the method, excluding sample
heterogeneity. In the case of many biomaterials, homogeneity can be
the limiting factor on the reproducibility of the total analytical
method. Because the reproducibility of the method itself will limit
the measurement of homogeneity, it is imperative that sets of real
samples with known homogeneity be generated. Once reproducibility
of the total method has been established, the second requirement is
to determine how close the result is to the true composition of the
material. While this difference can never be known exactly,
techniques are available to evaluate the amount of error. Most
often this is done by using basically different methods to determine
a specific constituent in a given material whose homogeneity has
been well-characterized. The term, basically different, relates to
the need for the underlying physico-chemical principles of the
method to be substantially different. If the different methods
agree, then confidence is generated that the methods provide a
concentration of a constituent that is close (within the

reproducibility limit) to the true value. Another approach is to design a series of experiments to measure directly the expected systematic error. Either of these methods is limited by the total reproducibility of the methods used.

It is necessary to assess the reproducibility of each of the measurement methods in any given analytical laboratory. One assessment procedure is through laboratory visits by qualified analytical chemists. Another assessment procedure is to evaluate the results of the laboratory method being used in order to analyze a well characterized reference material with respect to homogeneity. These materials, called performance evaluation standards, should be identical and indistinguishable from normal routine samples measured in the laboratory. At this point assistance to the laboratory can often result in improvement of the reproducibility. This assistance takes two forms; first, data are provided which show how the measurement reproducibility compares with historical data, and second, technical consultation can help the laboratory to improve the methodology.

Again, once reproducibility has been established, reference materials that are certified for absolute concentration can be used to assist the laboratory in eliminating any systematic error. Assistance to the laboratory through appropriate feedback of data, along with expert consultation, can help to eliminate these errors.

To establish ACQA throughout a large complex of laboratories, coordinating laboratories for ACQA (CLACQA) should be established. Their sphere of coordination should be limited to the number of laboratories that can be supported in all of the areas discussed below, which depends greatly upon the staff quality and quantity. These CLACQA's would have a number of responsibilities. First and foremost they must demonstrate exceptional proficiency in measuring the type of materials that the coordinated laboratories routinely analyze, and provide assistance to the coordinated laboratories on all matters of method development, and quality assurance. Second, they produce performance evaluation standards to be used by the coordinated laboratories. These materials are verified to be homogeneous within the limits achievable with routine samples of the same type and by the reproducibility of the analytical method. Third, a series of these standards is distributed among the coordinated laboratories, and each is given information as to their status with respect to past measurement reproducibility and how they compare to the composite of data obtained from other laboratories. If any laboratory has poor measurement quality with respect to the rest, the CLACQA provides assistance to that laboratory to improve the reproducibility of its measurements. Fourth, the CLACQA provides changes to the methods used in the coordinated laboratories to reduce any systematic errors that have been detected. As indicated above, this can be done by designing experiments to evaluate these errors or by using reference materials that have been certified for true composition.

Each CLACQA, along with their coordinated laboratories, should be specialists in the analysis of the trace elements in specific types of biomaterials. The complexity and uniqueness of certain types of biomaterials may require important adjustments to the methods in order to maximize the ACQA.

Finally, a central standards laboratory should be established to work with each of the CLACQA, by providing assistance in all areas of responsibility.

The successful manifestation of such a plan requires some type of sponsor, such as a government organization, large corporation, trade association, or some type of international organization [e.g., the International Atomic Energy Agency, which has considerable experience with the ACQA for trace elements in biomaterials (9)]. To improve the quality of these ultra-trace elemental measurements the emphasis must be placed upon interlaboratory cooperation. The need for a group of selected coordinating laboratories is paramount because the understanding of how to make quality concentration measurements requires dedicated effort.

Measurement Accuracy in Trace Element Analysis of Biomaterials at NIST. The IARD has developed a host of analytical methods which have been carefully designed to eliminate many sources of systematic error. The methods are as unique (basically different, see definition above) as possible so that comparisons of the methods give assurance of minimal statistical correlations. For example, most biomaterials are analyzed non-destructively by neutron activation analysis. When the biomaterial is dissolved in order to be analyzed by methods of demonstrated accuracy such as isotope dilution mass spectrometry and optical methods of analysis, great care is taken to evaluate potential errors in the dissolution step which is a medium of correlation between these two methods.

While the considerations for the preparation and certification of Standard Reference Materials are no different for biomaterials than any other material, these substances present some unusual difficulties. Many biomaterials cannot be reproducibly weighed. This is because they may contain deliquescent components or volatile oils which easily vaporize continuously over extended periods. In addition, care must be taken to assure that any trace elements bound to volatile species are not lost. Some of the biomaterials are very difficult to sample because of their inherent inhomogeneity. Biomaterials are subject to bacterial decomposition at a rate which is faster than most types of materials, and often sterilization tends to perturb the chemical structure of the material.

Because of these problems, not every biomaterial can be prepared to the level of accuracy that may be required. It should be understood that these special materials cannot serve as performance evaluation standards, and it remains the responsibility of the CLACQA to evaluate the systematic errors in a measurement method through use of the SRM. Consideration must be given by the expert analytical chemists in this field as to the type of materials that would be most useful. Also, consideration should be given to the type of biomaterials which should be made available as Standard Reference Materials. For example, does our recently prepared oyster tissue standard, SRM 1566a, suffice for all fishes? or crustaceans? We believe that through working with some appropriate forum which represents a broad cross section of the analytical community who analyze biomaterials (CLACQA's?), appropriate reference materials can be identified and made available.

Future Considerations. As indicated in the introduction, it is becoming clear that it is insufficient to know only the total elemental concentration of a trace element in a biomaterial. For example, it has been shown that the element zinc plays a very important role in the binding domains of DNA (10,11). Since we know that in almost every instance the element is coordinated with some bioorganic moiety, it is now necessary that we measure the concentration of such chemical compounds. There are profound problems with performing these types of analyses. In most cases the chemical compounds must be separated from the biomaterial in such a way that the concentration of the compound is unperturbed. If the compound is in thermodynamic balance with other compounds in the material, the act of extracting it(possibly with some solvent) may result in imbalance and destruction of all or part of the trace element-bound compound. Even if the material can be extracted successfully, the determination of the completeness of the extraction is very difficult because direct analysis of the analyte in the solid is currently infeasible. Moreover, even if the material is robust and some of the elemental compounds are extracted quantitatively, subsequent separation of the various fractions of pure compounds (e.g., by some type of chromatography) presents a much more difficult chemical analysis (12). The amount of element present in the various fractions is reduced, and sample contamination from the laboratory environment becomes a very severe problem. These difficulties need not impair progress in this field, but new concepts in experimental design, such as the use of stable isotopic tracers or neutron activation analysis, are needed to determine the sources and magnitude of systematic errors.

To this end, it is extremely important for the analytical chemist to work with those in the biomedical field as part of a team oriented research project. Too often the biomedical research project suffers from attempting to correlate inaccurate quantitative compositional data to clinical and epidemiological results.

Literature Cited

1. Greenberg, R. R.; Zeisler, R.; Malozowski, S. Anal. Chem. 1989, Vol. 220-28.
2. Iyengar, G. V. Biomedical, Environmental, Compositional and Methodological Aspects of Trace Elements - Volume 1 CRC Press: Boca Raton, FL, 1989.
3. Iyengar, G. V.; Woittiez, J. R. W. Proc. Trace Element Anal. Chem. in Medicine and Biology, 1988, pp 267-69.
4. Nielsen, F. H.; Uthus, E. O. In Biochemistry of the Essential Ultratrace Elements; Plenum Press: New York, 1984; pp 319-40.
5. Behne, D.; Iyengar, G. V. Analytiker Paschenbuch; Springer Verlhe: Berlin, 1986; pp. 237-80.
6. Aalbers, T. G.; Houtman, J. P. W.; Makkink, B. Trace Elements in Medicine 1988, 3, 114-19.

7. Banner, W. P.; DeCosse, J. J.; Tan, Q. H.; Zedeck, M. S. Carcinogenesis 1984, 5, 1543-46.

8. Iyengar, G. V.; Woittiez, J. R. W. Clin. Chem. 1988, 34, 474-81.

9. Parr, R. M. J. Radioanal. Chem. 1977, 39, 421-33.

10. Lee, M. S.; Gippert, G. P.; Soman, K. V.; Case, D. A.; Wright, P. E. Three-Dimensional Solution Structure of a Single Zinc Finger DNA-Binding Domain 1989, 245, 635-37.

11. Rajavashisth. T. B.; Taylor, A. K.; Andalibi, A.; Svenson, K. L.; Lusis, A. J. Science 1989, 245, 640-42.

12. Stone, S. F.; Zeisler, R.; Gordon, G. E. Science 1989, 245, 157-66.

RECEIVED July 16, 1990

Chapter 8

Evaluation and Improvement of Nutrient Composition Data for Trace Elements in Foods

Wayne R. Wolf

Nutrient Composition Laboratory, U.S. Department of Agriculture, Beltsville, MD 20705

Uses for trace element nutrient composition data in foods for dietetic, nutrition, and epidemiological purposes are greatly increasing. There is a large amount of presently available composition data and there is a strong demand for generating new data. Ability to assess the quality of reports and information stemming from these uses of the available composition data rests directly upon a good knowledge of their reliability. An accurate assessment of this reliability is needed. Due to the complexity and scope of the presently available data, this assessment requires a multifaceted evaluation process utilizing computerized technology. For the first step of this assessment, multidisciplinary criteria have been formulated, and an expert software system developed for evaluating the available published data on selenium content in foods. In order to assess reliability of calculations using different nutrient composition databases, a comparison of six USDA databases has been carried out. This comparison looked at data and calculations from these databases using a common list of foods representative of U.S. dietary intake.

The discussion in this paper will focus mainly on the general area of nutrient composition data, of which trace elements are an important part and serve as specific examples. Nutrient composition data in foods are used for a wide variety of dietetic, nutritional, and epidemiological purposes. These composition data are typically made use of in composition databases to estimate dietary nutrient intake via calculations based on food consumed. These calculated estimations of intake are used to formulate menus designed to meet some defined requirements for nutrient intakes, usually in institutional or medical situations. Calculated intakes are also used in nutrition research stud ies to define individual nutrient requirements. For example, studies are carried out to define the metabolism of the trace elements; including understanding the role and interactions trace elements have on the metabolism and utilization of other nutrients, both at the macro and trace levels; and to determine the bioavailability of the trace elements.

Although composition databases are sometimes used to calculate dietary exposure or intake of individuals, it is well recognized that the variability and uncertainty in these estimates can be very significant. Many of the variabilities and uncertainties in the calculated estimates are inherent in the data and in the use of composition databases.

<u>Sampling and Representative Data</u>

Composition data on the nutrient content of a specific food are not single static numbers, because the samples representing that food constantly change. In addition to analytical variation in generating the data, foods have natural variability from sample to sample depending upon a number of factors. These factors include origin (geographic changes in soil or climate), processing, distribution and genetic differences in variety or type. For example, in a recent study reporting the selenium content in bread, coefficients of variation among samples were approximately 50%(<u>1</u>).

Complete composition information on each and every variety of food sample is not available. Composition data in databases are a compilation of available information, usually merging data from several sources. The values are representative of a food or food type and not a specific sample. The data can only be as representative as the underlying sampling plans which generated the analyzed samples. Often the information available in the databases is insufficient to ascertain the representativeness of data in relation to the specific use. The data may represent every possibility from single samples to extensive sampling plans representative of the nation's food supply. Composition data in many databases are representative of fewer individual samples rather than many.

The "science" (in reality an art) of sampling the food supply for purposes of generating nutrient composition data is practiced only in a very few locations which are also sources of analytical data. An important example of a nationwide sampling plan is the US Food and Drug Administration's Total Diet Study Program (2). This nationwide monitoring program, which does include determination of a number of trace elements, will be discussed in more detail later in this paper. In order to have a truly representative nationwide sampling scheme, information on factors such as, marketing and product distribution must be statistically combined with the ability to procure, transport, and most importantly composite and subsample the collected food materials. The important role of representative sampling within a QA scheme for determining trace constituents in food has been summarized by Horwitz, et.al.(<u>3</u>).

In addition to variabilities in the composition databases themselves, estimations of the exact types and amounts of food consumed by individuals are a complex and difficult task(<u>4</u>). Thus, although concern with individual menu or meal planning is becoming more important, and more use of databases and dietary calculations are appearing, considerable caution should be reserved in interpreting information based upon calculated estimates of nutrient intake of individuals.

More often databases are used to calculate dietary intakes and exposures for population groups in order to gain information on public health concerns. Population surveys and population intakes are often combined with other population information on health statistics in the context of epidemiological studies. For example, dietary intake calculations are used to formulate dietary recommendations in terms of nutrient intake for health or toxic concerns for trace elements.

There are no nutritional requirements for specific foods, only for the nutrients supplied in foods. What people eat is defined in terms of kinds and amounts of foods. To get from food intake to nutrients and nutrient requirements requires translation from food to nutrients. This is done either infrequently by analyzing directly the individual foods consumed or most often, by calculation using nutrient composition databases. Thus the body of information used in making any decisions, statements or public policy about health or disease concerns relating to nutrition rests almost solely on a foundation of nutrient composition databases. The reliability of this information and of the resultant decisions or policy, are dependent directly upon the soundness of these databases.

Evaluation And Improving Quality Of Analytical Data

A whole body of metrological knowledge and published information describes ways to evaluate and improve individual analytical data(5-7). This knowledgewill not be further discussed in detail here but includes areas such as: improved and correct analytical methodology; quality control in practice of methodology; and validation and quality assurance of methodology through use of food reference materials.

As stated above, the quality of information used in public policy decisions relating to nutrition rests very strongly on the foundation of information generated by use of nutrient composition databases. These databases are composed of combined individual values usually from a number of sources. Sometimes databases contain imputed or estimated values in the absence of available analytical data. Ability to validate the information generated by these databases, rests directly upon our ability to assess the quality of the individual data contained in the databases.

As an approach to development of methodology for evaluating available composition data, criteria have been established to assess and quantitatively rate published data on selenium in foods(8). These criteria were developed in five categories: number of samples; analytical method; sample handling; sampling plan; and analytical quality control. Ratings based upon these criteria were assigned to data from each individual published study. A Quality Index (QI) derived from the ratings over all categories was assigned to each reported value. Criteria were defined for an acceptable QI and data with less than the acceptable QI were omitted from further consideration. The QI's for acceptable data were then summed over each food to determine a Confidence Code(CC) associated with the mean selenium content of that food. This CC was intended to indicate the relative degree of confidence the user can have in each selenium value. A confidence code of "a" means the user can have considerable confidence in the value. A confidence code of "c" means the user can have less confidence due to limited quantity and /or quality of the data. A computerized expert system has been developed(9) using these criteria and a Table of selenium values ranking the important food sources of selenium has been published(10). The criteria have been adapted to evaluation of copper content of foods and a Table of evaluated data published(11). The concepts and approach used in this evaluation have potential for extension to many other sets of nutrient composition data in addition to the potential for establishment of criteria for evaluation of proposed analytical methods.

Assessment Of Nutrient Composition databases

Evaluation of the confidence in the individual data contained in databases is the first step in establishing the reliability of use of these databases. Several additional questions must be asked in using calculations from a number of databases:

* How compatible are the various calculated results?
* How valid are these calculated results?

In order to gain some information about these questions, a study has been carried out to compare nutrient intake calculations generated from six US Department of Agriculture nutrient composition databases(12). The Nutrient Data Research Branch of the Human Nutrition Information Service(HNIS) in Hyattsville, MD, maintains and provides information on the nutrient composition of foods important to the American diet. HNIS maintains a computerized database which is utilized primarily for the Continuing Survey of Food Intake by Individuals(CSFII)(13). Release 4 of this database was used in this comparison. In addition each of the five Human Nutrition Research Centers (located in Grand Forks, ND; Boston, MS; Beltsville, MD; Houston, TX; and San Francisco, CA) maintains and utilizes a computerized nutrient composition database for calculations of dietary intakes for their research studies and mission oriented responsibilities. Each of these groups by virtue of location and mission would have interest in different nutrient data needs.

Although each of the groups is using an independent database, due to interactions in historical development there are strong ties among several. The HNIS, Grand Forks and Beltsville databases were developed originally with the Houston, Boston and San Francisco databases each developing respectively from one of the original. All of the databases are based on USDA standard reference tapes, with data additions and changes in user software developing in each center.

For scientific validity in comparing nutritional findings among these USDA centers, comparability of calculated results among their nutrient databases is required. Several specific goals of the study relating to evaluation of the databases were: estimate completeness of data in each database; identify and categorize nutrient data differences; and identify multiple data sources (12).

For this evaluation, each Center provided a calculated nutrient content of a representative daily diet based upon an identical list of 200 foods.   This list of foods was obtained from the Food and Drug Administrations Total Diet Study(TDS) program(2). In the TDS program, 200 individual foods and food items, representative of diets of the adult population, are collected in three cities within four regions of the United States. Four to five collections per year are obtained. The individual foods from each collection are sent to the FDA Kansas City Laboratory where they are prepared ready to eat, composited, and each of the 200 food item composites analyzed for a series of constituents for a nationwide monitoring program.   Determination of trace element content of these foods is one part of the FDA monitoring program(2). In addition a complete nutrient profile is being generated on composite samples of these foods as a supplement to a program being carried out in conjunction to the normal FDA-TDS Study. (Tanner, J.T.; Iyengar, G.V.; Wolf, W.R.; Fresenius J. Anal. Chem., In Press, 1990). Thus a large amount of validated analytical information will be available for this list of foods.   This list of foods was developed by FDA using extensive nationwide dietary survey data and is fully representative of foods most frequently consumed in the US diets.

For the USDA database evaluation, each center was asked to code these 200 foods and supply a data tape with the individual food composition information and calculation of total nutrient in a simulated composite based upon a given amount of each individual food. These tapes were forwarded to the Boston HNRC computer center and comparisons were carried out on 28 nutrients common to all six databases. As a first step for comparison, mean values, standard deviations and coefficients of variation(CV) were determined for the total intake across the six centers. Results of these comparisons have been reported(12). One immediately evident result was that not all databases had complete information on all 28 of the nutrients compared. For those nutrients where there was complete information in all databases, the CV's were in general lower than 10%. For example CV's for calories, protein, carbohydrate, total fat, cholesterol, riboflavin, niacin, vitamin C, iron, calcium, potassium and phosphorus were all lower than 5%. For the additional trace elements compared, zinc and copper had CV's over 20% and magnesium had 17%, all three showing missing data in several databases.

While the complete comparison of values for each of the trace elements in the individual foods is too lengthy for summation in this paper, there were significant differences seen in the data among the six centers. These differences were seen to be predominately due to incomplete data in several of the databases. A full evaluation of these results is underway.

In summary, for accurate estimation of individual nutrient intakes the actual foods consumed MUST be analyzed. For populations,   intake calculations from composition databases may be adequate, providing the data have been properly evaluated and validated.

<u>Literature Cited</u>

1. Holden, J. M.; Gebhardt, S., Davis, C.S., and Lurie, D.C. A Nationwide Study of the Selenium Levels and Variability in Bread Products, 74th Annual Meeting, <u>Fed. of Amer. Soc. Exper. Bio.</u>, Washington D.C., April 1990, (Abstract).

2. Pennington, J.A.T. Revision of Total Diet Study Food List and Diets, J. Am. Diet. Assoc; 1983, 82, 166-173.

3. Horwitz, W.; Kamps, L. R.; Boyer, K.W. Quality Assurance in the Analysis of Foods for Trace Constituents,   J. Assoc. Off. Anal. Chem. 1980, 63,1344-1354.

4. Human Nutrition Information Service, USDA: Food Intakes: Individuals in 48 States, Year 1977-78. Nationwide Food Consumption Survey 1977-78, Report No.I-1, 1983.

5. Wolf, W. R.; Ihnat, M. Evaluation of Available Certified Biological Reference Materials for Inorganic Nutrient Analysis, In Biological Reference Materials, Wolf, W.R., Ed.,John Wiley & Sons, New York, 1985,pp 89-105.

6. Uriano, G.A.; Cali, J. P. The Role of Reference Materials and Reference Methods in the Measurement Process, In Validation of the Measurement Process DeVoe, J. R., Ed., American Chemical Society 1977, Chapter 4.

7. Wolf WR, Quality Assurance for Trace Element Analysis, In Trace Elements in Human and Animal Nutrition, 5th Ed., Mertz, W., Ed., Vol 1, Acad. Press, 1987 pp 57-78.

8. Holden, J. M.; Schubert, A. S.; Wolf, W.R.; Beecher, G.R. A System for Evaluating the Quality of Published Nutrient Composition Data: Selenium , A test Case. In Food Composition Data: A Users Perspective., Rand, W.,;Windham, C. T.; Wyse, B.; Young, V.; Eds., Food and Nutrition Bulletin, Suppl. 12, 177-193, 1987

9. Bigwood, D.W.; Heller, S.R.; Wolf, W.R.; Schubert, A.S.; Holden, J.M. SELEX, An `Expert System For Evaluating Published Data on Selenium in Foods Anal. Chem. Acta., 1987, 200, 411-419.

10. Schubert, A.S.; Holden, J.M.; Wolf, W.R. Selenium Content of a Core Group of Foods Based on a Critical Evaluation of Published Analytical Data, J. Of Am. Diet. Assoc 1987, 87, 285-289.

11. Lurie, D.G.; Holden, J.M.; Schubert, A.S.; Wolf W. R.; Miller-Ihli, N.J. The Copper Content of Foods Based on a Critical Evaluation of Published Analytical Data, J. Food Composition and Analysis; 1989, 2, 298-316.

12. Scura, L.; Wolf, W.R.; Rand, W. Comparison of Six Nutrient Databases Within the U.S. Department of Agriculture, Proc. 13th Nutrient Databank Conference; Framingham, Mass. May 1988, pp125-142.

13. U.S. Department of Agriculture, Human Nutrition Information Service, 1986. USDA Nutrient database for Individual Intake Surveys, Rel. 2, Springfield Va: National Tech. Information Service. Acc. No. PB86-206299/HBF. Computer Tape.

RECEIVED August 14, 1990

Chapter 9

Trace Metal Analysis and Quality Assurance in Clinical Chemistry

J. Savory, M. G. Savory, and M. R. Wills

Departments of Pathology, Biochemistry, and Internal Medicine, Health Sciences Center, University of Virginia, Charlottesville, VA 22908

The establishment of a trace element analysis laboratory requires many considerations. Specimen collection must be controlled and detailed protocols established. Sources of contamination in analysis are serious problems although some relatively inexpensive devices can help enormously. Of the instrumental techniques used for the analysis, electrothermal atomic absorption spectrometry has the widest applications. Sample preparation for analysis varies according to the biological material, element in question and instrumental technique. Minimizing the number of manipulations reduces the risk of contamination. Localization of trace elements in tissue requires special techniques which include laser microprobe, electron probe and electron energy loss microanalysis. Conventional clinical chemistry tools such as the use of standard reference materials, and internal and external quality assurance programs, are essential if satisfactory results are to be obtained in the long-term.

The field of metal toxicology over the years has had to cope with many serious problems in order to advance the knowledge base. Foremost among these problems has been the development of suitable analytical methods having the necessary degree of precision, accuracy, and sensitivity. Since many of the inorganic analytes are present in body fluids and tissues in the parts per million and often only parts per billion range, it has taken a considerable degree of refinement to solve these analytical problems. Atomic spectroscopy has contributed most to providing satisfactory analytical methods, although the application of good analytical technique to avoid contamination has been of major importance.

This review summarizes some of the more important considerations in the establishment of a trace metal clinical laboratory.

0097–6156/91/0445–0113$06.00/0

Specimen Collection

Many different types of specimens are submitted for trace element
measurements. The most common are biological fluids usually serum
(or plasma) and urine, although there is sometimes interest in
cerebrospinal fluid. Tissue analysis is also of importance and the
laboratory must be equipped to analyze a wide variety of tissues such
as bone, brain, muscle, liver, etc. Hair analysis has gained some
interest but it should be discouraged in most instances since
contamination from the environment is a major problem. Recently,
because of the high level of interest in aluminum toxicity in
hemodialysis patients, dialysis water and dialysate solutions are
commonly analyzed. General aspects of specimen collection,
processing and storage have been reviewed by Aitio and Jarivisalo (1)
and also by Seiler (2).

    The present authors recommend for trace metal measurements such
as aluminum, copper and zinc that an acid-washed plastic syringe be
used with a stainless steel needle and that the blood be transferred
to a polypropylene tube (Falcon, Oxnard CA) for processing and
storage. An alternate approach which is satisfactory for aluminum is
to use a plastic collection tube containing uncontaminated lithium
heparinate as an anticoagulant. Even greater care must be taken for
ultra trace analysis such as the measurement of chromium, cobalt or
nickel in serum. The blood collection technique must not allow the
specimen to come into contact with a metal needle, and to circumvent
this problem we collect our specimen into an acid-washed
polypropylene test tube using a needle with a teflon or polyethylene
intravenous catheter (3).

    Procedures for the collection and storage of urine and fecal
specimens have been developed in our laboratory (4). Twenty-four
hour urine specimens are collected in plastic containers (Scientific
Products, McGraw Park, IL) and an aliquot transferred and stored at
4°C in a polypropylene tube (Falcon, Oxnard, CA). Fecal specimens
are collected directly into plastic bags, weighed and frozen.

Sources of Contamination in Analysis

Every item used during analysis is a possible source of trace metal
contamination. To be considered are such items as glass and
plasticware, pipette tips, collection tubes, sample cups, purity of
reagents, standards, acids, water and the working environment.
Contamination of glass and plasticware is removed by acid washing and
all acids should be ultrapure grade. Techniques for preparing
exceptionally high purity acids have been reported (5-7). Water
should produce a resistivity of at least 18 megohms and continuous
monitoring is necessary. The room chosen for analysis should have
limited access to ensure a clean working environment and powder-free
gloves must be worn at all times. Sample preparation should be
carried out in an environmental laminar flow hood. We use a filter
unit (122 x 61 cm) suspended from the ceiling with plastic sheeting
enclosed to bench level. This filter unit (MAC 10, Envirco,
Hagerstown, MD) is inexpensive, convenient and provides class 100

air. Several units can be used together to provide more bench space having a clean air environment.

Analytical Techniques

The older chemical and physicochemical methods in general have been replaced by more specific and sensitive new techniques. X-ray fluorescence, neutron activation, atomic spectroscopic, electrochemical and, recently, mass spectrometric techniques have all been applied to the measurement of trace elements in biological specimens. Each of these approaches has its own inherent advantages and disadvantages. However, for most clinical chemistry applications, atomic absorption spectrometry has been most widely used and can be recommended as a reliable analytical technique. Flame atomization methods perform well for some of the more commonly measured elements such as calcium, magnesium, zinc and copper, but for trace and ultratrace analysis electrothermal atomic absorption spectrometry (EAAS) is the method of choice. Advantages of this technique include: (i) sample pretreatment usually can be eliminated, (ii) sample requirements are small (2-100 μl), (iii) graphite furnaces are capable of attaining the high temperature needed to form ground state atoms, and (iv) the atoms stay in the light path for a longer time than flame atomization, resulting in increased concentration and thus greater sensitivity.

There are many factors to consider in choosing parameters for EAAS. The use of pyrolytically coated graphite tubes to minimize reactions between the analyte and the tube is an important consideration and such tubes used together with a pyrolytic graphite platform are recommended for optimal results. However, throughput and other financial considerations sometimes force the analyst to choose alternate approaches. Thus, regional laboratories providing aluminum measurements for a large number of renal dialysis units are often forced to dispense with the use of the platform in order to perform the large number of analyses requested. Such a compromise does not appear to jeopardize patient care since the quality of the results can be maintained at a high level provided other analytical principles are not violated. Other considerations in establishing a procedure for the measurement of a trace element are the use of matrix modification, background correction and sample preparation. These factors especially as they pertain to aluminum measurements have been reviewed by the present authors (4) and by Slavin (8).

The use of EAAS is not ideal for all applications but, as mentioned earlier, there are distinct advantages over other methods. Neutron activation analysis is very sensitive but is a complex technique requiring access to a nuclear reactor. Also for biological analyses, the elements sodium, chlorine and bromine can cause problems of masking, often making a separation step necessary. X-ray fluorescence is specific but lacks sensitivity for many analytes. Inductively coupled plasma emission has advantages for very refractory elements but often sensitivity is less than EAAS. Inductively coupled plasma interfaced with mass spectrometry (ICP/MS) is relatively new as a commercially available technique although it was first described over a decade ago. ICP/MS is a multielement

technique which provides the specificity inherent in isotope dilution
mass spectrometry and which also allows elemental bioavailability to
be studied using stable isotopes. Biomedical applications of ICP/MS
have been reviewed recently by Delves (9).

Sample Preparation for Analysis

A wide variety of sample preparation methods have been used, the
level of complexity being dependent on the final analytical technique
used.
    For EAAS of biological fluids we use the following four
procedures depending on the analyte: (i) dilution and direct
analysis with external calibration, (ii) direct analysis with
standard additions, (iii) digestion and extraction and (iv) protein
precipitation.
    The first of these procedures is the most straightforward and
should be used whenever possible.   Ideally the standard curve is
constructed using aqueous standards.   However if the sample matrix
causes problems then matrix based calibrators may be used.   Some
workers (10) have used this approach for serum aluminum by preparing
standards in a serum pool containing a minimal amount of endogenous
aluminum. Standard additions can be used to minimize matrix effects
but this approach inherently is imprecise since multiple analytical
measurements, each with its own imprecisions, are made to achieve the
final result. A valuable technique for the determination of nickel,
and aluminum in serum is to precipitate proteins using a small amount
of ultrapure concentrated nitric acid (11,12). The final preparation
which is relatively protein free has markedly reduced matrix effects.
    Urine samples similarly may be simply diluted prior to EAAS
measurement (10), although acid digestion is often necessary (Rains,
T.C., NIST, personal communication, 1988). Analysis of feces is more
complicated and the procedure we use is as follows (13).   Frozen
specimens are thawed and weighed in the original plastic container
used for collection. Deionized water is added (1 ml per 2 g feces)
and the sample is homogenized on a paint shaker in a sealed paint
can.   A 10-ml aliquot is ashed, dissolved in dilute HNO$_3$, and
analyzed by EAAS.
    Soft tissue samples, such as brain, liver, or muscle, must be
homogenized before processing, and this can be accomplished easily by
pummelling the tissue in a "Stomacher" blender (Fisher Scientific
Company, Pittsburgh, PA 15219). Deionized water (5 ml) is added to
the bag with the tissue. The sealed bag is placed in the blender and
blended for 5-15 min, which completely homogenizes the sample.   The
homogenate can then be processed for analysis.
    Microwave digestion of tissue specimens is a new technique
which provides an excellent means of preparing a sample for EAAS
without risking excessive contamination (14).   We use a microwave
digestion bomb (Parr Instrument Company, Moline, IL 61265) which is
modified with a teflon liner with a smaller sample well than the
standard liner. Approximately 50 mg (100 mg maximum) of dried tissue
plus 1 ml of 50% nitric acid are added to the teflon well. The bomb
is sealed and placed in a standard microwave oven for approximately
1 minute. Elevated temperature and pressure promote rapid digestion

of the tissue, while the sealed teflon liner eliminates external
contamination and loss due to volatilization.   Upon cooling the
digest is ready for EAAS analysis.

Localization of Trace Elements in Tissues

There is extensive literature on elemental microanalysis but for the
purposes of this review we present only information on aluminum whose
presence in tissues, particularly bone, is diagnostic of aluminum
toxicity. A histochemical reaction for the demonstration of aluminum
in tissues was developed in 1955 (15) using ammonium aurintricarboxy-
late (aluminon) which forms a lake with a number of metallic ions;
aluminum produces a cherry red color.   This technique has been
validated by electron probe x-ray microanalysis (16).
    The scanning electron microscope has been used to detect
aluminum deposits by energy-dispersive x-ray analysis. Aluminum
deposits have been located within the glomerular basement membrane
(17). Using this same technique, Perl, et al. (18) examined brain
tissues from patients with amyotrophic lateral sclerosis and
parkinsonism-dementia, and located aluminum in neurofibrillary
tangle-bearing hippocampal neurons. Other workers also using this
technique have localized aluminum in bone tissues (19,20).
    An alternative approach to aluminum localization in tissues is
to use the new technique of laser microprobe mass spectrometry
(LAMMA). In this technique an intense laser beam is directed toward
a specific organelle and metals localized in that region are analyzed
in a mass spectrometer.   This technique has been used to localize
aluminum in the lysosomes of hepatocytes as well as Kupffer cells
from liver tissue of patients on chronic hemodialysis (21,22). We
recently have used LAMMA to study the ultrastructural localization of
aluminum in livers of aluminum maltol-treated rabbits (23). An even
newer technique is that of electron energy loss spectroscopy (EELS)
which provides exceptional resolution even though the original
promise of excellent sensitivity has not materialized.   The
principles of the technique are given in two excellent reviews (24,
25). We have applied this technique to the analysis of liver from
aluminum maltol-treated rabbits (26).
    A major problem with microanalysis of tissue is in the
processing of the specimen. Conventional chemical fixation as
developed for morphological studies, is not ideal for localizing
trace elements. Movement of metal ions within the cell must be a
serious consideration as tissues are left for several days in a
fixative. We presently use rapid freezing and freeze substitution as
a means of tissue processing, but these techniques are extremely
difficult.
    Although both the LAMMA and EELS require very expensive
sophisticated equipment, both are extremely powerful techniques for
studying metal localization in tissues. The sensitivity of the LAMMA
technique appears to be better than EELS but the resolution of the
latter techniques is better. Our understanding of the toxicity of
metals will be aided considerably by the use of such localization
techniques.

Standard Reference Materials and Quality Assurance

The availability of standard reference materials is of considerable importance in the application of reliable assays for trace and ultratrace metals. The National Institute of Standards and Technology (NIST, Washington, DC 20234) provides aqueous standard solutions, which contain 10 mg/mL of a standard in acid medium. In addition, NIST offers other certified reference materials for serum, urine and other biologic matrices. These materials are useful not only in evaluating accuracy but can also be used to calibrate instrumentation. The materials are certified for accuracy based on determinations by definitive methods. Aqueous standards are also available from commercial sources, one being Fisher Scientific Company. Dilution to working standards must incorporate an intermediate standard if accuracy is to be maintained.

Quality assurance in trace metal measurements follows established principles used in all areas of clinical chemistry. Internal quality control materials, usually serum from a human donor or animal source such as a cow or steer, at two or three different concentrations, should be used on a daily basis. The concentrations in these materials should coincide with various decision levels. Interlaboratory control materials also are an important part of any quality assurance scheme. Presently, such a program is conducted by the Robens Institute (University of Surrey, Guildford, Surrey, GU2 5XY, United Kingdom), for example. The institute mails samples monthly to laboratories for selected trace metal analyses. The data and statistical information of the 38 participating laboratories are then made available for review.

The clinical chemist inspired to perform trace element measurements must appreciate the subtleties needed to provide satisfactory results. Not only is the instrumental method selection, but also specimen collection and processing of samples are important. Strict adherence to basic principles of quantitative analysis is essential if the clinician and, of course, the patient is to benefit from the efforts of the laboratorian.

Literature Cited

1. Aitio, A.; Jarvisalo, J. Pure Appl. Chem. 1984, 56, 549-66.
2. Seiler, H.G. In: Handbook on Toxicity of Inorganic Compounds, Seiler, H.G.; Sigel, H., Eds.; New York: Marcel Dekker, Inc., 1988; pp 39-49.
3. Brown, S.S.; Nomoto, S., Stoeppler, M.; Sunderman, F.W., Jr. Pure Appl. Chem. 1981, 53, 773-81.
4. Savory, J.; Wills, M.R. In: Aluminum and Health A Critical Review, Gitelman, H.J., Ed.; New York: Marcel Dekker, Inc., 1988; pp 1-26.
5. Moody, J.R.; Beary, E.S. Talanta 1982, 29, 1003-10.
6. Mattinson, J.H. Anal. Chem. 1972, 44, 1715-16.
7. Maas, R.P.; Dressing, S.A. Anal. Chem. 1982, 55, 808-9.
8. Slavin, W. J. Anal. Atomic Spectrmet. 1986, 1, 281-5.
9. Delves, H.T. Chem. Brit. 1988, 24, 1009-12.
10. Leung, F.Y.; Henderson, A.R. Clin. Chem. 1982, 28, 2139-43.

11. Sunderman, F.W.; Crisostomo, C.; Reid, M.C.; Hopfer, S.M.; Nomoto, S. Ann. Clin. Lab. Sci. 1984, 14, 232-41.
12. Brown, S.; Bertholf, R.L.; Wills, M.R.; Savory, J. Clin. Chem. 1984, 30, 1216-18.
13. Brown, S.; Mendoza, N.; Bertholf, R.L.; Ross, R.; Wills, M.R.; Savory, J. Res. Commun. Chem. Pathol. Pharmacol. 1986, 53, 105-16.
14. Nicholson, J.R.P.; Savory, M.G.; Savory, J.; Wills, M.R. Clin. Chem. 1989, 35, 488-490.
15. Irwin, D.A. Arch. Ind. Health 1955, 12, 218-70.
16. Smith, P.S.; McClure, J. J. Clin. Pathol. 1982, 35, 1283-93.
17. Smith, D.M., Jr.; Pitcock, J.A.; Murphy, W.M. Am. J. Clin. Pathol. 1982, 77, 341-6.
18. Perl, D.P.; Gajdusek, D.C.; Garruto, R.M.; Yanagihara, R.T.; Gibbs, C.J., Jr. Science 1982, 217, 1053-5.
19. Boyce, B.F.; Elder, H.Y.; Fell, G.S.; Nicholson, W.A.P.; Smith, G.D.; Dempster, D.W.; Gray, C.C.; Boyle, I.T. Scanning Electron Microsc. 1981, 111, 29-37.
20. Cournot-Witmer, G.; Zingraff, J.; Plachot, J.J.; Escaig, F.; Lefevre, R.; Boumati, P.; Bourdeau, A.; Garabedian, M.; Galle, P.; Bourdon, R.; Drueke, T.; Balsan, S. Kidney Int. 1981, 20, 375-85.
21. De Broe, M.E.; Van de Vyver, F.L.; Bekaert, A.B.; D'Haese, P.; Paulus, G.J.; Visser, W.J.; Van Grieken, R.; de Wolff, F.A.; Verbueken, A.H. In: Trace Elements in Renal Insufficiency; Quellhorst, E.A.; Finke, K.; Fuchs, C. Eds.; Basel:Karger, 1983; pp 37-46.
22. Verbueken, A.H.; Van de Vyver, F.L.; Van Grieken, R.E.; Paulus, G.J.; Visser, E.F.; D'Haese, P.; De Broe, M.E. Clin. Chem. 1984, 30, 763-8.
23. Vandeputte, D.; Savory, J.; Van Grieken, R.E.; Jacob, W.A.; Bertholf, R.L.; Wills, M.R. Biomed. Environ. Mass Spectrom. 1989, 18, 598-602.
24. Isaacson, M. Scanning Electron Microsc. 1978, 1, 763-76.
25. Ottensmeyer, F.P.; Andrew, J.W. J. Ultrastructure Res. 1980, 72, 336-48.
26. Vandeputte, D.F.; Van Grieken, R.E.; Savory, J.; Wills, M.R.; Jacob, W.A. 1st European Workshop on Modern Developments and Applications in Microbeam Analysis (Abstract). Antwerp, Belgium, 8-10 March 1989.

RECEIVED July 16, 1990

Chapter 10

Accuracy and Precision of Trace Metal Determinations in Biological Fluids

Interlaboratory Comparison Program

Jean-Phillipe Weber

Centre de Toxicologie du Quebec, Le Centre Hospitalier de l'Université Laval, 2705 Boulevard Laurier, Quebec G1V 4G2, Canada

Since 1979, we have been conducting an interlaboratory comparison program for several toxic elements in blood and urine with a view to validate the accuracy and precision of toxicological trace element analyses. Presently, over 120 North American and European laboratories participate. Samples are prepared by pooling the specimens obtained from exposed workers and patients and are sent bimonthly to participants. The target values determined from the results of reference laboratories are used to evaluate proficiency of the participating laboratories. The participants are ranked according to their accuracy and reproducibility. Analytical performance has improved over time. Comparison of methods has enabled us to identify problems in the determination of several analytes, especially serum aluminum and urine mercury.

The accurate determination of trace metals in human biological fluids is not a trivial task. Many problems confront the analyst among which are contamination during the sampling and analytical process, and the complexity of the biological matrix. The absence of standardized methodology adds to the burden of the analyst who must validate his chosen method and ensure its reliability over time.

Within the laboratory, it is feasible to verify the reproducibility of results generated by a given analytical technique, simply by repetitive analysis of the same sample (e.g., a patient sample) over the course of time. The results, obtained within a preset window of tolerance will indicate whether the desired level of precision has been achieved.

This however does not address the question of the accuracy of the measurement, for which a representative reference standard of known concentration is needed. Aqueous standard solutions of metals, used to prepare calibration curves,are not suitable since they do not take into account matrix effects. Available reference materials, such as NIST (US National Institute of Standards and Technology) bovine liver are better but still have drawbacks, e.g., the concentration of the analyte is not necessarily in the desired range and the matrix is not identical, necessitating a different analytical procedure.

One possible solution is an interlaboratory comparison program in which participants analyze aliquots of the same representative sample. The results are used to estimate the true concentration or target value. By comparing their own results with the target value, participants can then assess their accuracy and precision. We have operated such a program for several toxic metals in human blood and urine over the course of the past eleven years. The data generated by participants have enabled us to identify problems in the determination of serum aluminum and urine mercury and thus to suggest alternative techniques.

0097–6156/91/0445–0120$06.00/0

PROGRAM DESCRIPTION

This program has been described elsewhere (1). It is based on the analysis of control samples by participants.   As shown in Fig. 1, the number of participants has increased steadily over the course of time to the present level of more than 120 laboratories. The international composition of the program is illustrated in Fig. 2.

Available analytes and matrices include lead and cadmium in blood; aluminum, copper, zinc and selenium in serum; and arsenic, cadmium, mercury, chromium and fluoride in urine.

Control samples are prepared either by pooling material obtained from exposed persons, or if temporarily unavailable, by adding a known quantity of the trace metal to normal human blood or urine.   These samples are thus very similar to the real samples. Blood and serum specimens testing positive for HIV and Australia antigen are rejected prior to pooling. Each sample is then divided into aliquots (usually about 5 ml for blood and 20 ml for urine). The aliquots are sent to participants who are thus allowed five weeks to analyze the sample and report the results to us. Laboratories are identified only by a code number in order to ensure confidentiality.   Each of these six annual runs includes three samples per analyte/matrix pair. Several samples are sent in duplicate over the course of a year in order to evaluate long-term reproducibility.

We compile results, and perform statistical calculations including determination of mean, standard deviation and median for all results and according to the analytical method used.

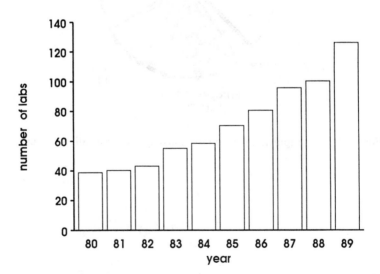

Figure 1. Participation in the program as a function of time (for all analytes and matrices).

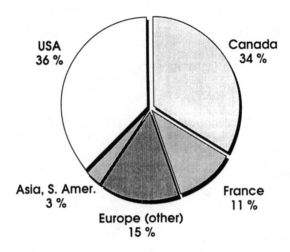

Figure 2. Geographical location of participants (for all analytes and matrices)

A target value is set, based on the results obtained by a subgroup of "reference" laboratories identified by their past reliable performance. Limits of acceptability are chosen at a high and low level for each analyte using criteria adapted from those developed by Yeoman [2] and Taylor and Briggs [3]. "Outer" limits are based on what is necessary for clinical purposes, and what can be achieved using currently available techniques (e.g., for blood lead ± 0.15 µmol/l at 0.5 µmol/l and ± 0.3 µmol/l at 3.0 µmol/l). Limits at intermediate concentrations are obtained by linear interpolation. More stringent "inner" limits allow discrimination between highly and moderately proficient laboratories. These criteria are shown in Table I. A report is sent back to the participants within two weeks.

The performance of each participant is evaluated annually using a scoring system which takes into account both accuracy and reproducibility. For accuracy, each result within the outer limits scores 1 point. An additional point is awarded if the result is also within the inner limits. Samples sent in duplicate are used to evaluate reproducibility. For each pair, the difference between the two results is used to compute a "reproducibility score", using the same criteria. An individual performance summary is prepared for each participant.

The data gathered during the course of the program are also used to evaluate the various analytical methods.

We have previously reported our findings for blood lead and cadmium, and urine arsenic and cadmium [1,4]. Performance trends for aluminum in serum and mercury in urine are discussed below.

ALUMINUM IN SERUM

The performance for aluminum in serum was examined over a three-year period (1985-1987). As shown in Table 1, only a few participants were able to attain a reasonable proficiency level (e.g. for 1987, only 26 % of laboratories scored over 75 %). In an attempt to shed some light on the causes of this situation, we sent a questionnaire to all laboratories participating for serum aluminum asking for detailed information on analytical methodology and laboratory practices, with the following topics being addressed : sample preparation procedures (sample volume used, use of digestion, complexation, deproteinization, dilution, etc.); labware used (glass or plastic, acid wash of material prior to use, etc.); instrumental techniques (type, instrument model, specific operation conditions); calibration (type of calibration such as aqueous standards, matrix-matched standards, standard additions and the use of standard reference materials; experience in using the method (number of years, samples/year).

The data obtained were compared with the observed performance to identify parameters linked to analytical proficiency. Performance for the year 1987 was quantified using the overall score obtained by each laboratory. Participants were then assigned to one of three groups as shown below:

Performance Category	Overall 1987 Score
good	> 75 %
average	50 - 75 %
poor	< 50 %

There were many common features in the analytical approach used by participants: all used graphite furnace atomic absorption spectrophotometry with 75 % preferring Perkin-Elmer instrumentation; a majority of participants used background correction (89 %, of which 56 % preferred the Zeeman system and 33 % chose the deuterium system), and pyrocoated graphite tubes (85 %, of which 60 % also used the pyrocoated platform); sample preparation was kept to a minimum, consisting mainly of dilution with water.

Table I. Acceptability Criteria

Substance	Target Value	Inner limit	Outer limit
Blood Lead	0.5	± 0.05	± 0.15
(μmol/l)	3.0	± 0.15	± 0.30
Urine Arsenic	0.7	± 0.1	± 0.3
(μmol/l)	3.3	± 0.3	± 0.5
Urine Mercury	50	± 7	± 15
(nmol/l)	1250	± 125	± 250
Urine Fluoride	50	± 5	± 10
(μmol/l)	525	± 25	± 50
Blood Cadmium	20	± 4.5	± 9
(nmol/l)	220	± 9	± 18
Urine cadmium	20	± 4.5	± 9
(nmol/l)	220	± 9	± 18
Serum Aluminum	1.0	± 0.2	± 0.4
(μmol/l)	4.0	± 0.4	± 0.8
Urine Chromium	40	± 8	± 15
(nmol/l)	400	± 18	± 36
Serum Selenium	0.75	± 0.06	± 0.12
(μmol/l)	2.00	± 0.10	± 0.20
Serum Copper	4.0	± 0.5	± 1.0
(μmol/l)	20.0	± 0.8	± 1.5
Serum Zinc	4.0	± 0.5	± 1.0
(μmol/l)	20.0	± 0.8	± 1.5

Notes :   1)   Limits at other concentrations are obtained by linear interpolation.
           2)   Results within the outer limits are considered acceptable.

However, differences in methodology existed in the following areas : acid rinse of laboratory material (wash all containers and pipettes, wash only micropipette tips, wash only containers, wash nothing); the dilution factor varied between 2 and 10-fold; the calibration methods varied among aqueous standards (30 % of participants), matrix-matched standards (26 %) and the method of standard additions (44 %); use of reference material (60 % of participants); variations in furnace temperature program (times and temperatures) were numerous; participants' experience in using the technique varied between 6 months and 5 years.

Major differences in analytical technique were found between "good" laboratories and others in the following areas:

PREWASHING PROCEDURE: "Good" performers tended to either acid-wash both pipette tips and containers (50%), or do no prewashing (25%). In contrast "poor" performers always carried out at least a partial wash of labware, especially containers. This would seem to indicate that prewashing with a dilute acid solution can actually do more harm than good. Whether this is due to contamination of the acid used for cleaning or simply a consequence of additional handling is unclear. However, even somewhat contaminated acid would not contaminate containers as long as the subsequent water rinse was done with pure water. It would probably be a better idea to determine the necessity of cleaning the labware by running blanks on a sample from each new lot.

CALIBRATION PROCEDURE: Three different calibration methods were used: aqueous standards, matrix-matched standards and standard additions. As seen in Fig. 3, the "good" performers used all three methods in roughly equal proportion (38, 38, 24% respectively). Surprisingly, all "poor" performers used the method of standard additions. This would suggest that more complicated techniques do not guarantee better results; in fact the opposite seems true.

EXPERIENCE: The picture for this parameter is more murky. However all "good" performers had been using their analytical method for at least one year. No pattern could be distinguished for "poor" performers. A certain amount of experience thus seems necessary to achieve a satisfactory degree of proficiency. Experience is however no guarantee of proficiency.

For other parameters, no relationship was observed. Methodological differences did not appear to influence performance. These included the type of graphite tube used (pyro-coated or not, with or without platform) and the background correction scheme (deuterium or Zeeman correctors). In this latter case, "good" performers were more frequent users of deuterium (50%) than Zeeman (40%). This probably reflects their greater experience and consequently older instruments which are more likely to be fitted with deuterium correctors.

Our general conclusion is that simpler methods tend to yield more accurate and reproducible results. Complexity increases the likelihood of contamination in the case of trace metal determination. It would be a worthwhile exercise for each participant to examine his or her analytical method critically to assess the necessity of each step.

Of course such a survey can only take into account objective parameters, and may not explain why two laboratories using apparently identical methods obtain divergent results. Conclusions must therefore be made in the light of this reality.

MERCURY IN URINE

Traditionally, the determination of mercury in urine has been performed by cold vapor atomic absorption spectrophotometry (CVAAS). The sample is usually pretreated by acid digestion, in order to generate Hg (II) which is then reduced to its elemental form using an appropriate reductant (stannous chloride or sodium borohydride). However, it had been reported as early as 1977 (5) that the digestion step was unnecessary. During the past few years, participants in the program have increasingly switched to a non-digestion cold vapor technique as shown in Fig. 4.

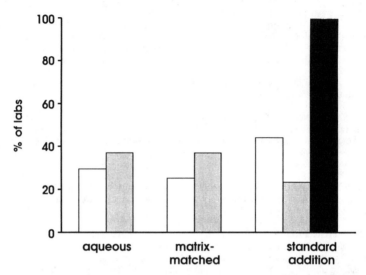

Figure 3. Serum Aluminum : Calibration methods used by participants, according to their performance category (□: overall, ▨: "good" performers, ■ : "poor" performers)

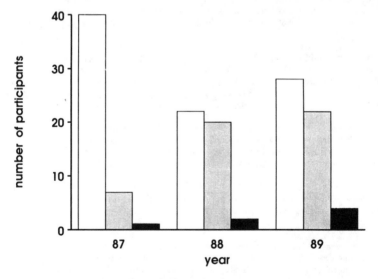

Figure 4. Urine Mercury : Analytical methods used by participants as a function of time. (□: cold vapor with digestion, ▨: cold vapor without digestion, ■: gold film analyzer)

Recently, a new instrumental technique, the gold film analyzer (Jerome Instruments, Jerome, AZ) has emerged on the marketplace. As in the CVAAS technique, elemental mercury vapor is generated. The mercury is then adsorbed onto a gold film. The electrical conductivity of the resulting amalgam varies with the amount of mercury present.

We have examined the performance of participants for the three-year period, 1987-1989, in using CVAAS methods with or without sample digestion. Due to the small number of laboratories using the gold film method, no comparison was made for this technique.

As seen in Fig. 5, participants using the digestion procedure initially scored much higher. The performance of laboratories which did not use a digestion procedure improved gradually. By the end of 1989, 70% of these laboratories were able to generate acceptable results. The reasons for this improvement remain to be explained. However the necessity of digesting urine samples prior to the reduction of mercury can be questioned. Nevertheless, laboratories using a digestion procedure still performed somewhat better with 80% obtaining acceptable results.

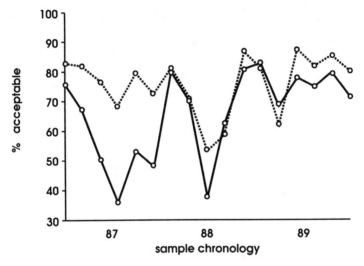

Figure 5. Urine Mercury : Observed performance over time according to the analytical method used (--- O --- : cold vapor with digestion, —O— : cold vapor without digestion)

CONCLUSION

Participation in this interlaboratory comparison program has provided feedback to participants, enabling them, when necessary, to revise their analytical methods and operating procedures. Data obtained over the duration of the program have allowed us to evaluate the different methods used by participants and to suggest modifications in specific instances.

For both aluminum in serum and mercury in urine it seems acceptable results can be obtained with only limited sample pretreatment. Additional steps in sample preparation do not necessarily improve performance.

ACKNOWLEDGMENTS

The financial support of the Institut de Recherche en Santé et Sécurité du Travail du Québec is gratefully acknowledged. I thank Alain Beaudet, Suzanne Morin, Sergine Lapointe and Jacinthe Larochelle for their able technical assistance and Claudette Blais for her secretarial and administrative work.

Literature Cited

1. Weber, J.P. Sci. Total Environ. 1988, 71, 111-23.

2. Yeoman, W.B., In Analytical Techniques for Heavy Metals in Biological Fluids; Occupational and Environmental Commission of the European Communities Joint Research Centre, Ed., ISPRA, Italy. 1980.

3. Taylor, A.; Briggs, R.J. J. Anal. At. Spectromet. 1986, 1, 391-94.

4. Savoie, J.Y.; Weber, J.P. In Chemical Toxicology and Clinical Chemistry of Metals; Brown, S.S.; and Savory, J. Eds.; Academic Press : New York, 1983, pp 77-80.

5. Ebbestadt, V.; Gunderson, G.; Torgrimsen, T.A. At. Absorp. Newslett. 1977, 14, 142-43.

RECEIVED August 6, 1990

DETERMINATION

Chapter 11

Trace Elements in Biological Fluids
Determination by Stabilized Temperature Platform Furnace Atomic Absorption Spectrometry

K. S. Subramanian

Environmental Health Centre, Health and Welfare Canada, Tunney's
Pasture, Ottawa, Ontario K1A 0L2, Canada

The application of stabilized temperature platform
furnace atomic absorption spectrometry (STPFAAS) to
the determination of Al, As, Au, Be, Cd, Cu, Mn, Pb,
Pt, Sb, Se and Tl in various biological fluids is
critically reviewed. Al, Se, Pb and Cd have received
the greatest attention. The biological fluids
analyzed include whole blood, serum, plasma,
erythrocytes, urine, cerebrospinal fluid and
intraocular fluid. Although some workers have sought
to make the STPF methods simpler and faster by using
direct or dilution methods, the original
recommendation to use matrix modifiers still seems
valid. Matrix interferences have been considerably
minimized (but, not eliminated) under STPF conditions
compared to conventional GFAAS approach. STPF only
partially overcomes the temporal and spatial
temperature variations found in the commercial non-
isothermal atomizers. As a result, total isothermal
conditions are not maintained within the stabilized
temperature platform furnace and problems with the
interference-free, precise and accurate measurement of
trace elements in biological fluids still persist.

It is now well recognized that ultratrace amounts of many
metals play a vital role in several biochemical, clinical,
nutritional, toxicological, and environmental and occupational
health problems. Changes in the concentration of both
beneficial and harmful trace metals in human body fluids are
observed in a variety of pathological conditions (1).
Therefore, the determination of trace metals in biological
fluids such as blood (the term 'blood' is used here to denote
whole blood, erythrocytes, serum and plasma), urine, semen and
cerebrospinal fluid can assist in the diagnosis of metabolic

0097–6156/91/0445–0130$08.00/0

disorders; confirm environmental and occupational exposure to
toxic metals; be useful in monitoring a patient's response to
treatment; give valuable information about the net intestinal
absorption and net retention of elements; and be helpful in
studies of synthetic oral diets and parenteral nutrition (2).

The impetus for the impressive developments in biomedical
trace element research came with the introduction of
ultrasensitive analytical techniques such as graphite-furnace
atomic absorption spectrometry (GFAAS), inductively-coupled
plasma atomic emission spectrometry and mass spectrometry (ICP-
AES and ICP-MS), isotope-dilution mass spectrometry (IDMS), and
neutron activation analysis (NAA). The high sensitivity (10^{-13}-
10^{-11} g absolute), scope (possibility of determining at least 40
elements), the small sample size required (µl volumes), the
comparatively low cost of apparatus and low cost of analysis,
the relative speed of measurement, ease of operation and the
potential for direct analysis have made GFAAS the most popular
technique for the routine determination of trace metals in
biological fluids (3-5).

Table I provides a list of the elements that have been
determined in blood and urine to date by GFAAS, their "normal"
physiological concentrations and GFAAS sensitivities (6,7).
Note that at least 27 elements have been determined and that
GFAAS possesses adequate sensitivity for the direct
determination of even "normal" levels of many of the elements
listed in Table I. Reliable data for the "normal"
physiological levels of trace elements in biological fluids
other than blood and urine are not available at present.

Despite its popularity, GFAAS is not without problems.
Although papers on the GFAAS determination of trace elements in
biological fluids have appeared since 1971, publications prior
to 1980 are now only of historical interest. The pre-1980
Massman-type commercial non-isothermal AA spectrometers were
designed to process steady state signals from the flame and
were too slow (time constant > 1 s) to faithfully follow the
fast, transient signals from the furnace (8,9). The slow
detection, the vaporization of the sample from the hot tube
surface into a cooler gaseous atmosphere and the virtual use of
peak height measurements resulted in severe interferences,
especially in the analysis of complex matrices such as
biological samples. The pre-1980 attempts to overcome or
minimize these matrix interferences were, in general,
unsatisfactory (3).

Since 1981, Slavin and colleagues (8-15) have sought to
minimize the spectral and non-spectral interferences that
plague GFAAS by the optimization of a set of mutually dependent
furnace parameters. The cardinal features of this modern
furnace technology, referred to by them as the stabilized
temperature platform furnace or STPF, have been described
elsewhere (3), but are briefly delineated below. STPF consists
of: (i) a pyrolytic graphite platform to facilitate atomization
under isothermal conditions; (ii) fast electronics (time
constant < 20 ms) to minimize signal distortion; (iii)
pyrolytically coated graphite tube to minimize interaction of
the analyte (e.g., carbide formation) with the graphite surface
and to prevent entrapment of the analyte or matrix within its
porous structure; (iv) use of peak area rather than peak height
to minimize dependence on the nature of the sample matrix; (v)
accurate establishment of the baseline of the absorbance

Table I.   Elements Determined in Blood and Urine by Graphite
Furnace Atomic Absorption Spectrometry[a]

Element	Blood Reference Range (µg/l)[b]	Urine Reference Range (µg/l)[b]	GFAAS Sensitivity (pg/0.0044 A.s)[c]
Ag	≤ 1	-[d]	1.4
Al	< 10	-	10.0
As	≤ 1	-	17.0
Au	-	-	13.0
Be	-	-	1.0
Bi	-	-	24.0
Cd	0.3-1.2	≤ 1	0.4
Co	0.1-0.3	≤ 1	7.0
Cr	0.04-0.39	0.2-0.5	3.3
Cu	815-1370	30-60	8.0
Fe	800-1200	100-150	5.0
Ga	-	-	14.0
Li	-	-	1.4
Mn	0.3-1.0	-	2.2
Mo	0.3-1.0	-	9.0
Ni	0.1-1.0	-	13.0
Pb	90-150	-	12.0
Pd	-	-	24.0
Pt	-	-	115.0
Sb	≤ 0.5	-	38.0
Se	50-120	25-50	30.0
Si	-	-	40.0
Sn	-	-	23.0
Te	-	-	15.0
Tl	-	-	17.0
V	-	-	40.0
Zn	800-1100	400-600	0.5

[a]Reliable reference ranges are not available for physiological
fluids other than blood and urine.
[b]The values given are the most probable range for humans not
exposed to these elements particularly through occupation and
smoking. The reference ranges for blood, taken from references 6
and 7, refer to serum or plasma except for Cd and Pb. The values
for Cd and Pb refer to human whole blood. The urine values, taken
from reference 6, are highly debatable.
[c]The sensitivity values, expressed as characteristic mass and
obtained under STPF conditions with Zeeman background correction
(except for Pt and V for which wall atomization was used), are
taken from reference 8. The values refer to aqueous standards as
data for biological fluids under STPF conditions are not available
at present. The authors of ref. 8, however, claim that the
characteristic mass value given for the various analytes should
not differ by more than 20% for different operators, instruments,
matrices, etc. To convert the characteristic mass value into
concentration unit in µg/l, divide it by the value of the
microlitre volume injected.
[d]Signifies non-availability of data.

profile immediately before and after atomization; (vi) rapid
heating of the order of 1500-2000°C just prior to atomization
to ensure maintenance of isothermal conditions; (vii)
interruption of gas flow during atomization to minimize
disruption of the thermal equilibrium within the furnace;
(viii) use of appropriate matrix modifiers to thermally
stabilize the analyte and prevent its preatomization loss; and
(ix) background correction, preferably Zeeman, to compensate
for the rapidly changing background signals. Application of
the STPF technology together with the use of modern data
handling systems has the potential to make the graphite furnace
an almost interference-free analytical technique; it may even
be possible to use simple aqueous calibration standards.
However, it is important to be aware that complete freedom from
interference and use of simple calibration may not be possible
if the STPF conditions are compromised. Thus, a number of
workers have used the peak height mode of measurement, but the
equality of the peak height absorbance values in standards and
samples requires that the rate of analyte vaporization be
independent of matrix, a condition rarely achieved.
    The present review will focus on the various STPF-AAS
methods that have been developed to date for determining trace
elements in biological fluids beginning with aluminum.

ALUMINUM

The role of aluminum in the pathogenesis of a number of
clinical disorders such as Alzheimer's disease and dialysis
encephalopathy has been receiving increasing attention lately
(16).
    Published reference intervals for Al in serum by convention-
al GFAAS prior to 1981 show wide variations (2-35 µg/l), which
indicate difficulties such as sample spluttering during drying,
carbonaceous residue build-up, volatility losses and matrix
interferences in determining this element (17). Recent studies
show that use of STPF conditions with Zeeman background correction permits reliable determination of Al in biofluids, and
Table II summarizes the STPF methods available for analyzing
serum and urine (18,19,21-23). STPF methods for the
determination of Al in other physiological fluids are yet to
appear.
    Background correction is essential; in its absence, the
error obtained in the evaluation of Al levels in blood could be
as high as 25 µg/l (17). However, at the 309.3 nm line,
neither the deuterium nor the tungsten lamp is as bright as the
hollow cathode lamp. Therefore balancing the two beams requires
reducing the hollow cathode lamp current and hence the signal
to noise ratio. This problem can be solved by using the 396.2
nm line which is only 20-24% less sensitive than the 309.3 nm
line; also, at this wavelength the intensities of the hollow
cathode lamp and the tungsten lamp are better matched without
the need to reduce the optimum operating hollow cathode lamp
current (20). The loss in Al sensitivity at the 396.2 nm line
can be easily compensated for by increasing the volume of
injection, especially when measuring < 10 µg/l Al. However,
Zeeman background correction provides favorable signal to noise
ratios at both the Al lines, and gives better detection limits
and precisions at low concentrations (22).

Table II. STPF-AAS Methods and Parameters for the Determination of Aluminum in
Biological Fluids

Leung and Henderson (18)
Method: 1 + 1 dilution of
serum with
0.1% Mg(NO₃)₂ and
0.2% Triton X-100.
Injection Vol: 10 μl
Sensitivity: 11 pg/0.0044 A.s
Linearity: 0–500 μg/l
Calibration: Matrix-match

Instrumentation: P-E 5000/HGA 500/D₂

Temp°(C):	250	1500	2500	2700	20
Ramp (s):	1	1	0	1	1
Hold (s):	45	45	6	3	10 / 5

Bettinelli et al. (19)
Method: 1 + 1 dilution of
serum with
0.2% Mg(NO₃)₂
Injection Vol: 10 μl
Sensitivity:
10.5 pg/0.0044 pg/A.s
Linearity: 0–150 μg/l
Calibration: Standard Additions

Instrumentation: As above

Temp°(C):	80	130	500	1500	1500	2400	2600	20
Ramp (s):	1	30	30	1	15	0	1	1
Hold (s):	4	25	55	25	30	6	6	20
Int.Alt.:			O₂					
ml/min):			50					

Gardiner and Stoeppler (21)
Method: 1 + 1 dilution of
plasma with
0.1% Triton X-100
and 1 mM HNO₃
Injection Vol: 10 μl
Sensitivity: 10.6 pg/0.0044 A.s
Calibration: Standard Additions

Instrumentation: P-E 4000/HGA 500/D₂

Temp°(C):	100	1400	2400	2600	20
Ramp (s):	5	5	0	1	1
Hold(s):	30	20	4	3	30

Slavin (22)
Method: 1 + 1 dilution of
serum with
0.2% Mg(NO₃)₂
Injection Vol: 10 µl
Sensitivity:
11 pg/0.0044 A.s
Linearity: 0-200 µg/l
Calibration: aqueous in modifier

Instrumentation: P-E Z-5000/HGA 500

Temp°(C):	130	600	600	1700	2400	2600	20
Ramp (s):	1	30	1	15	1	1	1
Hold (s):	45	55	25	25	6	6	20
Int.alt.		O₂					
(ml/min)		50					

Faqioli et al. (23)
Method: dilution of 400 µl
of serum with 50 µl
1% Mg(NO₃)₂ and
50 µl 1% Triton X-100
Sensitivity: 10 pg/0.0044 A.s

Instrumentation: P-E 603/HGA 500/D₂

Temp°(C):	110	500	1500	1500	2400	2600	20
Ramp (s):	15	30	15	1	0	1	1
Hold (s):	50	30	30	4	8	3	20
Ar (ml/min):	0		0				

General Parameters: Hollow cathode lamp current, 20 mA; Wavelength, 309.3 nm;
Spectral bandwidth, 0.7 nm; Argon flow rate was generally set at
300 ml/min except at atomization during which the flow rate was set at
set at 0 ml/min.

The highest possible atomization temperature is needed for
maximum Al sensitivity. At high atomization temperatures,
however, the furnace is vulnerable to oxidation. Many workers
sought to overcome this problem by using metal (Mo, Ta, Th,
Zr)-coated tubes (22). Recent studies show that pyrocoated
tubes and platforms are equally effective; they can attain high
temperatures and longer life, and also minimize formation of
carbide and permeation of the injected sample into the pores of
the graphite thereby improving sensitivity (21,22). The non-
pyrocoated tubes should be avoided as their use degrades the
precision of Al determination (21,22).
     In spite of using the platform, however, Gardiner et al.
(20) found that a significant portion of Al was lost before
isothermality was attained. This problem has been overcome by
using magnesium nitrate as a matrix modifier which delays
atomization until isothermality is attained (18,19,22-24).
Also, magnesium nitrate stabilizes Al to higher temperatures
(1400-1700°C) thereby facilitating the complete removal of
organic matter and the conversion of Al to a stable chemical
form preventing its loss as a volatile chloride.
     The main reason for recommending the STPF method for Al is
that the peak area (integrated absorbance, A.s) is independent
of the rate of vaporization of Al and hence independent of
matrix interference from the chloride salts of Ca, Mg, Fe and
Zn (25). In the peak height mode, there was interference from
these salts. The furnace should be heated rapidly to obtain the
highest sensitivity for Al and this calls for instruments with
short time constants to avoid signal distortion and to minimize
interference. For example, the Perkin-Elmer model 5000 AA
provides a threefold improvement in sensitivity compared to the
slower models 603 and 2380 (22). Only an instrument with a
sufficiently short time constant (~20 ms) would provide an
accurate Al signal at low concentrations. Finally use of argon
rather than nitrogen is preferred as the purge gas because
nitrogen reacts with Al to form aluminum nitride and produces
variable results (22).

CADMIUM

The determination of Cd in whole blood and urine may be useful
in monitoring occupational exposure while its determination in
plasma or serum may indicate cadmium toxicity before
irreversible changes have occurred. Cadmium concentrations in
the whole blood and serum of healthy individuals are < 1 μg/l
and 0.1 μg/l, respectively (26). To detect such low levels,
highly sensitive techniques permitting direct determination are
desirable. GFAAS with STPF and Zeeman background correction
offers such a possibility.
     Table III shows the STPF methods published to date for Cd in
biological fluids (27-30). Yin et al. (27) found it necessary
to dilute the blood fivefold with a 0.05% Triton X-100 solution
in order to facilitate pipetting and to keep the background
absorption at a reasonable level. They used a matrix modifier
composed of palladium nitrate and ammonium nitrate. The sample
and matrix modifier could not be mixed prior to injection
because of the precipitation of proteins. The introduction of
the modifier should precede that of the diluted blood sample to
the furnace for obtaining reliable results; when the blood
sample was injected first followed by the modifier, the two

solutions did not mix properly resulting in some Cd loss during
the ashing stage. Oxygen had to be introduced as an ashing aid
in stages 2 and 3 of the temperature program (Table III) in
order to reduce background absorbance from 0.6 to 0.15, and to
eliminate any variation in sensitivity. Under these
conditions, as low as 0.4 µg/l of Cd could be measured, but
even this low value may not allow determination of "normal"
levels. For serum, it was possible to use a dilution factor of
2.5; this permitted measurement of Cd to 0.2 µg/l. Lum and
Edgar (28) found it unnecessary to use matrix modification.
They simply diluted the blood 10-fold with high-purity water
and deposited an aliquot of the diluted sample into the
platform. Multiple drying stages with ramping had to be used
to prevent bumping, boiling and splattering of samples. The Cd
peak was well resolved in time from the background peak, and
there was no interference, enabling use of acidic standards for
calibration. Zeeman background correction was required since
the background signal for blood was beyond the compensation
capability of the deuterium background corrector even at 10-
fold dilution.

Non-STPF methods for determining Cd in urine are fraught
with problems of contamination; variable loss of Cd during the
char step; matrix interference requiring the method of standard
additions; and high background-to-analyte ratio (3,29,30).
Pruszkowska et al. (29) sought to overcome these problems, and
improve the ruggedness of Cd measurement in urine by using STPF
technology involving matrix modification with diammonium
hydrogenphosphate and nitric acid. Residual matrix
interferences were still present and the authors had to use
calibration standards prepared in a solution of $(NH_4)_2HPO_4$, NaCl
and HNO_3. Zeeman background correction was mandatory because
the background signal of even a 5-fold diluted urine was 0.7-
2.0 A, depending on the sample.

The phosphate is often contaminated with Cd and requires
extensive purification (30). To avoid this problem, McAughey
and Smith (30) proposed direct injection of urine after it is
diluted 1+1 with water. They were able to temporally separate
the Cd signal from the background signal by using an
atomization temperature of 800°C under maximum power heating.
At such a low atomization temperature the background was within
the compensating abilities of the deuterium background
corrector and sensitivity for Cd was comparable to the Zeeman
AA method of Pruszkowsaka et al. (27) as can be seen from Table
III. However, matrix interferences could not be completely
eliminated under these conditions and the method of standard
additions was mandatory for obtaining satisfactory results
especially because of the large matrix variation among urine
specimens. Note also that the method of selective
volatilization used by these authors calls for the precise
control of the furnace temperature program (29). Yin et al.
(27) used a modifier composed of palladium nitrate and ammonium
nitrate to facilitate separation of analyte from matrix and to
minimize non-specific absorption. Reduction of background was
aided by a 3-fold dilution; use of NH_4NO_3; and an atomization
temperature of 1300°C. Under these conditions, up to 0.1 µg/l
Cd could be determined in urine based on matrix-free aqueous
standards.

Table III. STPF-AAS Methods and Parameters for the Determination of Cadmium in Biological Fluids

Yin et al. (27)

Method: 5-fold dilution of whole blood (2.5-fold for serum and 3-fold for urine) with 0.05% Triton X-100. Inject 10 µl modifier composed of 50 µg Pd(NO₃)₂ and 500 µg NH₄NO₃ followed by 10 µl sample solution.

Sensitivity: 0.45 pg/0.0044 A.s
Linearity: 0-100 pg
Calibration: aqueous in modifier

Instrumentation: P-E Z-3030/HGA 600

Temp°(C):	90	120	400	800	1700	2500	20
Ramp (s):	1	1	40	10	0	1	1
Hold (s):	20	20	10	20	4	4	5
Int.Alt. (ml/min)	O_2 300	O_2 300					

Lum and Edgar (28)

Method: Direct injection of whole blood or urine after 5- to 10-fold dilution with water.

Sensitivity: 0.3 pg/0.0044 A.s
Calibration: Aqueous

Instrumentation: Hitachi Zeeman Z-800

Temp°(C):	90–120	120–150	150–600	600–600	2000–2000	2500–2500
Time (s):	15	30	10	30	7	3

Pruszkowska et al. (29)

Method: 20 µl injection of 5-fold diluted urine followed by 5 µl injection of 4% (NH₄)₂HPO₄ and 15% HNO₃.

Instrumentation: P-E Z-5000/HGA 500

Temp°(C):	140	700	1600	2600	20
Ramp (s):	1	1	0	1	1
Hold (s):	60	45	5	6	20

Sensitivity: 0.35 pg/0.0044 A.s
Calibration: Aqueous

McAughey and Smith (30)
Method: 2-fold dilution of
         urine with water
Sensitivity:
0.4 pg/0.0044 A.s
Linearity: 0-4 µg/l
Calibration: Standard Additions

Instrumentation:	P-E 4000/HGA 400/D_2				
Temp°(C):	150	300	800	2500	20
Ramp (s):	25	15	0	2	1
Hold (s):	15	60	9	5	20

General Parameters: EDL lamp, 5 W; Wavelength, 228.8 nm; Spectral Bandwidth, 0.7 nm; Argon flow rate, 300 ml/min throughout except during atomization when it was interrupted.

LEAD

The concentration of Pb in blood reflects recent exposure and
is extensively used throughout the world for the biological
monitoring of populations exposed to this metal (31). The
concentration of Pb in urine may not be a reliable test of
exposure, but some scientists regard levels of 80 µg and 150 µg
of Pb in 24 h urine as evidence of an upper limit for tolerable
exposure and of possible clinical manifestations, respectively
(31).
     GFAAS is the method of choice for the determination of Pb in
blood, but the status of conventional GFAAS methodology is poor
(32). Recent attempts to overcome matrix interferences, and to
improve the accuracy and precision of blood-Pb determination
under STPF conditions has been the subject of a critical review
(32). Suffice it to say that the STPF method for Pb in blood
is not totally fool-proof. Thus the use of maximum power
heating (an important STPF feature) in the presence of the
commonly used ammonium phosphate-nitric acid-Triton X-100
modifier gave poor precision (average increase of 2-3%
coefficient of variation), higher detection limit and high
background absorption (33). Therefore, a 1-s ramp and a purge
gas flow of 20 ml/min were used during atomization to lower the
detection limit and the background signal, respectively. The
low background signal of 0.05 A.s obtained under these
conditions was well within the allowable limits of the
deuterium background corrector, and obviated the need for
Zeeman background correction. At a purge gas flow of 0 ml/min,
the background was about 0.2 A which was at the limit of
accurate correction by the deuterium system. Also, these
authors thoroughly validated their method by analyzing the
National Bureau of Standards Blood-Pb reference material (SRM
955), by using IDMS as an independent technique and by
participation in interlaboratory programs. Bruhn et al. (34)
also used the above modifier. However, they found carbonaceous
residue build-up within the furnace at the optimum atomization
temperature of 1500°C used under maximum power heating
conditions. These authors used peak height since the use of
peak area gave erratic results. The use of peak height,
however, necessitated calibration by the method of additions
since its slope (26 pg/0.0044 A.s) differed by about 20% with
respect to the aqueous calibration slope (20 pg/0.0044 A.s).
Shuttler and Delves (35) used a complex modifier composed of
NH₃, NH₄H₂PO₄, (NH₄)₂H₂EDTA and HNO₃, oxygen ashing and a 20-fold
dilution in order to attain an accuracy of 2-3% at 100 µg/l Pb.
The oxygen ashing was required to reduce residue formation; to
aid ashing; and to facilitate efficient atomization. However,
the use of oxygen caused complete removal of the pyrolytic
coating and virtual degradation of the platform within about 30
firings because of the rapid formation and removal of CO. This
problem was solved by desorbing the oxygen prior to
atomization. Since desorption occurred only at 950°C, it was
necessary to use phosphate as a modifier to prevent the loss of
Pb. At the 20-fold dilution used by these workers, the non-
specific signal was < 0.05 A, which was clearly within the
compensation capabilities of the deuterium background
corrector. An important finding by these scientists (36) was
that the within-batch and between-batch variations in platforms
and tubes could lead to variable sensitivity (20-26 pg/0.0044

A.s compared to the expected value of 14-17 pg/0.0044 A.s), and
decreased appearance time (0.0-0.2 s compared to the expected
value of 0.4 s). The problem was traced to poor thermal
contact between the platform and the interior grooves of the
pyrolytic tube resulting in loss of isothermality within the
furnace (37). The significant change in sensitivity indicated
a serious problem affecting atom formation. The variable
appearance times suggested that the rate at which Pb atoms were
produced varied from one platform-tube combination to another;
these were caused by differences in the heating rate of the
platform. Also the optimum drying conditions varied
considerably among different lots of tubes and platforms, and
each lot of platform-tube combination had to be evaluated to
ensure drying of the blood specimens without spattering
(33,36). Thus, the use of the grooves to hold the platform in
place was found to be unsatisfactory and these authors (35,36)
resorted to manual positioning of the platform in non-grooved
tubes. However, considerable care was required to ensure
reproducible positioning. The problem has also been
demonstrated for the determination of Cd in urine (37).

Conventional GFAAS methods for the determination of urine-Pb
are susceptible to variable matrix effects not only from one
urine specimen to another but also with sequential specimens
from the same donor. Recently, STPF methods were developed for
determining urinary Pb involving either $(NH_4)_2HPO_4$ - HNO_3 (38),
or La - HNO_3 (39) modification, Zeeman background correction
and a 1 + 1 dilution. Problems were encountered, however. The
background was 1.5 A.s with the phosphate method when the gas
flow was interrupted during atomization as required by STPF.
Such a high background value was near the limit of the accurate
correction capability of the Zeeman system. An internal purge
gas flow rate of 20 ml/min had to be introduced during
atomization to reduce the background signal to 0.25-0.30 A.s,
which was well within the compensation capability of the Zeeman
background corrector (38). With the La - HNO_3 method, the
background at 0 ml/min was 0.5 A.s, which was correctable by
Zeeman, but not by continuum background correctors such as
deuterium. Also, both the methods required the use of matrix-
matched calibration plots owing to the presence of residual
matrix effects. The La - HNO_3 method in which the gas flow
could be interrupted during atomization gave better sensitivity
(15 pg/0.0044 A.s) than the phosphate method (20 pg/0.0044
A.s), which used a flow of 20 ml/min.

SELENIUM

The concentration of Se in serum and urine is one of the
indicators of Se status, and Se deficiency has been linked with
cancer and cardiovascular diseases (40,41). These links
indicate the value of this biological marker in epidemiological
studies and the need to develop fast, sensitive and accurate
methods for the determination of selenium in body fluids.

The determination of Se in biological fluids by conventional
GFAAS is subject to problems such as preatomization losses,
spectral interferences, chemical interference and formation of
carbonaceous deposit inside the furnace (3-5). Spectral
interference in the determination of Se occurs in the presence
of the decomposition products of iron salts and phosphate when
continuum background correctors are used. Since blood is rich

in iron and urine is rich in phosphate, the spectral
interference makes measurement of Se in these and other
biological fluids difficult under conventional GFAAS
conditions. For example, Fig. 1 clearly shows that the signal
distortion and baseline offset for serum samples with the
deuterium background corrector give low recoveries irrespective
of whether peak height or peak area is used (42). This
spectral interference has been significantly minimized under
STPF conditions with Zeeman background correction.

Table IV gives the STPF methods available for determining Se
in biological fluids (42-50). The use of a variety of matrix
modifiers suggests problems with STPF methodology. The
modifiers used include Ag, Cu, Fe, Mg, Ni, Pd and Pt either
alone or in combination. Use of either Cu alone or Fe alone
was not satisfactory, but the mixed modifier, Cu-Fe, in the
presence of Zeeman background correction was claimed to yield
reliable results even by using simple calibration (42). Morisi
et al. (48) found otherwise. They obtained Se values of only
88.3-94.6 µg/l for the National Bureau of Standards (now
National Institute of Standards and Technology or NIST) human
serum reference material (SRM 909) compared to the certified
value of 106 µg/l when aqueous calibration was used. This low
recovery could be ascribed Se loss at the charring temperature
of 1000°C used by Welz et al. (42). On the other hand, at char
temperatures < 1000°C, it was difficult to control Se recovery
because of the high background absorption. Thus, this method
may not be suitable for work of the highest accuracy. Matrix
modification with nickel caused a 25% suppression of the Se
signal in serum when aqueous calibration was used (42), but
Paschal and Kimberly (44) found Ni modification to be
satisfactory when Zeeman background correction, ramp
atomization (i.e., no maximum power heating), and matrix-
matched calibration plots were used. Lewis et al. (45)
reported that when STPF conditions were not compromised and
Zeeman background correction was used, accurate results could
be obtained for Se in plasma with a Ni-Mg modifier and matrix-
free calibration. They validated the accuracy of their method
by comparison with IDMS. Eckerlin et al. (47) found it
impossible to use this mixed modifier as it caused rapid
settling and precipitation of whole blood samples resulting in
poor reproducibility. Neve and Molle (46) used a Cu-Mg
modifier, Zeeman background correction and oxygen ashing to
improve the accuracy and precision of serum-Se determination.
In the absence of oxygen ashing, the recovery was poor (~85%
for a serum sample containing 93 µg Se/l). Oxygen ashing led
to complete removal of organic matter and better removal of
carbonaceous residue. However, it was important to use the
lowest possible ashing temperature in the presence of oxygen to
prolong tube and platform life. Although they used a second
ashing step, the temperature used was not sufficient to desorb
the oxygen completely. As a result, residual matrix
interferences were still present and they had to use the method
of standard additions to obtain reliable results. Morisi et
al. (48) found that oxygen alone could not eliminate the
carbonaceous crust. They also found it necessary to add Ag to
the Cu-Mg modifier in order to completely remove the carbon
deposit. Under these conditions, they could obtain reliable
values for Se in serum when matrix-matched calibration was

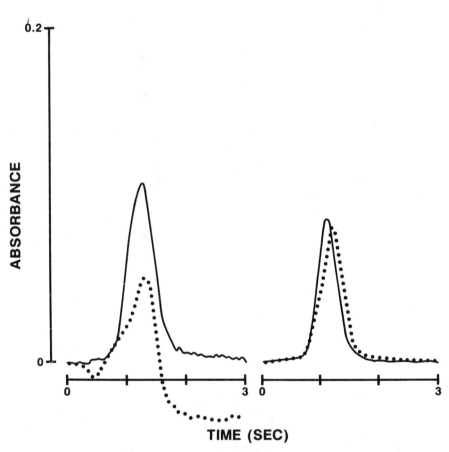

Figure 1.   Determination of selenium in serum by STPF-
AAS.   In the presence of deuterium background
correction, the spectral interference causes
a signal distortion (left) which is not seen
with the Zeeman system (right).   Dotted line:
serum sample; solid line: reference solution.
(Reproduced with permission from Ref. 42.
Copyright 1983 Springer-Verlag.)

Table IV. STPF-AAS Methods and Parameters for the Determination of Selenium in Biological Fluids

Welz et al. (42)[a]

Method: 1 + 2 dilution of serum with 0.2% Triton X-100. Injection of 15 μl of diluted sample followed by 10 μl of modifier composed of 0.125% Cu and 0.05% Fe in 0.2 M HNO_3

Calibration: Aqueous

Instrumentation: P-E Z-5000/HGA 500

Temp°(C):	90	140	1000	1000	2100	2650	20
Ramp (s):	1	10	20	1	0	1	1
Hold (s):	40	10	20	4	3	3	10
Ar (ml/min):				0	0		

Bauslaugh et al. (43)

Method: Injection of 10 μl serum followed by 10 μl of modifier composed of 0.1% Ni + 2.5% Pt + 3 M HNO_3

Instrumentation: P-E 5000/HGA 500/D_2

Temp°(C):	180	220	450	1000	1300	1300	2600	20
Ramp (s):	15	5	5	5	10	1	0	1
Hold (s):	15	20	5	5	5	4	6	20
Ar (ml/min):						10	10	

Paschal and Kimberly(44)

Method: Injection of 10 μl of serum diluted 1 + 1 with water followed by 5 μl of 0.24% $Ni(NO_3)_2$

Sensitivity: 34 pg/0.0044 A.s

Calibration: Matrix-matched

Instrumentation: P-E Z-5000/HGA 500

Temp°(C):	180	1200	2400
Ramp (s):	10	10	1
Hold (s):	35	20	5

Lewis et al. (45)
Method: 1 + 1 dilution of plasma
   with modifier composed
   of 1% Ni(NO₃)₂ and
   1% Mg(NO₃)₂

Instrumentation:	P-E Z-5000/HGA 500				
Temp°(C):	150	1000	2100	2700	20
Ramp (s):	60	1	1	0	1
Hold (s):	10	20	4	4	14

Neve and Molle (46)
Method: 1 + 3 dilution of serum
   with modifier composed of
   0.05% Cu(NO₃)₂ + 0.15%
   Mg(NO₃)₂ + 0.15%
   Triton X-100 in 0.2%
   HNO₃
Injection Vol: 15 µl
Sensitivity: 40 pg/0.0044 A.s
Calibration: Standard Additions

Instrumentation:	P-E Z-3030/HGA 600						
Temp°(C):	130	500	500	1350	2200	2600	20
Ramp (s):	1	10	1	1	1	1	
Hold (s):	30	30	20	10	3	5	5
Int. Alt.:			O₂				
(ml/min):			300				

Eckerlin et al. (47)
Method: 10-fold dilution of whole
   blood with a modifier
   of 0.09% PdCl₂ in 0.25%
   HCl + 0.1% Triton X-100
   + 1.25% Ni(NO₃)₂
Injection Vol: 20 µl
Sensitivity: 32.6 pg/0.0044 A.s
Calibration: Matrix-matched

Instrumentation:	P-E Z-5000/HGA 500						
Temp°(C):	80	120	500	500	750	2200	2700	20
Ramp (s):	10	1	15	1	1	0	1	1
Hold (s):	30	20	45	30	30	6	6	20
Int. Alt.:				Air				
(ml/min):				300				

Continued on next page

Table IV. Continued

Morisi et al. (48)
Method: 15 µl injection of 1 + 2 diluted serum with 0.2% Triton X-100 followed by 15 ul modifier composed of 0.1% $AgNO_3$ + 0.2% $Cu(NO_3)_2$ + 0.2% $Mg(NO_3)_2$ + 0.4% HNO_3; Method of Standard Additions

Instrumentation: P-E Z-5000/HGA 500

Temp°(C):	120	500	1200	1200	2200	2650	20
Ramp (s):	80	15	1	10	0	1	1
Hold (s):	20	20	30	20	5	5	10
Ar (ml/min):							

Carnrick et al. (49)
Method: 10 µl injection of urine diluted 5-fold with water followed by 10 µl modifier composed of 60 µg Ni, 25 ug Mg, 3% HNO_3
Sensitivity: 30 pg/0.0044 A.s

Instrumentation: P-E Z-5000/HGA 500

Temp°(C):	160	800	2000	2600	2600	20
Ramp (s):	1	20	1	0	1	1
Hold (s):	60	45	20	8	6	20

Neve et al. (50)
Method: 0.3 ml seminal plasma + 0.7 ml modifier as in ref. 46
Injection Vol: 15 µl
Sensitivity: 25-30 pg/0.0044 A.s
Calibration: Standard Additions

Instrumentation: as in ref. 46

Temp°(C):	130	450	1250	2200	2600	
Ramp (s):	1	10	1	0	1	5
Hold (s):	30	30	20	10	3	
Int.Alt.:			O_2			
(ml/min):			300			

General Parameters: EDL lamp, 6 W; Wavelength, 196 nm; Spectral bandwidth, 2 nm; Argon flow rate, 300 ml/min except during atomization at which it was 0 ml/min.

[a]In a later paper (Welz, B.; Melcher, M.; Schlemmer, G. In Trace Element Analytical Chemistry in Biology and Medicine; Bratter, P.; Schramel, P., Eds.; Walter de Gruyter: Berlin, 1984; Vol. 3, pp 207-215.), the authors used a Cu-Mg modifier. No reasons were given for replacing the Fe with Mg.

used. The method did not work with whole blood samples.
Eckerlin et al. (47) used a Pd-Ni modifier, a 10-fold dilution,
air ashing and Zeeman background correction to determine Se in
whole blood samples. Air ashing could not eliminate the
carbonaceous build-up completely. Also Pd was not effective in
stabilizing the volatile organoselenium compounds likely to be
found in blood. Thus, their method was less reliable.
Bauslaugh et al. (43) sought to overcome the spectral
interference using a Pt-Ni modifier. In the presence of
platinum, the absorption signal of Fe was completely separated
in time from that of Se and thus helped remove the spectral
interference. The increased appearance time of Fe was perhaps
due to the formation of a Pt-Fe alloy from which the Fe
volatilised more slowly than the Se from a Pt-Se alloy or
selenide. The addition of Pt also decreased the
overcompensation effect of phosphate by enhancing the formation
of P atoms (43). However, Pt had no stabilizing effect. Ni,
on the other hand, stabilized both the organic and inorganic
selenium species, but was ineffective for removing the spectral
interference. Thus, a matrix modifier composed of Ni and Pt
gave interference-free determination of Se in blood and serum
samples. Under these conditions, even deuterium background
correction was possible.

Only one STPF paper seems to have been published to date on
determining Se in urine (49) attesting to the difficulty
involved in its measurement. Se in aqueous solution was stable
up to 1200°C in presence of Ni, but in a 5-fold diluted urine
it was stable only up to 900°C. There was nearly a 40% loss of
Se from urine at 1200°C. The cause of this loss was traced to
the presence of Na_2SO_4 in the urine matrix. Modifiers such as
Ag, Mo or Cu instead of Ni were not effective in preventing the
loss. Also the amount lost was variable from one urine sample
to another. Temperatures ≤ 700°C were not usable because of
the high background (1.5 A) in spite of using Ni and 6% HNO_3.
The use of a Ni-Mg modifier, however, was effective in reducing
the loss of Se at temperatures ≤ 900°C. An optimum ashing
temperature of 800°C was chosen. The background signal at
800°C was between 0.8-0.5 depending on the urine sample. Such
a high background value was correctable by the Zeeman system,
but not by the deuterium lamp. Thus the use of Ni-Mg matrix
modification, Zeeman background correction, dilution factors ≥
5, and careful control of the temperature program was required
to reliably determine Se in urine with respect to matrix-free
calibration. Nevertheless the loss of volatile Se species at
temperatures as low as 200°C is still a possibility (49).

Se deficiency is associated with poor reproductive function
including sperm quality. Therefore, Neve et al. (50)
determined Se in the seminal plasma by the same Cu-Mg matrix
modification method they had earlier developed for measuring
this element in blood serum (46). Zeeman background correction
was necessary as there was spectral interference from the
phosphate present in seminal plasma. Also the method of
standard additions was required as calibration with aqueous
standards produced inaccurate results despite careful
optimization of analytical conditions.

OTHER ELEMENTS

Table V lists the STPF methods available for Sb (51), As (52-54), Be (55), Cu (56,57), Au (58), Mn (59), Pt (60,61), and Tl (62,63) in various biological fluids.
To determine the non-essential toxic element, antimony, Constantini et al. (51) diluted whole blood samples 10-fold and urine samples 2-fold with water followed by Ni matrix modification. The 10-fold dilution was required to minimize the formation of carbonaceous deposit, but it limited detection of Sb to levels > 200 µg/l. To obtain better sensitivity (40 µg/l), a digestion procedure involving 2-fold dilution was used and even this approach was inadequate to detect the < 5 µg/l levels expected in the blood and urine of populations not exposed to Sb. The determination of arsenic in body fluids such as blood and urine may serve as a monitor of recent exposure (64). The problems involved in the conventional GFAAS measurement of As in biological fluids are essentially the same as described earlier for Se. Therefore, a STPF method was developed for determining As in whole blood involving 5-fold dilution, nitric acid deproteinization, nickel fortification and deuterium background correction (52). Deproteinization overcame problems of carbonaceous residue build-up. However, the inorganic blood matrix interferences could not be completely eliminated at the optimum charring temperature of 1200°C and matrix-matched calibration was mandatory especially at dilution factors < 20. In other words, a dilution factor of at least 20 was required to eliminate the chemical interference and facilitate use of matrix-free calibration. Even at 5-fold dilution the sensitivity of the method was only 3 µg/l, and was not low enough to permit measurement of reference values; it may, however be sufficient for screening exposed individuals.
    Paschal et al. (53) determined total-As in urine by adapting the Ni-Mg modification - STPF method developed for determining urine-Se by Carnrick et al. (49). Zeeman background correction was essential as overcompensation occurred with the deuterium corrector due to structured background interference from phosphate. The sensitivity (23 pg/0.0044 A.s) of the method was not sufficient to permit determination of "normal" levels. The method of standard additions was essential to obtain reliable results. The authors validated their method by the analysis of standard reference materials (urine SRM from NBS with an As value of 480 µg/l), and by participation in interlaboratory studies.
    Subramanian (54) showed that the phosphate and sulfate salts of Ca and Mg, which are likely constituents in urine, interfered strongly in the direct determination of urinary arsenic in spite of using STPF conditions and Pd or Pd-ascorbic acid modification. Since the phosphate and sulfate concentrations would be different in different urine samples or even in the urine sample of the same individual collected at different times, it is clear that the direct STPF method would be fraught with matrix problems and even the method of additions may not be satisfactory. Further, the poor sensitivity of the direct method facilitated determination of As only at levels ≥ 20 µg/l as no signals could be detected even at 5 µg/l. The interference was most likely spectral in nature and perhaps could have been overcome by the use of

Table V. STPF-AAS Methods and Parameters for the Determination of Some Other Elements in Biological Fluids

Antimony: Constantini et al. (51)
Method: 10-fold dilution of whole
        blood or 2-fold dilution of
        urine with 0.2% Ni
Injection Vol: 10 µl
Linearity: 0-400 µg/l

Instrumentation: P-E 430/HGA 500/D_2

Temp°(C):	90	130	700	1100	1100	2500	2700	20
Ramp (s):	15	30	60	30	1	0	1	1
Hold (s):	4	10	30	10	5	7	4	20

Hollow cathode lamp current, 20 mA; Wavelength, 217.6 nm; Spectral bandwidth, 0.2 nm

Arsenic: Subramanian (52)
Method: Dilution of 200 µl whole
        blood with 680 µl water.
        Deproteinization with
        100 µl 50% HNO_3. Addition
        of 20 µl 5% Ni $(NO_3)_2$

Instrumentation: P-E 5000/HGA 400/ D_2

Temp°(C):	200	2500
Ramp (s):	40	0
Hold (s):	30	6

EDL, 10 W; Wavelength, 193.7 nm; Spectral Injection Vol: 10 µl
bandwidth, 0.7 nm

Sensitivity: 0.61 µg/l (diluted)
             3.1 µg/l (undiluted)
Linearity: 0-40 µg/l
Calibration: Matrix-matched

Arsenic: Paschal et al. (53)
Method: 1 + 7 dilution of urine
        with a modifier composed
        of 0.6% Ni(NO₃)₂, 0.6%
        Mg(NO₃)₂, 0.1% Triton X-100
        and 2.3% HNO_3
Sensitivity: 23 pg/0.0044 A.s
Calibration: Standard Additions

Instrumentation: P-E Z-5000/HGA 500

Temp°(C):	140	1400	2400	20
Ramp (s):	5	5	0	10
Hold (s):	20	20	5	10

EDL, 8 W; Wavelength, 193.7 nm; Spectral
bandwidth, 0.7 nm

Continued on next page

Table V. Continued

Beryllium: Paschal and Bailey (55)
Method: 1 + 3 dilution of urine
with a modifier composed
0.25% $Mg(NO_3)_2$ plus 0.1%
Triton X-100 plus 1% HNO_3
Volume: 20 µl
Sensitivity: 1.7 pg/0.0044 A.s
Linearity: 0–4 µg/l
Calibration: Aqueous with modifier

Instrumentation: P-E Z-5000/HGA 500

Temp°(C):	180	1400	2400
Ramp (s):	5	5	0
Hold (s):	30	10	4

HCL, 20 mA; Wavelength, 239.4 nm; Spectral bandwidth, 0.7 nm

Copper: McGahan and Bito (56)
Method: 1 + 39 dilution of plasma
with 0.2% HNO_3
Volume: 20 µl
Sensitivity: 0.3 µg/l (diluted)
Calibration: Aqueous
Note: Same instrumental parameters
were used for the determination
of Cu in cerebrospinal fluid by
direct injection.

Instrumentation: P-E 4000/HGA 400/D_2

Temp°(C):	250	350	1000	2100	2600	20
Ramp (s):	5	5	5	0	1	5
Hold (s):	20	20	3	5	5	20

Copper: Carelli et al. (57)
Method: 1 + 99 dilution of red
cells with water
Volume: 20 µl
Sensitivity: 0.4 µg/l (diluted)
Calibration: Standard Additions

Instrumentation: P-E 4000/HGA 500/ D_2

Temp°(C):	100	1000	2250	2700	20
Ramp (s):	30	10	0	1	1
Hold (s):	35	25	4	3	15

HCL, 20 mA; Wavelength, 324.7 nm; Spectral bandpass, 0.7 nm

Gold: Matthews and McGahan (58)
Method: Injection of 15 µl of plasma or intraocular fluid followed by 5 µl of 2% cysteine
Sensitivity: 10.2 pg/0.0044 A.s

Instrumentation: P-E 3030/HGA 400/Zeeman

Temp°(C):	250	350	1000	1800	2650	20
Ramp (s):	15	10	10	0	1	1
Hold (s):	20	20	40	5	3	30

Ar (ml/min): 30
Wavelength, 242.8 nm; Spectral bandwidth, 0.7 nm

Manganese: Paschal and Bailey (59)
Method: 1 + 1 dilution of serum with 0.5% Triton X-100
Volume: 20 µl
Sensitivity: 3.2 pg/0.0044 A.s
Linearity: 0-12 µg/l
Calibration: Aqueous

Instrumentation: P-E Z-5000/HGA 500

Temp°(C):	180	1400	2400	20
Ramp (s):	5	5	1	1
Hold (s):	25	15	4	4

HCL, 30 mA; Wavelength, 279.5 nm; Spectral bandwidth, 0.7 nm

Platinum: McGahan and Tyczkowska (60)
Method: Direct injection of 10 µl of plasma or urine
Sensitivity: 116 pg/0.0044 A.s
Linearity: 0-800 µg/l

Instrumentation: P-E 4000/HGA 400/D_2

Temp°(C):	150	350	1500	2650	20
Ramp (s):	10	10	10	0	1
Hold (s):					30

Ar (ml/min):
Wavelength, 265.9 nm; Spectral bandwidth, 0.7 nm

Thallium: Paschal and Bailey (62)
Method: 1 + 1 dilution of urine with a modifier composed of 1% $Mg(NO_3)_2$, 4% HNO_3 and 0.01% Triton X-100
Volume: 20 µl
Sensitivity: 35 pg/0.0044 A.s
Calibration: Matrix-matched standards

Instrumentation: P-E Z-5000/ HGA 500

Temp°(C):	180	650	2000	20
Ramp (s):	5	5	0	1
Hold (s):	55	20	5	4

EDL, 7 W; Wavelength, 276.8 nm; Spectral bandwidth, 0.2 nm

Zeeman background correction.   The author lacked this facility.
He eliminated the interference by extracting the As from an
acidified (9 M HCl) urine sample using toluene followed by
stripping of the toluene layer with Ni-HNO$_3$.   The As present in
the Ni-modified solution was determined under STPF conditions.
The extraction procedure permitted selective determination of
As(III), and the combined determination of As(III), As(V),
monomethylarsenic acid (MMA) and dimethylarsenic acid (DMA).
The various arsenic species could be detected down to 1 µg/l at
a concentration factor of 1.   Using concentration factors of 10
each in the toluene and the back extraction steps, sensitivity
could be reduced to 0.01 µg/l permitting detection of As even
in individuals not exposed to this element.
      Paschal and Bailey (55) determined Beryllium in urine under
STPF conditions involving a 1 + 3 dilution with a magnesium
nitrate-Triton X-100-nitric acid modifier.   Deuterium
background corrector was found to be adequate as the non-atomic
absorbance was only 0.08-0.10 A.s and there was no spectral
interference.   Accurate determination required the method of
standard additions.   Copper is essential for the functioning of
many enzymes and yet at high concentrations it can be toxic.
Thus, it is important to study the metabolism of this metal in
body fluids.   McGahan and Bito (56) directly analyzed Cu in
blood plasma (40-fold dilution with 0.2% nitric acid),
cerebrospinal fluid or CSF (direct injection), and intraocular
fluid (direct injection) under STPF conditions with deuterium
background correction.   Careful optimization of pyrolysis
temperature was required for overcoming interference from NaCl.
Optimum pyrolysis temperature for the three fluids was found to
be 1000°C at which the interference as well as the loss of Cu
was negligible.   A double drying step was employed to reduce
splattering of the viscous intraocular and cerebrospinal
fluids.   Carelli et al. (57) determined Cu in red cells by
diluting a 100-mg sample to 10 ml with water and injecting a
10-µl aliquot upon the platform.   Despite the 100-fold
dilution, the integrated absorbance signal of Cu in the red
blood cells was enhanced by nearly 32% with respect to aqueous
standards indicating matrix interferences.   Matrix modification
might have eliminated the problem.
      The measurement of gold in body fluids is useful in
elucidating its mechanism of action as a therapeutic agent in
the treatment of rheumatoid arthritis (RA), and in controlling
its level in the blood of RA patients in order to minimize its
side effects.   Matthews and McGahan (58) determined Au in blood
plasma or IOF by injecting a 15-µl aliquot followed by
injection of a 5-µl solution of 20 mg/l cysteine as a modifier.
The use of cysteine caused a build-up of carbonaceous residue
in spite of using an ashing temperature of 1000°C, and this
could lead to erratic results.
      Biological monitoring of manganese is important in
occupational exposure as chronic Mn absorption can lead to
tardive dyskinesia and other health problems.   Paschal and
Bailey (59) diluted serum samples 1 + 1 with 0.5% Triton X-100
and determined Mn by STPF without matrix modification and
maximum power atomization.   Even without the use of a modifier,
ashing temperatures of 1150-1400°C could be easily used without
Mn loss.   The temperatures were more than sufficient to remove
the organic matter.   Zeeman background correction was preferred

as the deuterium corrector was not totally effective at the 279.5 nm resonance line.

Studies on the distribution of the popular anticancer drug, cis-dichlorodiammineplatinum, in blood is essential for understanding the mechanism of its therapeutic action and also its toxic side effects. <u>Platinum</u> in serum and urine has therefore been determined; samples were injected directly under STPF conditions with Zeeman background correction. An ashing temperature of 1500°C was used to remove organic matter and inorganic matrix salts. At this temperature there was little background absorption and it was possible to use even deuterium background correction. Delves and Shuttler (<u>61</u>), however, found loss of Pt at ashing temperatures ≥ 1100°C and recommended ashing at 1000°C. The use of this temperature led to build-up of carbonaceous residue and made reliable determination difficult. The authors (<u>61</u>) overcame this problem and obtained reliable data using oxygen ashing and the mixed modifiers employed with their blood-Pb work (<u>35</u>). Nevertheless, direct measurement of Pt in blood and urine was possible only at levels > 50 µg/l because of the poor GFAAS sensitivity of 100 pg for Pt. This restricts measurement to gross concentrations at therapeutic levels and precludes characterization of the various Pt-binding species that may be present in body fluids.

Environmental and occupational exposure to Tl can cause gastrointestinal and cardiac disorders, and neurotic effects. Tl can be monitored by analyzing samples of blood, urine or hair. Paschal and Bailey (<u>62</u>) determined Tl in urine by a 1 + 1 dilution with a modifier composed of $Mg(NO_3)_2$, HNO_3 and Triton X-100. Grobenski et al. (<u>63</u>) preferred the $NH_4H_2PO_4$-HNO_3-Triton X-100 modifier. In both cases, the background was ≥ 1.5 A.s and the use of Zeeman was therefore essential as the background is beyond the deuterium background correction capability.

CONCLUSION

At least 12 elements have been determined in the various biological fluids notably blood and urine using STPF-AAS. Among the elements determined, Al has received the greatest attention in terms of the number of methodology papers [8 papers] followed by Se [7], Pb [5], Cd [4], As [3], Cu, Pt, Tl [2 each], Au, Be, Mn and Sb [1 each]. STPF is not effective for metals such as Co, Mo, Ni, Ti and V because the appearance temperatures for these metals are too close to the maximum temperature of the furnace (<u>10</u>). The biological fluids analyzed include blood, urine, cerebrospinal fluid and intraocular fluid. STPF methods are yet to appear for other metals, and for biological fluids such as amniotic fluid, bile, breath, hip-joint fluid, saliva, semen, sweat and synovial fluid.

Although matrix modification is one of the STPF features, some workers have attempted direct injection of either the raw sample or the sample after simple dilution with water or Triton X-100. The injection of the raw sample usually gives rise to: (i) foaming, frothing and spattering of samples requiring careful attention to drying steps (e.g., use of multiple drying steps); (ii) build-up of carbonaceous residue, which may

require for its removal oxygen ashing followed by a desorption
step to minimize oxidation of the pyrocoated tube and platform;
(iii) high background absorbance requiring the use of either
Zeeman or Smith-Hieftje background correction; in some cases,
such as the analysis of urine samples, even these background
correctors may be inadequate; and (iv) residual matrix effects
requiring use of the method of standard additions or matrix-
matched calibration. For these reasons, the direct injection
of samples onto the platform may not be desirable even with the
state-of-the-art STPF instrumentation.

Dilution methods may be feasible for metals such as Cu, Fe
and Zn because they are found in high levels in biological
fluids; also the GFAAS sensitivity of these metals is
excellent. The remaining elements, with the possible exception
of Pb, are present at ≤ μg/l levels in biofluids making simple
dilutions virtually impractical. Under these conditions, the
background will be high; there will still be build-up of the
carbonaceous crust; and it will be difficult to ascertain the
precise charring temperature and time required for maximum
matrix loss and minimum analyte loss especially in the case of
volatile elements such as As, Cd, Pb, Sb, Se, Zn, etc. At the
optimum permissible ashing temperature (usually ≤ 500°C) for
these elements, the organic components of the biological matrix
will only be partially destroyed resulting in smoke production
during atomization and consequent increase in background
absorption. On the other hand, use of temperatures at which
the organic matter could be completely ashed results in the
volatile loss of the analytes.

It is true that some workers sought to make the STPF methods
simpler and faster by the use of direct or dilution methods,
but the original recommendation to use modifiers still seems
valid. The addition of modifiers to samples permits removal of
all the organic matter and bulk of the inorganic matrix prior
to atomization without loss of the analyte. Also, the higher
thermal stability resulting from the addition of modifiers
facilitates temporal resolution of the analyte peak from the
non-analyte peak. The combination of rapid heating and
atomization from a platform ensures near-isothermal conditions
in the furnace when the sample is vaporized and helps reduce
matrix effects. The use of integrated absorbance measurement
may further minimize the effect of the matrix concomitants on
the accuracy of the analysis. The rationale for the
application of efficient background correction systems such as
Zeeman has been succinctly summarized by Carnrick and Slavin
(65). Also, detailed guidelines for the development of
analytical methods and for optimizing the instrumental
parameters under STPF conditions have been documented in an
excellent paper by Slavin et al. (12).

No doubt matrix interferences have been considerably
minimized under STPF conditions. Yet, STPF is not without
problems. The L'vov platform placed especially in a grooved
tube (as is the case at present) only partially overcomes the
temporal and spatial temperature variations found in the
commercial atomizers (37). As a result, a temperature gradient
may still exist between the tube and platform nullifying total
isothermal conditions within the furnace. This can lead to
poor analytical accuracy, precision and sensitivity even when
using integrated absorbance measurement as has been
demonstrated with the measurement of Pb in blood (37), and Cd

in urine (38). Also, the transfer or adaptation of methods
developed using one make of instrument to another make is
difficult even under STPF conditions probably because the
temperature program is not always transferable between
instruments. Thus stringent quality assurance of graphite
tubes, platforms, and the instruments themselves are required.
Slavin et al. (66) have recently introduced a quality control
protocol (QCP) based on the characteristic mass and Zeeman
ratio for aqueous solutions of Ag, Cr and Cu under improvised
STPF conditions without matrix modification and pyrolysis
steps. The validity of this QCP for real samples and under
non-compromised STPF conditions remains to be tested. In
summary, precise and accurate STPF-AAS determinations of trace
elements in biological fluids are still difficult. The nuances
of stabilized temperature platform furnace atomization seem to
be eluding us.

LITERATURE CITED

1. Mckenzie, H.A.; Smythe, L.E. In Quantitative Trace
    Analysis of Biological Materials; McKenzie, H.A.; Smythe,
    L.E., Eds.; Elsevier: Amsterdam, 1988; Chapter 1.
2. Iyengar, G.V. Elemental Analysis of Biological Systems;
    CRC Press, Boca Raton, FL, 1989; Vol.1, Chapter 2.
3. Subramanian, K.S. Prog. Anal. Spectrosc. 1988, 11, 511-
    608.
4. Subramanian, K.S. Prog. Anal. Spectrosc. 1986, 9, 237-
    334.
5. Subramanian, K.S. At. Spectrosc. 1988, 9, 169-78.
6. Iyengar, G.V. Elemental Analysis of Biological Systems;
    CRC Press, Boca Raton, FL, 1989, Vol.1, Chapter 9.
7. Versieck, J.; Cornelis, R. Trace Elements in Human Plasma
    or Serum; CRC Press, Boca Raton, FL, 1989; Chapter 4.
8. Slavin, W; Carnrick, G.R. At. Spectrosc. 1985, 6, 157-60.
9. Slavin, W; Manning, D.C.; Carnrick, G.R. J. Anal. At.
    Spectrom. 1988, 3, 13-19.
10. Slavin, W; Manning, D.C.; Carnrick, G.R. At. Spectrosc.
    1981, 2, 137-45.
11. Slavin, W.; Carnrick, G.R. Spectrochim. Acta 1984, 39B,
    271-82.
12. Slavin, W.; Carnrick, G.R.; Manning, D.C.; Pruszkowska,
    E. At. Spectrosc. 1983, 4, 69-86.
13. Slavin, W. Graphite Furnace AAS, A Source Book; Perkin-
    Elmer Corporation: Ridgefield, CT, 1984.
14. Burnett, W.B.; Bohler, W.; Carnrick, G.R.; Slavin, W.
    Spectrochim. Acta 1985, 40B, 1689-1703.
15. Carnrick, G.R.; Slavin, W. Am. Lab. 1988, 20, 88-98.
16. Krishnan, S.S.; McLachlan, D.R.; Krishnan, B.; Penton,
    S.S.A.; Harrison, J.F. Sci. Total Environ. 1988, 71, 59-
    64.
17. Frech,W.; Cedergren, A.; Cederberg, C.; Vessman, J. Clin.
    Chem. 1982, 28, 2259-63.
18. Leung, F.Y.; Henderson, A.R. Clin. Chem. 1982, 28, 2139-
    42.
19. Bettinelli, M.; Baroni, U.; Fontana, F.; Poisetti, P.
    Analyst (London) 1985, 110, 19-22.
20. Gardiner, P.E.; Stoeppler, M.; Nurnberg, H. Analyst
    (London) 1985, 110, 611-17.

21. Gardiner, P.E.; Stoeppler, M. J. Anal. At. Spectrom.
    1987, 2, 401-04.
22. Slavin, W. J. Anal. At. Spectrom. 1986, 1, 281-85.
23. Fagioli, F.; Locatelli, C.; Gilli, P. Analyst (London)
    1987, 112, 1229-32.
24. Savory, J.; Wills, M.R. In Aluminum and its Role in
    Biology; Sigel, H.; Sigel, A., Eds.; Marcel Dekker: New
    York, 1988; Chapter 10.
25. Sanz-Medel, A.; Roza, R.R.; Alonso, R.G.; Valllina, A.N.;
    Canata, J. J. Anal. At. Spectrom. 1987, 2, 177-83.
26. Black, M.M.; Fell, G.S.; Ottaway, J.M. J. Anal. At.
    Spectrom. 1986, 1, 369-72.
27. Yin, X.; Schlemmer, G,; Welz, B. Anal. Chem. 1987, 59,
    1462-66.
28. Lum, K.R.; Edgar, D.G. Internat. J. Environ. Anal. 1988,
    33, 13-21.
29. Pruszkowska, E.; Carnrick, G.R,; Slavin, W. Clin. Chem.
    1983, 29, 477-80.
30. McAughey, J.J.; Smith, N.J. Anal. Chim. Acta 1984, 156,
    129-37.
31. Subramanian, K.S. In Quantitative Trace Analysis of
    Biological Materials; McKenzie, H.A.; Smythe, L.E., Eds.;
    Elsevier: Amsterdam, 1988; Chapter 34.
32. Subramanian, K.S. Sci. Total Environ. 1989, 89, 237-50.
33. Miller, D.T.; Paschal, D.C.; Gunter, E.W.; Stroud, P.E.;
    D'Angelo, J. Analyst (London) 1987, 112, 1701-04.
34. Bruhn, C.G.; Piwonka, J.M.; Jerardino, M.O.; Navarrete,
    G.M.; Maturana, P.C. Anal. Chim. Acta 1987, 198, 113-23.
35. Shuttler, I.L.; Delves, H.T. Analyst (London) 1986, 111,
    651-56.
36. Shuttler, I.L.; Delves, H.T. J.Anal. At. Spectrom. 1987,
    2, 171-76.
37. Shuttler, I.L.; Delves, H.T. J.Anal. At. Spectrom. 1988,
    3, 145-49.
38. Paschal, D.C.; Kimberly, M.M. At. Spectrosc. 1985, 5,
    134-36.
39. Lian, L. Spectrochim. Acta 1986, 41B, 1131-35.
40. Kok, F.J.; Hofman, A.; Vanderbroucke, J.P.; Valkenburg,
    H.A. Internat. J. Epidemiol. 1985, 14, 335.
41. Clark, L.C. Fed. Am. Soc. Exp. Biol. 1985, 44, 2584-89.
42. Welz, B.; Melcher, M.; Schlemmer, G. Fresenius' Z. Anal.
    Chem. 1983, 316, 271-76.
43. Bauslaugh, J.; Radziuk, B.; Saeed, K.; Thomassen, Y.
    Anal. Chim. Acta 1984, 165, 149-57.
44. Paschal, D.C.; Kimberly, M.M. At. Spectrosc. 1986, 7, 75-
    78.
45. Lewis, S.A.; Hardison, N.W.; Veillon, C. Anal. Chem.
    1986, 58, 1272-73.
46. Neve, J.; Molle, L. Acta Pharmacol. Toxicol. (Suppl.
    VII). 1986, 59, 606-09.
47. Eckerlin, R.H.; Hoult, D.W.; Carnrick, G.R. At.
    Spectrosc. 1987, 8, 64-66.
48. Morisi, G.; Patriarca, M.; Menotti, A. Clin. Chem. 1988,
    34, 127-30.
49. Carnrick, G.R.; Manning, D.C.; Slavin, W. Analyst
    (London) 1983, 108, 1297-1312.
50. Neve, J.; Chamart, S.; Trigaux, P.; Vertongeu, F. At.
    Spectrosc. 1987, 8, 167-69.

51. Constantini, S.; Giordano, R.; Ruzzica, M.; Benedetti, F. Analyst (London) 1985, 110, 1355-59.
52. Subramanian, K.S. J. Anal. At. Spectrom. 1988, 3, 111-14.
53. Paschal, D.C.; Kimberly, M.M.; Bailey, G.G. Anal. Chim. Acta 1986, 181, 179-86.
54. Subramanian, K.S. Can. J. Spectrosc. 1988, 33, 173-81.
55. Paschal, D.C.; Bailey, G.G. At. Spectrosc. 1986, 7, 1-3.
56. McGahan, M.C.; Bito, L.Z. Anal. Biochem. 1983, 135, 186-92.
57. Carelli, G.; Bergamaschi, A.; Altavista, M.C. At. Spectrosc. 1984, 5, 46-50.
58. Matthews, D.O.; Mcgahan, M.C. Spectrochim. Acta 1987, 42B, 909-13.
59. Paschal, D.C.; Bailey, G.G. At. Spectrosc. 1987, 8, 150-52.
60. McGahan, M.C.; Tyczkowska, K. Spectrochim. Acta 1987, 42B, 665-68.
61. Delves, H.T.; Shuttler. I.L. In Biochemical Mechanisms of Platinum Antitumour Drugs; McBrien, D.C.H.; Slater, T.F., Eds.; IRL Press, Oxford, 1986; pp 329-46.
62. Paschal, D.C.; Bailey, G.G. J. Anal. Toxicol. 1986, 10, 252-54.
63. Grobenski, Z.; Lehman, R.; Radzuik, B.; Voellkopf, U. At. Spectrosc. 1986, 7, 61-63.
64. Subramanian, K.S. In Quantitative Trace Analysis of Biological Materials; McKenzie, H.A.; Smythe, L.E., Eds.; Elsevier: Amsterdam, 1988; Chapter 33.
65. Carnrick, G.R.; Slavin, W. Am. Lab. 1989, 21, 90-95.
66. Slavin, W.; Manning, D.C.; Carnrick, G.R. Talanta 1989, 36, 171-78.

RECEIVED July 16, 1990

Chapter 12

Inductively Coupled and Other Plasma Sources

Determination and Speciation of Trace Elements in Biomedical Applications

Ramon M. Barnes

Department of Chemistry, University of Massachusetts, Lederle Graduate Research Center Towers, Amherst, MA 01003–0035

An appraisal of currently popular electrical plasma sources, including the inductively coupled plasma (ICP), microwave induced plasma (MIP), and direct current plasma (DCP), employed with atomic and mass spectrometry is presented for the determination and speciation of trace elements in the biomedical field. The advantages and limitations of the methods based on these techniques, problems of interferences and their resolution, elements that can and cannot be determined relative to their reference values in various biological materials, the current status, and future developments are discussed. Interfacing liquid, gas, and supercritical fluid chromatography with these plasma detectors also is examined. Practical illustrations are given based upon a survey of recent literature.

The application of state-of-the-art electrical plasmas as sources for the atomic and mass spectrochemical analysis of biological and biomedical materials has grown with the development of inductively coupled plasma (ICP), direct current plasma (DCP), and microwave induced plasma (MIP) discharges (1). Major and minor elements and some trace elements are determined routinely in biological fluids and tissues (e.g. hair) by ICP spectrometry. Not only can individual or groups of metals and non-metals be determined, but the unique chemical form or complex of some can be identified and quantified when appropriate separation methods utilize plasma detection systems. Numerous demonstrations of these applications have been published, and commercial instrumentation now is readily available.

Low power, flowing inert gas plasma discharges at atmospheric pressure complement and supplement flame and furnace atomic absorption spectrophotometry (AAS) (Fig. 1), because plasmas are most often employed in emission or mass spectrometry. Except perhaps for the emission determination of alkali metals, flame spectroscopy has been replaced by ICP or DCP atomic emission spectrometry (AES) for determination of major and minor

0097–6156/91/0445–0158$06.75/0

elements in biological materials (2). For trace and ultratrace determinations, electrothermal vaporization AAS (ETV-AAS), characteristically performed with a graphite furnace, is employed commonly but only for a relatively few elements in a sample. Challenging ETV-AAS powers of detection is ICP mass spectrometry (ICP-MS); furthermore, its additional capability for rapid sequential multi-element and isotopic analyses makes it the technique of choice for trace element analysis (3).

For these spectrochemical sources proper sample collection and handling, sample preparation, and sample introduction represent the major limitations in the universal application to biological materials analysis (4). Considerable effort has been expended recently to improve sample introduction techniques for furnace and plasma sources in order to extend the versatility of the techniques, increase the analysis speed, or minimize sample preparation and manipulation steps. Techniques including direct gas or gaseous product introduction, micro-volume solution techniques, slurry nebulization, and direct or ablated solid sampling are currently under development. Analysis accuracy, however, is restricted more often than not by the quality of the sample obtained.

The essential problems in the elemental analysis of biological materials today are *determinations of ultra-trace concentration levels* and *speciation of chemical forms*. These problems are compounded by the requirement for *in situ* micro-distribution of elements and species (5). Chemical speciation can best be achieved with plasma discharges sources (Fig. 2) when separation techniques, such as ion chromatography (IC), high performance liquid chromatography (HPLC), gas chromatography (GC), supercritical fluid chromatography (SFC), or chemical reactions, such as hydride generation or chlorination, forming volatile metal species are applied. Highly successful GC-MIP methods, in fact, are the major examples, and commercial GC-MIP or similar detectors have only become available during the past year. An IC-ICP-AES system also was introduced commercially in 1989, and gaseous hydride generation arrangements have been available for AAS and AES for years. HPLC- and SCF- systems are used only in research laboratories. In total these plasma speciation arrangements are dwarfed by the commercial GC-MS popularity, however.

ELEMENT CONDITIONS AND SCOPE. In a review of trace metal analysis for biocompatibility testing and metal biotolerance studies, Michel (6) identified prerequisite conditions for adequate investigations, which can be extended to all trace metal analysis studies of biological materials:
1. Multi-element techniques are preferred so that element concentration ratios and the possibility of detecting unexpected effects can be evaluated.
2. High accuracy trace-metal analysis is essential to minimize bias and flaws found in earlier analytical data.
3. Validity of the analysis results is established by quality control of the analysis scheme.
4. Extremely low detection limits are essential to measure the very low normal concentrations of some metals in many body tissues and fluids.
5. Low analytical blank values, indicating freedom from contamination, guarantee minimum systematic error.
6. Spatial distribution of trace elements can provide quantitative description of trace elements at microscopic and cellular scales, and
7. Speciation provides information on the chemical form of trace metals.

Although the reasons for determining trace metals in biological and biomedical samples may differ with the specific study and application, the chemical action and concentrations of trace elements in organisms generally

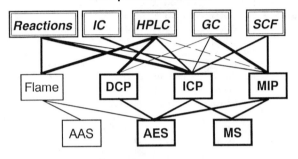

Figure 1. Sample introduction and signal detection arrangement for popular spectroscopic sources.

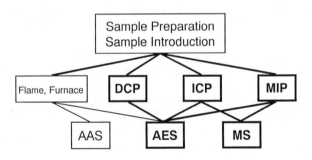

Figure 2. Interactions between chemical reaction and/or separation approaches and spectroscopic sources to achieve elemental speciation.

defines their typical classification as essential, toxic, or accidental, the latter comprising elements without any known physiological action. Another category is non-essential therapeutic elements (*e.g.*, Al, Au, B, Bi, Co, Ge, Li, Pt). Trace elements in living organisms can be chemically oncogenic and/or mutagenic as distinguished from toxic elements by their dose-response relationships (1). These responses may also depend upon the chemical form of the element. For example, Michel composed a priority list of 18 elements (Al, Ti, Cr, Fe, Co, Ni, Cu, Zn, Zr, Nb, Mo, Pd, Ag, Sn, Ta, Au, Hg, and Pb) after considering the composition of medical implants, the importance of the biological action, the feasibility of analysis, and the pathological potency in studies of biocompatibility testing. Schramel (7) indicated that trace element research in the biomedical field is primarily directed toward the known essential trace elements (Co, Cr, Cu, F, Fe, I, Mn, Mo, Ni, Si, Sr, V, and Zn), whereas for toxological investigations heavy metals (Cd, Hg, Pb, Tl) are of primary interest.

ICP SPECTROMETRY. The inductively coupled plasma exhibits both attributes and limitations for elemental analysis. Among the advantages are included the following:
1. Good excitation and ionization of metals and metalloids leading to excellent analysis sensitivity, precision, and accuracy.
2. Good thermal energy and solvent loading of the plasma to desolvate and destroy the analyte matrix resulting in simplified sample introduction and few matrix effects and large linear dynamic concentration range.
3. Modest electrical power requirements (<1 kW) resulting in compact instrumentation competitively priced and available from numerous vendors.
   Limitations of ICP spectrometry include the following:
1. Poor excitation of elements with high ionization potentials (*e.g.*, F) which limits their sensitivity.
2. No direct chemical information is provided owing to the complete destruction of chemical bonds of the analyte, thus requiring supplementation of the ICP source with a system providing separation based upon chemical form.
3. For emission spectrometry spectral interferences among elements is the major limitation, and high-resolution spectrometers are recommended for samples with spectrally complex matrices.
4. The plasma can be seriously perturbed and even extinguished if overloaded with volatile solvents especially organics.
   The benefits of employing ICP analysis for biological materials include the following (8):
1. Accurate multi-element data can be obtained from small quantities (<0.1 g) of a single sample under one set of experimental operating conditions.
2. Minimal sample handling eliminates sample modification or dilution and reduces contamination.
3. Statistical analysis can be performed with a common method database to reduce errors introduced from diverse methods.
4. Data can be obtained on a large number of samples in a minimum time owing to the analysis speed and insignificant measurement errors from instrument drift.
5. A few, multi-element standards are used, and
6. Extensive samples study can be performed owing to the versatility and speed of analysis.

DIRECT CURRENT PLASMA. The DCP also exhibits both virtues and shortcomings for elemental analysis. Among the advantages are the following:

1. Good excitation of metals
2. Good thermal energy and volatile solvent loading
3. Low generator power requirement (<750 W), and
4. Simple, compact equipment.
   In the sole commercial system available the small emitting zone of the DCP (i.e., 1 mm diameter) is optically matched to a high spectral-resolution, echelle spectrometer, which leads to minimal spectral interferences. Major disadvantages of the DCP in general are the severe matrix effects and the lack of competitive DCP equipment. A comparison of the performance characteristics atomic spectroscopy techniques including AAS, ICP, and DCP was made by Routh (9), who evaluated 12 general performance standards that ranged from sample throughput to operating costs. Overall, the DCP and ICP received equally high scores; whereas, graphite furnace AAS was rated worse owing to its limited sample throughput, precision, dynamic range, interferences, element range, qualitative analysis capability, and ease of use. The determination of trace elements in blood and other biological fluids with graphite furnace AAS was reviewed by Subramanian (10, 11).

   A direct comparison of some experimentally measured analytical properties of the DCP and ICP reported by Leis et al. (12) indicated a slight advantage in favor of the ICP based upon the limits of detection of 27 spectral lines of 20 elements. Owing to the convenient coupling of the DCP with an echelle spectrometer, the analysis of materials with line-rich emission spectra by DCP is practical and is a major field of application. Liquid samples containing high salt concentrations and even slurries can be analyzed readily by DCP-AES without possibility of nebulizer blockage owing to the wide nebulizer tube and robust nebulizer (13). For practical analysis of samples with variable alkali content, calibration procedures must be performed carefully to avoid systematic errors arising from effects alkalis on other analyte signals.

   Schramel (7) reviewed ICP and DCP-AES parameters and practical aspects as well as applications for trace element analysis in biomedical and environmental samples. The normal concentrations of heavy metals in body fluids and organs cannot be determined by plasma emission spectroscopy owing to their very low concentration levels. Mineral elements like Ca, K, Mg, Na, and P can be determined in body fluids. In DCP-AES the determination of Al and Li is about a factor of 10 more sensitive than with ICP-AES.

   In a comparison of DCP and ICP-AES for the determination of copper in blood serum, Bamiro et al. (14) found it necessary to calibrate with matrix matched standards for DCP-AES while for ICP-AES the sample was diluted ten-fold with distilled water and analyzed using standard solutions for calibration. The relative precision of the ICP method was about 2%.

ICP-MS. The ICP with mass spectrometric detection exhibits unique features and limitations for elemental and isotopic analyses beyond those observed with ICP-AES (15, 16, 17) Among the advantages are the following:
1. Ionization of metals and most metalloids with ionization potentials below 9.6 eV is uniform which results in similar sensitivity and powers of detection for elements throughout the periodic table.
2. High sensitivity inherent in mass spectrometry is achieved.
3. Isotope analysis for all elements with more than one stable nuclide is possible.
4. Rapid semi-quantitative analysis is achieved through a combination a fast measurement with a quadrupole mass spectrometer, relatively simple spectra with like sensitivity, and sophisticated computer software.
   As with ICP-AES the discharge provides good thermal energy for sample decomposition and reasonable solvent tolerance at a low generator power

requirement (<1.5 kW). Limitations include some matrix effects resulting from the mass spectrometer (*i.e.*, space charge effects in the ion optics), poorer signal stability than ICP-AES thus making an internal reference essential, poor ionization of elements with very high ionization potential (*e.g.*, F), and no direct chemical information is produced.

Practical topics of current interest or areas of active development in ICP-MS related to the analysis of biological materials include identifying isobaric overlap and spectral interferences; characterizing sample matrix effects; improving signal stability and long-term drift; extending the number of commercially available instrument manufacturers; reducing the capital, operating, and maintenance costs; developing measurement capabilities for multi-element transient signals; improving the ionization efficiency of elements with high ionization potential; extending the number of applications; and combining methods such as sample introduction techniques to provide chemical information.

STATUS OF ICP MULTI-ELEMENT ANALYSES. The status of ICP spectrometric analysis in clinical and medical laboratories for multi-element analysis is reviewed annually (18, 19, 20). Excerpts from these reviews during the past few years reproduced below are revealing:

Elemental Analysis with Plasma Discharge Sources

Developments in Multi-element Analysis: 1988
"...ICP-AES, particularly with simultaneous instruments, will likely take its place in the clinical laboratory alongside the existing multi-channel analysers now in common use. In the near future...ICP-AES will replace FAAS completely and ETA-AAS partly. The capabilities of the technique have been clearly shown....
" The analysis of *body tissues* is an area where multi-element capability of ICP-AES has really made its mark." (18)

Developments in Multi-element Analysis: 1989
"The ability of *inductively coupled plasma atomic emission spectrometry* to handle a wide range of samples with minimum interferences is still being demonstrated. Applications are limited to those elements present at higher concentrations, *e.g.*, Fe and Cu in serum, B in rat tissues, Ba, Ca, Mg, and Sr in foetal bones, Pt in serum, ultrafiltrate, B, S, P in tissues, a range of elements in hair, and trace elements in children's urine and faeces to study their ingestion of soil or dirt. So successful is the technique that, in contrast to others such as ETAAS and ICP-MS, description of the methods is of little value as there is nothing unusual. More attention is likely to be paid to the problems of dissolution of the sample or pre-concentration techniques." (19)

Inductively coupled plasma mass spectrometry: 1988
"An appraisal of the value of this technique to real analysis in the clinical and biological field is difficult. Up to the end of this review year [1987], there were still no refereed papers to review; all references here are to conference reports.
"The ability of the technique to measure different *isotopes* is being explored.
"A further important use of ICP-MS is for *speciation*." (18)

Inductively coupled plasma mass spectrometry: 1989
"Exponents of ICP-MS in biological analysis are apparently too busy
collecting data for the next ICP-MS conference to find time to actually
submit their work for *publication*. This review year [1988] there are
eight published papers to report and 15 conference reports to discuss;
last year [1987] there were no refereed papers and 19 conference reports. The published papers are all of a high standard and give a good
insight of the potential and problems of the technique, but what has
happened to all that accumulated knowledge in former years' conference papers?" (19)

Inductively coupled plasma mass spectrometry: 1990
"The *growth of ICP-MS as an analytical technique* is evident from the
increase in published and conference papers. ...It is becoming clear,
however, that despite the claims for ICP-MS as a multi-element technique, practising analysts are not using this facility so frequently as the
ability of the technique to measure isotopic composition or to act as a
sensitive direct analytical technique for one to two elements in speciation studies or total element determination." (20).

Analysis of Clinical and Biological Materials: 1988
"Multi-element analysis by ICP-AES seems to have become more firmly
established. Its ability to determine a large number of elements in digested samples with a minimum further sample pre-treatment and minimum interference is good news for the analyst, but it makes the description of the methods...rather boring. Progress with ICP-MS seems
slow. There are a number of conference abstracts to report, but...no reports of formally published applications in the field have been received." (18)

Conclusions: 1988
"Present research applications indicate a likely revolution in laboratories carrying out clinical and biological analyses. ...the simultaneous
multi-channel ICP-AES spectrometer is likely to become a feature of
the clinical laboratory, taking its place amongst the other multi-channel analysers already used routinely. Initially the application will be to
take over the determination of the common elements currently determined by FAAS or colorimetry.... What will be the impact of the ease of
tissue analysis by this technique which has been demonstrated in so
many papers? Will every autopsy lung sample be screened for evidence
of occupational exposure to toxic elements...?"
"When ICP-MS becomes a routine analytical tool, the vast range of data
that could be produced raises further questions about the control of the
quality and the use to which so much of the data could be put. The skill
of the clinical analyst may well then be more in the ability to sift out
what information is important and relevant. The misuse of the straightforward analysis of hair by ICP-AES is a lesson in trying to seek relevance in the measurements that are made." (18)

Conclusions: 1989
"Studies so far on ICP-MS as a multi-element technique suggest that for
elements of clinical interest, the relationship to ICP-AES will be rather
like that of ETAAS to FAAS. Both ICP-AES and FAAS can determine elements in rather complicated matrices with minimal interference but
with a lack of sensitivity. Inductively coupled plasma mass spectrome-

try has good sensitivity but suffers from interferences, requiring skill in chemical manipulation and pre-treatment to avoid interference, as is true for ETAAS. The value of ICP-MS may be more in measuring elements of mass number greater than 80 so that future years will see studies of more exotic elements." (19)

Conclusions: 1990
"...The number of laboratories so far attempting to use ICP-MS for more routine clinical analysis seems small. ...It is apparent also that the attraction of the ICP-MS technique is not only its multi-element capability but its ability to act as a sensitive technique for difficult elements, such as the uranium group metals; its sensitivity as a detector in speciation studies and its ability to measure accurately different isotopes for studies in nutrition of the uptake of trace elements and for identification of the source of environmental exposure to toxic elements." (20).

A survey of plasma atomic spectrometry and comparison with other methods for inorganic trace element analysis including absorption spectrophotometry and laser techniques by Broekaert (21) reviewed developmental prospects and trends with each of these sources.

APPLICABILITY OF ICP SPECTROMETRY IN HEALTH

ICP spectrometry is a well established, essential tool for element-oriented problems in human health and environmental protection, according to Caroli (1). The main classes of samples found amenable for ICP-AES in health and environmental studies are biological materials (*e.g.*, serum, plasma, urine, milk, cerebrospinal fluid, tissues, hair, nails); waters (*e.g.*, fresh, marine, precipitation, waste); plant tissues; airborne particulates; soils; sediments; fossil fuels; and pharmaceuticals. When these groups were examined qualitatively, Caroli (1) classified 25 elements of particular clinical and environmental interest as being optically, sufficiently, or insufficiently sensitive when determined by ICP-AES. For example, ICP-AES detection limits for Ti, Tl, and V were universally inadequate to permit their routine determination in any of these classes of materials without significant chemical preconcentration. Despite these sensitivity limitations, Caroli et al. (1) described the analysis of human lung tissues (22), hair, aluminum in connection with neurotoxicity, and platinum-based antitumor agents (23). In the latter example, 0.1 ml sample volumes were introduced into the ICP by means of an electrothermal vaporizer.

Maessen (24) comprehensively scrutinized early ICP-AES applications to biological and clinical materials including sampling details, sample digestion procedures, and analytical practice such as sample introduction and examples of analytical techniques. A tabulation of ICP-AES biological analysis applications through 1985 was presented along with discussion of the relative merits of ICP-AES.

Thompson (25) had previously examined the capabilities of ICP-AES and emphasized the difficulties and suitable laboratory approaches, including multi-element solvent extraction, involved in attempting the determination of ultratrace amounts of elements in similar natural materials (*e.g.*, geological, plant, and water samples). The need exists for either element enrichment or better measurement sensitivity for ultratrace analyses. Schramel (7) also described which biomedical and environmental samples (*i.e.*, whole blood, milk, amniotic fluid, bile, blood serum, tissues, urine) could and could not be analyzed in practice for selected elements by plasma emission spectrometry. For most tissues, for example, only the mineral elements and trace elements Cu, Fe, Mn, and Zn can be determined directly (26). On the other hand, the multi-element character of ICP and DCP-AES compensates for the relatively small total number of essential trace elements that can be detected.

Today a feasible instrumental approach for improved sensitivity is to employ ICP-MS rather than ICP-AES as demonstrated recently by Templeton et al. (27, 28). They emphasized that the significant improvement with ICP-MS over the sensitivity obtained with conventional ICP-AES provides novel simultaneous multi-element capabilities for more than 30 elements (Ag, Al, As, B, Ba, Ca, Cd, Co, Cs, Cu, Fe, Hg, Li, Mg, Mn, Mo, Ni, P, Pb, Pd, Pt, Sb, Sn, Sr, Tl, V, and Zn) of biological interest in liver tissue and serum. Among the peculiar considerations related to ICP-MS determinations of biological materials, they discussed

1. Overlap of signals from argon and other background species with analyte ions (e.g., $^{40}Ar^+$ on $^{40}Ca^+$, $^{40}Ar^{12}Cl^+$ on $^{52}Cr^+$, $^{40}Ar^{16}O^+$ on $^{56}Fe^+$, and $^{40}Ar_2^+$ on $^{80}Se^+$) which generally requires selection of alternative and often less sensitive nuclides.
2. Overlap of nuclides with the same mass, isobaric interferences (e.g. ^{87}Rb and ^{87}Sr).
3. Spectral overlap arising from unwanted ions including doubly charged ions, oxides (e.g., $^{79}Br^{16}O$ on ^{95}Mo), hydroxides, and argon dimers with metals, e.g. $^{40}Ar^{23}Na^+$ on ^{63}Cu.
4. Mass-interferent matrix effects resulting from the composition of biological materials, such as the chloride in biological materials causing $^{40}Ar^{35}Cl^+$ overlap on ^{75}As.
5. Matrix effects from the total composition of the sample as a result of space-charge effects in the mass spectrometer leading to reductions in the ion current.
6. Deposition of salts and corrosion by acids at the MS sampling interface limits the concentration and composition of sample nebulized.

They documented ICP-MS detection limits, short- (10 hour) and long-term (3 month) precision, and accuracy with standard biological reference materials. They found that P, Mg, and Ca were present in sufficient concentration in most tissues and fluids to permit ICP-MS determination; ^{44}Ca was used in place of ^{40}Ca. K measurement was difficult owing to spectral interference of major isotopes ^{39}K and ^{41}K. Other alkali metals were determined without interferences. High Fe and Zn concentrations made their measurement easy, and the ^{57}Fe was used instead of the ^{56}Fe. Owing to spectral interferences ^{51}V and ^{52}Cr were not reported. For heavy metals Cd, Hg, Pb, and Bi were considered ideally suited for ICP-MS owing to the low background above 80 atomic mass units. Monoisotopic ^{27}Al and ^{75}As experienced spectral interference limiting their sensitivity. Rare earth elements also were uniquely suited for ICP-MS measurement.

Using ICP-MS, Paudyn et al. (28) also examined trace element contamination in several blood sampling and storage devices. In our experience, ICP-MS is a critical tool for this type of testing, and careful controls over working procedures, laboratory environment (i.e., a clean room), and reagent blanks are essential for ultra-trace analyses.

Protein solutions, urine, and bovine liver were analyzed by Lyon et al. (29) in an evaluation of the multi-element capability of ICP-MS compared to ICP-AES and furnace AAS. The polyatomic interferences arising from the biological matrices were considered, and ICP-MS required less sample preparation than the other methods.

In another example, Vanhoe et al. (30) determined seven ultratrace elements (Co, Cu, Cs, Fe, Mo, Rb, and Zn) in human serum by ICP-MS. Polyatomic ion interferences, memory effects, and accuracy were examined in detail. The ICP-MS results showed good agreement with certified values.

Durrant and Ward (31) determined 18 elements by ICP-MS in pooled breast milk samples, and their ranges along with 38 other elements were established.

Igarashi et al. (32) determined Th and U concentrations in six typical biological standard reference materials, including liver, mussel, chlorella, bone, and lung.

In establishing the amount of dirt ingested by normal children (33), we measured eight elements (Al, Ba, Mn, Si, Ti, V, Y, and Zr) by ICP-AES and ICP-MS in biological (feces, urine), environmental (dust and soil), and food samples for 63 children and 6 adults (34, 35).

The determination of mineral intake normally is established by measuring the dietary intake, and Shiraishi et al. (36), for example, measured 11 elements in duplicate diets samples in Japan by ICP-AES. The applications of ICP-MS in food science was described in detail by Dean et al. (37).

ANALYTICAL DATA ASSESSMENT

Trace element analysis of biological samples often can be inaccurate, and establishing how much error is involved is critical in application of plasma sources for analysis of biological materials (4, 25). Veillon (38) described sources of analytical error resulting from inappropriate sampling, storage, sample preparation, analysis, and reference materials. Ramsey et al. (39) concluded that the quality of analysis typically produced by ICP-AES on practical samples, although acceptable for most applications, was poorer than that normally reported using naive tests, since these tests may not include errors in sample preparation or long-term instrumental errors with practical samples. Parameters of precision and bias were examined for ICP-AES. They recommended an analytical quality assurance scheme as the most practical way to achieve a true assessment of data quality for routine plasma spectrometric analysis.

PROTEIN CONCENTRATIONS AND STOICHIOMETRIES BY ICP-AES DETERMINATION OF SULFUR

The concentrations of several micromolar solutions of proteins with known sulfur contents were determined by ICP-AES employing the S I 180.73 nm emission (40). Accurate agreement was obtained with conventional methods of protein determination, and the ICP-AES determination of sulfur is a dependable means of measuring microquantities of simply diluted buffered proteins pumped directly into the nebulizer. For several metalloproteins containing Fe, Cd, Zn, and Cu, the multi-element ICP-AES determination provided an excellent means of determining stoichiometries of bound metal ions, since both metal and protein determinations could be performed on a single sample.

SEMI-QUANTITATIVE ANALYSIS

Two properties of ICP-MS provide a unique analytical capability for semi-quantitative analysis (41). First, the sequential MS scan is rapid so that the entire range of useful elements can be examined quickly. Secondly, the element response sensitivity is almost uniform throughout the periodic table.

168   BIOLOGICAL TRACE ELEMENT RESEARCH

This results in the identification of potential elements in an unknown sample
and an estimation of unknown concentrations with errors of less than ±30%.
Commercial instrument software consists of pre-calibrated or established re-
sponses for each element. In identifying the significant peaks in the unknown sample spectrum, the interactive software permits the ICP-MS operator choices in the level of confidence of the identification and quantification.
For example, the semi-quantitative analysis of NIST (National Institute of
Standards and Technology; formerly National Bureau of Standards) 1577
Bovine Liver are within 20% of the certified concentrations (Fig. 3) except for
As, Cr, and Hg, which owing to the sample dilution placed their solution
concentrations near their detection limit. Nevertheless, their concentrations
were within a factor of 2 to 3. When used in conjunction with spatial sampling devices like laser ablation, the semi-quantitative mode might provide a
very rapid screening capability and spatial evaluation for biological materials.

A refinement of this approach was recently described by Polk et al.
(42), who developed a new procedure for automatic interpretation of ICP-MS
spectra. With standard solutions to calibrate the response factors, the semi-
quantitative analysis of concentrations presumably can be accurate to
within 2% for solutions and 5% for solid samples.

The reliable semi-quantitative analysis of biomedical specimens can
be expected to improve the practical outlook for clinical multi-element analysis (cf. 18 - 20).

STABLE ISOTOPE RATIO ICP-MS

Many unresolved questions exist concerning bioavailability and mineral
metabolism in human and animal nutrition, and the utilization of ICP-MS
has stimulated the study of mineral metabolism by stable isotope labelling
(43). Because of methodological and ethical limitations, stable isotope labeling is preferable to radiotracer administration, but the main problem associated with stable isotope labelling is quantification. ICP-MS has proven to be
as reliable as thermal ionization MS and Fast Atom Bombardment MS but
faster, easier to use, and of higher sensitivity for stable isotope ratio determinations.

To date B, Br (44), Cu (45), Cr (46), Fe (47, 48), Li (49), Mg (50), Os, Pb
(51, 52), Se (53), Sr (54), U, and Zn isotope ratios have been examined in
various materials including whole blood and red blood cells, feces, milk,
serum and plasma, tissues, and urine (43). Details of practical considerations and applications are given by Janhorbani and Ting (43), who have been
pioneers in the field.

In the ICP-MS determination of iron absorption in normal women (47),
the introduction of samples with an electrothermal vaporizer (ETV) instead of
the normal nebulizer-spray chamber reduced the interference of some polyatomic ions. In the absence of water, the peaks at ^{54}Fe and ^{56}Fe corresponding to $^{40}Ar^{14}N^+$ and $^{40}Ar^{16}O^+$ were reduced almost to background
level. Based upon these measurements the time change of Fe ratio enrichment in serum after oral and intravenous ^{57}Fe administration was followed.

We have evaluated the effect of pectins on dietary iron utilization in
normal rats by examining the ^{58}Fe to ^{57}Fe ratios in rat blood after oral administration of enriched ^{58}Fe (55). ICP-MS was employed to measure iron
isotope ratios in rat blood, feces, and food, while total iron was determined
by ICP-AES. Blood samples were completely decomposed in a high-pressure
digestion device, and the accuracy of sample preparation was evaluated with
standard reference blood. The isotope ratio precision was better than 1%. El-

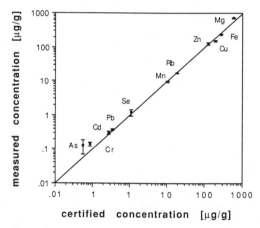

Figure 3. Comparison of results for ICP-MS semi-quantitative analysis (measured concentration) of 11 elements in NIST SRM 1577 Bovine Liver and certified values. Diagonal line is theoretical correlation.

evated ^{58}Fe to ^{57}Fe ratios in rat blood were achieved after 2 days of administration and remained for about 12 days.

The unique variation of lead isotope abundance in nature permits identification of sources of lead traceable to the original mine. This feature has been utilized in identifying sources of childhood lead poisoning, when differences in the ingested source exist (51, 52, 56). In our tests, lead isotope ratios were determined routinely in about five minutes by ICP-MS in blood, dust, paint, soil, and water.

ISOTOPE DILUTION ICP-MS

Isotope dilution is an accepted analytical technique for the definitive determination of trace elements (57). Spiked or isotope dilution analyses are employed to establish the amount of an element present in a sample by addition of a known quantity of a spike isotope to the sample followed by measuring the resultant isotope ratio. From the change in isotope ratio(s) before and after the spike and from the isotopic composition of the spike, the unknown concentration of the element in the sample can be calculated. Multi-element isotope dilution mass spectrometry also is possible. The limitations and precautions to be considered were discussed by Russ (49), and a summary of ICP-MS isotope dilution analyses was given by Houk and Thompson (15).

Recently Lasztity et al. (58) examined on-line isotope dilution ICP-MS as part of a study of lead content of biological (blood, milk), food, and other reference materials.

SPECIATION

The identification of specific metal containing compounds by the use of plasma spectrometry is powerful when combined with appropriate sample separation and introduction techniques like gas, liquid or supercritical fluid chromatography (Fig. 2). Other separation techniques like electrophoresis, ultrafiltration, or dialysis have been used much less often with plasma detection than chromatography. Combined techniques for chemical species identification in biological materials were reviewed by Gardiner (59, 60), Ebdon et al. (61, 62), and Uden (63, 64).

The advantages of chromatographic detection by plasma spectrometry include the following:

1. Many metals and non-metals can be monitored directly in the eluent of the chromatographic system. With high-speed sequential or simultaneous multi-element spectrometers numerous elements can be determined together.
2. Partial separation or imperfect chromatography with non-ideal elution or incomplete chromatographic resolution can be used owing to the generally high selectivity of the spectroscopic detector. With partial separations less time is required, species are likely to remain unaltered, and sample requirements are less severe.
3. High elemental sensitivity and selectivity can be achieved with good resolution spectrometers and background correction. Selectivity of an element against carbon can be 10^4 to 1 (cf. Table I).
4. Plasma detectors are generally compatible with chromatographic systems with respect to eluent flow rates and solvent loading. Organic solvent compatibility, however, may require thermostated nebulizer chambers and some modification of the plasma operating conditions.

Table I. Detection Limits, Selectivity, and Dynamic Range of Elements for Gas Chromatography - Microwave Induced Plasma Atomic Emission Spectrometry[a]

Wavelength	Detection Limit pg/s (pg)	Selectivity vs. Carbon[b]	Dynamic Range
H 486.1	4[c]		5,000
486.1	1		2,000
486.1	16 (45)	74	500
656.3	7.5 (22)	160	500
656.3	2[c]		
D 656.1	4[c]	50 (H)	10,000
656.1	2	781 (H)	500
656.1	7.4 (20)	194	500
B 249.8	3.6 (27)	9,300	500
C 193.1	1[c]		20,000
193.1	0.3		15,000
247.9	2.7 (12)	(1)	1,000
247.9	4[c] (15 second order)		20,000
N 174.2	50[c]	2,000	20,000
174.2	3	25,000	8,000
O 777.2	120[c]	10,000	5,000
777.2	50	20,000	12,000
F 685.6	40	31,000	
685.6	31	9,000	1,000
685.6	50[c]	20,000	2,000
685.6	8.5	3,500	
685.6	180 (510)	11,400	500
685.6	20 (64)	573	500
685.6	4.8	1,060	5,000
Al 396.2	5.0 (19)	3,900	500
Si 251.6	3	31,000	
251.6	85[c]	1,000	2,000
251.6	9.3 (18)	1,600	500
P 177.5	1	50,000	
177.5	1[c]	5,000	1,000
253.6	3.3 (56)	10,000	500
S 180.7	0.9	125,000	
	2[c]	8,000	10,000
	1	60,000	20,000
545.4	8.4	25,000	
545.4	52 (140)	4,600	500
Cl 479.5	30	71,000	
479.5	40[c]	3,000	10,000
479.5	15	25,000	8,000
479.5	86 (310)	1,500	500
481.0	16	2,400	
V 268.8	10 (26)	57,000	100
Cr 267.7	7.5 (19)	108,000	1,000
Mn 257.6	1.6 (7.7)	110,000	1,000
Fe 259.9	0.3 (0.9)	280,000	1,000
Co 240.7	6.2 (18)	190,000	1,000
Ni 231.6	2.6 (5.9)	6,500	1,000
Ge 265.1	1.3 (3.9)	76,000	1,000

Continued on next page

Table I. Continued

Wavelength		Detection Limit pg/s (pg)	Selectivity vs. Carbon[b]	Dynamic Range
As	228.8	6.5 (155)	47,000	500
Se	204.0	5.3 (56)	11,000	1,000
Br	470.5	10	1,400	
	470.5	33 (106)	300	500
	478.6	**60**	**19,000**	
	478.6	**50**	**9,500**	**1,000**
	478.6	**60[c]**	**2,000**	**1,000**
	478.6	34 (106)	600	500
Nb	288.3	69 (335)	32,000	100
Mo	281.6	5.5 (25)	24,000	500
Ru	240.3	7.8 (35)	134,000	1,000
Sn	284.0	1.6 (6.1)	360,000	1,000
	303.4	**1-100**		
I	206.2	21 (56)	5,000	500
	184.4	**1-100**		
W	255.5	51 (646)	5,500	500
Os	225.6	6.3 (34)	50,000	1000
Hg	**253.7**	**0.1**	**600,000**	
	253.7	0.6 (60)	77,000	1,000
	184.9	**1-100**		
Pb	283.3	0.17 (0.71)	25,000	1,000
	405.8	2.3 (7.2)	200,000	1,000
	405.8	**1-100**		

[a]Detection limit, selectivity, and dynamic range data taken from publications or manufacturer's literature. Updated values are given in (85). Information not available for blank entries.
[b]Selectivity is the peak area response per mole of analyte element divided by the peak area response of the background element per mole of that element (64).
[c]HP 5921A AED specifications; **values in boldface obtained with HP AED or its prototype.**

Among the limitations are physical and chemical interferences caused
by matrix constituents depending upon the separation method and plasma
detection combination; inadequate analytical sensitivity and sample dilution;
and extraneous trace metal contamination. Plasma spectrometric detectors,
in contrast to flame or furnace AAS systems, exhibit lower matrix interferences and longer linear dynamic concentration ranges (59). Biological sample collection, pretreatment, and storage must preserve the integrity of the
species (4, 60).

In general, species from gas chromatography are most effectively detected with MIP-AES operated in helium at atmospheric pressure. A commercial instrument (Hewlett Packard) became available in 1989. Liquid
chromatography is best handled with DCP-AES, ICP-AES, or ICP-MS. Although most arrangements are laboratory combinations of commercial
equipment, a commercial ion chromatograph ICP-AES system (Dionex -
Thermo Jarrell Ash) became available recently (65). Aqueous eluants are
accommodated directly, while organic eluants (*i.e.*, normal phase chromatography with hexane or methyl isobutyl ketone mobile phases) require extra
care in nebulizer and spray chamber design to minimize solvent overloading
of the plasma. Since DCP has the highest and MIP has the lowest organic
solvent loading capacity, the utility of detectors is DCP > ICP >> MIP. Although potentially useful, furnace AAS detection for liquid chromatography
is somewhat limited by the delay between sample analysis; it is a batch
rather than an on-line detector.

In their recent review of plasma instrumentation, Marshall et al. (66)
commented: "The area of *coupled chromatographic techniques* for the automatic introduction of sample into atomic spectroscopic instruments is now
maturing as an analytical technique ...." They continued: "... a plethora of
applications reporting the *coupling of GC to MIP* for the detection of mainly
non-metallic elements has appeared. This large number of papers, in excess
of 20, illustrates the popularity of this methodology for GC elemental analysis." Furthermore, they commented: "The introduction of supercritical fluid
chromatography (SFC) has led to the *inevitable interfacing of SCF to MIP*.
...The interfacing of SFC to ICP-AES also illustrates how this new form of
chromatography is gradually becoming popular as a sample introduction
procedure.... The coupling of liquid chromatographic systems to MIP has
long posed many problems. However, the recent use of a large volume He
MIP has indicated a possible way forward for this form of sample introduction."

LIQUID CHROMATOGRAPHY

Separation of biological molecules can be achieved with one or a combination of chromatographic modes: size exclusion, normal phase, reversed-
phase, paired-ion reverse-phase, and ion exchange.

We recently described the determination of calcium tetracycline compounds using reverse-phase HPLC-ICP-AES (67). The binding of tetracycline,
oxytetracycline, and chlortetracycline to calcium was evaluated with a PLRR-
1 polystyrene-divinylbenzene column as a function of pH from 3.5 to 8. In
order to accommodate the organic mobile phase (82% 10 mM phosphate
buffer and 18% acetonitrile) oxygen was added to the outer argon gas flow to
reduce carbon molecular band emission spectra. Calcium complexation was
verified between pH 6 and 7.5, and binding by the three tetracyclines followed their order of basicity.

Beauchemin et al. (68) separated and determined on line arsenic
species (*i.e.*, As(III), As(V), monomethylarsonic acid, dimethylarsinic acid,

and arsenobetaine) with ion pairing and ion exchange HPLC-ICP-MS in a dogfish muscle reference material. Heitkemper et al. (69) also separated and quantified four arsenic compounds (i.e., As(III), As(V), monomethylarsonate, dimethylarsinate) found in urine by HPLC-ICP-MS. Limits of detection were below 100 pg for each of these species. The determination of As(III) was complicated by co-elution of chlorine-containing species and the formation of $^{40}Ar^{35}Cl^+$.

Bushee et al. (70) determined a mercury-containing antimicrobial agent (thimerosal) in vaccines and toxoides with HPLC-ICP-MS by monitoring the ^{202}Hg signal. A similar method with post-column cold mercury vapor generation. ICP-MS was applied to the separation of methylmercuric acetate, ethylmercuric chloride, and mercuric chloride, and the determination of methylmercury in tuna (71). Post-column derivatization was not necessary for the determination of trace mercury in environmental samples owing the superior sensitivity of ICP-MS relative to DCP-AES, ICP-AES, or AAS.

Suyani et al. (72) separated and determined mixtures of up to five organotin compounds by means of micellar liquid chromatography and ICP-MS. Sub-μg/g level detection limits were reached by avoiding the hydro-organic mobile phase typical of HPLC.

ION CHROMATOGRAPHY. Ion chromatography (IC), used to separate charged species in complex mixtures, employs a primarily aqueous mobile phase that is easily introduced in an ICP at flow rates up to 2 ml/min. The mobile phase may contain numerous eluent ions leading to elevated background with conventional conductivity detectors. The plasma detector effectively separates the response of the solute ions from the constant background of the mobile phase.

We have determined the drug Didronel (1-hydroxyethylidene-1,1-disodium diphosphonic acid, HEDP) used for treatment of metabolic bone diseases by IC-ICP-AES based upon specific detection of the P I 213.649 nm emission with a commercial sequential ICP-AES system (73). Although several gas and liquid chromatographic analyses of HEDP had been reported, many diphosphonates including HEDP lack the functional groups readily detected by conventional liquid chromatographic detectors. With a retention time of 3.7 minutes and a mobile phase (7.9 mM nitric acid) flow rate of 1 ml/min, linear calibration curves, injection precision of 2%, and a limit of detection of 250 ng of phosphorus injected were obtained. The technique exhibited no matrix effects and provided a direct, simple, routine analysis that could not be achieved by other chromatographic approaches.

SIZE EXCLUSION CHROMATOGRAPHY. As discussed by Gardiner (59, 60, 74) size exclusion chromatography (SEC) provides advantageous separations based upon molecular radius of the species ideally with no chemical interactions with the separation media. The pH and ionic strength of the buffer can be selected to correspond with that of the sample. SEC is effective for the separation of biological macro-molecules such as proteins with minimum sample preparation. Operating conditions are generally compatible with plasma detection systems. Practical difficulties, however, include trace metal contamination (e.g., Cu, Fe, Zn) of the column and buffer; decomposition of labile species on contact with the metallic components of the column; sample dilution on the column, which may lead to a shift in the chemical equilibrium or may require sensitive detection; irreversible interaction of some species with charged surfaces on the separation medium resulting in loss of species from the sample.

Applications of SEC with plasma detection have included two major groups: protein-bound metals (metalloproteins) and inorganic metals and complexes. The former have included determination of Cd (75); Cu, Fe, and Zn with albumin and Fe with ferritin (1); Zn in human plasma and serum, Al-binding proteins, Cd species and metallothionein in animal organs (76); Ca, Cu, Fe, Mn, Mg, and Zn in human milk and cow's milk formulas (77); Se proteins in blood serum, red blood cell lysate, and breast milk (78); and Pt methionine complexes in plasma. For the second category, analyses have included Au drug metabolites in blood; Cu, Mg, Fe, and Zn in seminal fluids and serum; lead alkyls and As species in urine and blood; and Cr species in soil.

As a result of the high ICP-MS sensitivity, the use of SEC-ICP-MS permits studies of speciation (*e.g.*, of Cd or Pb) at normal background levels rather than abnormally high levels (75) or eliminates the need for on-line post-column hydride generation (78). For example, we recently demonstrated for the SEC-ICP-MS of lead-binding proteins in human and rat serum and red blood cells that the limit of detection of lead was not significantly degraded in the protein fraction (150 ng/l) compared to that in the SEC buffer (90 ng/l) when the latter was measured directly (79). Detection of lead-binding proteins would not be possible by SEC-ICP-AES owing to the poor limit of detection of 42 µg/l.

GAS CHROMATOGRAPHY WITH MIP DETECTION

The major plasma spectrometric detector for gas chromatography is the MIP (64). MIP is most economical to operate owing to its low flow rates and power requirements. The high ionization potential of helium provides high energy excitation especially for non-metals and halogens with high ionization potential. The spectral background emission is low, and the equipment is simple and compact. On the other hand, low-power MIP discharges have low tolerance to solvent loading, and exhibit low thermal energy owing to the non local thermodynamic equilibrium of the plasma. Severe matrix effects can occur, and until recently, no commercial equipment was available. Among the unique capabilities of the GC-MIP system is multi-element detection and empirical formula determination of resolved components. Two major types of MIP arrangements are popular: the Beenakker cavity introduced in 1976 and the Surfatron surface wave arrangement.

Most biological materials are not sufficiently volatile for direct injection GC-plasma detection. Extraction or chemical conversion to volatile forms often is necessary. Element-specific derivatization for GC-MIP has been described (80, 81). Thus, most applications described so far for GC-MIP-AES relate to relatively volatile environmental or energy related materials (64). Hydride forming elements and those forming volatile alkyl derivates include As, Pb, Hg, Sn, Se, Sb, Ge, Tl among others.

As an example, Tsunoda et al. (82) attempted to determine dimethylselenide in dolphin liver by GC-MIP after collection on a carbon molecular sieve of the gas evolved upon heating the sample in an oven. The column packing was then extracted with methanol, and an aliquot was injected into the GC. In another example, Hagen et al. (83) characterized fluorine containing metabolites in rat plasma after dosing with 1H, 1H, 2H, 2H-perfluorodecanol by monitoring the F I 685.6 nm emission from a GC-MIP system.

The first integrated GC-MIP-AES system was introduced in 1989 by Hewlett-Packard (84, 85, 86). The reported performance properties of GC-MIP-AES are summarized from a variety of sources in Table I. The long-term success of this GC-MIP-AES system with biological materials will depend upon development of effective derivatization schemes. On the other hand, the capability to establish molecular formulae for molecules with molecular weight ranging from 200 to 400 (64, 87) is an added motivation.

Atmospheric-pressure helium MIP discharges recently have been evaluated as ion sources for MS and as detectors of halogenated compounds in GC eluents (88).

MIP detection for liquid chromatography has been achieved only with arrangements that incorporate solvent evaporation and/or high-power MIP discharges. In a pioneering example, Kawaguchi et al. (89, 90) employed a tantalum filament vaporizer in the determination of zinc metalloenzymes after SEC separation. For high-power operation MIP discharges utilize 150 - 500 W and require special water-cooled cavities and torches (91), which would seem to compromise the convenience of low-power MIP operation.

MIP and ICP detection for supercritical fluid chromatography is only in its infancy (92, 93, 94), and no reports currently exist for the application to biological materials. This situation may be expected to change as the application of SFC in the biological field develops. Both liquid and supercritical fluid chromatography are as likely as gas chromatography to become the prime techniques for biological materials, since considerable development remains to be done in derivatizing compounds for gas chromatography.

CONCLUSION

Comprehensive discussions of the status and trends of plasma spectrometry for trace element determinations (5, 21, 95, 96) compare and contrast analytical features of plasma, flame, and furnace spectrochemical sources. In this review, some specific conclusions can be reached about the relative merits of plasma sources for the analysis of biological materials. First, ICP spectrometry, especially ICP-MS, is or will shortly become the prevailing method of choice for ultratrace, isotope dilution stable isotope ratio, and liquid chromatographic species analyses. The semi-quantitative survey capabilities provided by ICP-MS will play a considerable role in rapid screening of many biological materials, and previously unknown multi-element relationships in biological systems may become apparent. Probably the application of laser ablation ICP-MS of biological solids and tissues will provide multi-element, semi-quantitative spatial profiles. DCP-AES along with ICP-AES will continue to serve a supporting role for many practical, routine analyses of major and minor elements in biological materials.

Element speciation obtained with gas chromatography will continue to be dominated by GC-MIP detection for the next few years, especially now that a sophisticated, complete instrument is commercially available. When applied to biological materials, effort will be required to obtain volatile compounds suitable for GC, which in turn may stimulate development of SCF extraction and chromatography with either an ICP or MIP detection. The determination of biological macromolecules by SEC with ICP-MS detection will lead to some fundamental stable isotope tagging studies in medical and nutritional fields.

The existing possibilities outnumber the literature reports, and for elemental analysis of biological materials, plasma spectroscopy is here to stay.

ACKNOWLEDGMENTS

Preparation of this manuscript was supported in part by the *ICP Information Newsletter*.

LITERATURE CITED

1. Caroli, S. Spectrochim. Acta 1988, 43B, 371 - 80.
2. Nixon, D.E.; Moyer, T.P.; Johnson, P.; McCall, J.T.; Ness, A.B.; Fjerstad, W.H.; Wehde, M.B. Clin. Chem. 1986, 32, 1660 - 65.
3. Paschal, D.C. Spectrochim. Acta 1989, 44B, 1229 - 36.
4. Behne, D. Fresenius Z. Anal. Chem. 1989, 335, 802 - 5.
5. Toelg, G. Fresenius Z. Anal. Chem. 1988, 331, 226 - 35.
6. Michel, R. CRC Crit. Rev. Biocompatibility 1987, 3, 235 - 317.
7. Schramel, P. Spectrochim. Acta 1988, 43B, 881 - 96.
8. Williams, T.R.; Van Doren, J.B.; Smith B.R.; Mc Elvany, S.W.; Zink, H. Am. Biotech. Lab. 1986, 4(5), 52, 54, 56-7.
9. Routh, M.W. Spectroscopy (Eugene) 1987, 2, 45 - 52.
10. Subramanian, K.S. Prog. Anal. Spectrosc. 1986, 9, 237 - 334.
11. Subramanian, K.S. Prog. Anal. Spectrosc. 1988, 11, 511 - 608.
12. Leis, F.; Broekaert, J.A.C.; Waechter, H. Fresenius Z. Anal. Chem. 1989, 333, 2 - 5.
13. Ebdon, L.; Parry, H.G.M. Microchem. J. 1990, 41, 219 - 26.
14. Bamiro, F.O.; Littlejohn, D.; Marshall, J. J. Anal. At. Spectrom. 1988, 3, 279 - 84.
15. Houk, R.S.; Thompson, J.J., Mass Spectrom. Rev. 1988, 7, 425 - 61.
16. Hieftje, G.M.; Vickers, G.H. Anal. Chim. Acta 1989, 216, 1 - 24.
17. Gray, A.L. In Applications of Inductively Coupled Plasma Mass Spectrometry; Date, A.R., Gray, A.L., Eds.; Blackie: Glasgow, 1989; Chapter 1.
18. Brown, A.A.; Halls, D.J.; Taylor, A. J. Anal. At. Spectrom. 1988, 3, 45R - 77R.
19. Brown, A.A.; Halls, D.J.; Taylor, A. J. Anal. At. Spectrom. 1989, 4, 47R - 88R.
20. Crews, H.M.; Halls, D.J.; Taylor, A. J. Anal. At. Spectrom. 1989, 5, 75 - 129R.
21. Broekaert, J.A.C. Anal. Chim. Acta 1987, 196, 1 - 21.
22. Alimonti, A.; Coni, E.; Caroli, S.; Sabbioni, E.; Nicolaou, G.E.; Pietra, R. J. Anal. At. Spectrom. 1989, 4, 577 - 80.
23. Alimonti, A.; Petrucci, F.; Dominici, C.; Caroli, S. J. Trace Elem. Electrolytes Health Dis. 1987, 1, 79-83.
24. Maessen, F.J.M.J. In Inductively Coupled Plasma Emission Spectroscopy. Part II: Applications and Fundamentals; Boumans, P.W.J.M., Ed.; Wiley-Interscience: New York, 1987; Chapter 5.
25. Thompson, M. Analyst 1985, 110, 443 - 49.
26. Schramel, P.; Lill, G.; Hasse, S.; Klose, B.-J. Biol. Trace Elem. Res. 1988, 16, 67 - 75.
27. Temptleton, D.M.; Paudyn, A.; Baines, A.D. Biol. Trace Elem. Res. 1989, 22, 17 - 33.
28. Paudyn, A.; Templeton, D.M.; Baines, A.D. Sci. Total Env. 1989, 89, 343 - 52.
29. Lyon, T.D.B.; Fell, G.S.; Hutton, R.C.; Eaton, A.N. J. Anal. At. Spectrom. 1988, 3, 265 - 71.
30. Vanhoe, H.; Vandecasteele, C.; Dams, R. Anal. Chem. 1989, 61, 1851 - 57.
31. Durrant, S.F.; Ward, N.I. J. Micronutrient Anal. 1989, 111 - 26.
32. Igarashi, Y.; Kawamura, H.; Shiraishi, K.; Takaku, Y. J. Anal. At. Spectrom. 1989, 4, 571 - 76.

33. Calabrese, E.J.; Barnes, R.M.; Stanek III, E.J.; Pastides, H.; Gilbert, C.E.; Veneman, P.; Wang, X.; Lasztity, A.; Kostecki, P.T. Reg. Toxicol. Pharm. 1989, 10, 123 - 37.
34. Wang, X.; Lasztity, A.; Vicizian, M.; Israel, Y.; Barnes, R.M., J. Anal. At. Spectrom. 1989, 4, 727 - 35.
35. Lasztity, A.; Wang, X.; Vicizian, M.; Israel, Y.; Barnes, R.M., J. Anal. At. Spectrom. 1989, 4, 737 - 42.
36. Shiraishi, K.; Yoshimizu, K.; Tanaka, G.; Kawamura, H. Health Phys. 1989, 57, 551 - 57.
37. Dean, J.R.; Crews, H.M.; Ebdon, L. In Applications of Inductively Coupled Plasma Mass Spectrometry; Date, A.R., Gray, A.L., Eds.; Blackie: Glasgow, 1989; Chapter 6.
38. Veillon, C. Anal. Chem. 1986, 58, 851A - 66A.
39. Ramsey, M.H.; Thompson, M.; Banerjee, E.K. Anal. Proc. (London) 1987, 24, 260 - 65.
40. Bongers, J.; Walton, C.D.; Richardson, D.E.; Bell, J.U. Anal. Chem. 1988, 60, 2683 - 86.
41. Amarasiriwardena, C.J.; Gercken, B.; Argentine, M.D.; Barnes, R.M., J. Anal. At. Spectrom. 1990, 5, 000 - 000.
42. Polk, D.; Zarycky, J.; Ediger, R. Abstract of Papers, The Pittsburgh Conference in New York, 1990, paper 85.
43. Janghorbani, M.; Ting, B.T.G. In Applications of Inductively Coupled Plasma Mass Spectrometry; Date, A.R., Gray, A.L., Eds.; Blackie: Glasgow, 1989; Chapter 5.
44. Janghorbani, M.; Davis, T.A.; Ting, B.T.G. Analyst (London), 1988, 113, 405 - 11.
45. Ting, B.T.G.; Janghorbani, M. J. Anal. At. Spectrom. 1989, 4, 325 - 36.
46. Dever, M.; Hausler, D.W.; Smith, J.E. J. Anal. At. Spectrom. 1989, 4, 361 - 63.
47. Whittaker, P.G.; Lind, T.; Williams, J.G.; Gray, A.L. Analytst (London), 1989, 114, 675 - 78.
48. Woodhead, J.C.; Drulis, J.M.; Rogers, R.R.; Ziegler, E.E.; Stumbo, P.J.; Janhorbani, M.; Ting, B.T.G.; Fomon, S.J. Pediatr. Res. 1988, 23, 495 - 99.
49. Sun, X.F.; Ting, B.T.G.; Zeisel, S.H.; Janghorbani, M. Analyst (London) 1987, 112, 1223 - 28.
50. Schuette, S.; Vereault, D.; Ting, B.T.G.; Janghorbani, M. Analyst (London) 1988, 113, 1837 - 42.
51. Delves, H.T.; Campbell, M.J. J. Anal. At. Spectrom. 1988, 3, 343 - 48.
52. Campbell, M.J.; Delves, H.T. J. Anal. At. Spectrom. 1989, 4, 235 - 36.
53. Ting, B.T.; Moores, C.S.; Janghorbani, M. Analyst (London), 1989, 114, 667 - 74.
54. Dalgarno, B.G.; Brown, R.M.; Pcikford, C.J. Biomed. Environ. Mass Spectrom. 1988, 16, 377 - 80.
55. Amarasiriwardena, C.J.; Barnes, R.M.; Kim, M.; Atallah, M.T., 1990 Winter Conference on Plasma Spectrochemistry St. Petersburg, FL; ICP Information Newsletter: Amherst, MA, 1990; Paper WP30, p. 156.
56. Viczian, M.; Lasztity, A.; Barnes, R.M. J. Anal. At. Spectrom. 1990, 5, 000 - 000.
57. Russ III, G.P. In Applications of Inductively Coupled Plasma Mass Spectrometry; Date, A.R., Gray, A.L., Eds.; Blackie: Glasgow, 1989; Chapter 4.
58. Lasztity, A.; Viczian, M.; Wang, X.; Barnes, R.M. J. Anal. At. Spectrom. 1989, 4, 761 - 66.
59. Gardiner, P.E. J. Anal. At. Spectrom. 1988, 3, 163 - 68.

60. Gardiner, P.H.E. In Topics in Current Chemistry, Vol. 141, Springer-Verlag: Berlin, 1987; pp 147 - 74.
61. Ebdon, L.; Hill, S.; Ward, R.W. Analyst (London) 1986, 111, 1113 - 38.
62. Ebdon, L.; Hill, S.; Ward, R.W. Analyst (London) 1986, 112, 1 - 16.
63. Uden, P.C.; Yoo, Y.; Wang, T.; Cheng, Z. J. Chromatog. 1989, 468, 319 - 28.
64. Uden, P.C. In Environmental Analysis Using Chromatography Interfaced with Atomic Spectroscopy; Harrison, R.W.; Rapsomanikis, S., Eds.; Ellis Horwood Ltd.: Chichester, UK, 1989; Chapter 4, pp 96 - 125.
65. Siriraks, A.; Kingston, H.M.; Riviello, J.M. Anal. Chem. 1990, 62, 1185 - 93.
66. Marshall, J.; Haswell, S.J.; Hill, S.J. J. Anal. At. Spectrom. 1989, 4 111R - 136R.
67. Joseph, M.J.; Vecchiarelli, J.A.; Barnes, R.M.; Uden, P.C. Abstract Book 1989 Pittsburgh Conference, Atlanta, GA; Paper 1649.
68. Beauchemin, D.; Siu, K.W.M.; McLaren, J.W.; Berman, S.S. J. Anal. At. Spectrom. 1989, 4 285 - 89.
69. Heitkemper, D.; Creed, J.; Caruso, J.; Fricke, F.L. J. Anal. At. Spectrom. 1989, 4 279 - 84.
70. Bushee, D.S.; Moody, J.R.; May, J.C. J. Anal. At. Spectrom. 1989, 4 773 - 75.
71. Bushee, D.S. Analyst (London), 1988, 113, 1167 - 70.
72. Suyani, H.; Heitkemper, D.; Creed, J.; Caruso, J. Appl. Spectrosc. 1989, 43, 962 - 67.
73. Forbes, K.A.; Vecchiarelli, J.F.; Uden, P.C.; Barnes, R.M. In Advances in Ion Chromatography; Jandik, P.; Cassidy, R.M., Eds.; Century International: Franklin, MA, 1989; Vol. 1, pp 487 - 502.
74. Gardiner, P.E.; Braetter, P.; Gercken, B.; Tomiak, A. J. Anal. At. Spectrom. 1987, 2 375 - 78.
75. Dean, J.R.; Munro, S.; Ebdon, L.; Crews, H.M.; Massey, R.C. J. Anal. At. Spectrom. 1987, 2 607 - 10.
76. Crews, H.M.; Dean, J.R., Ebdon, L.; Massey, R.C. Analyst (London) 1989, 114, 895 - 99.
77. Brätter, P.; Gercken, B.; Roesick, U.; Tomiak, A. In Trace Element Analytical Chemistry in Medicine and Biology; Brätter, P.; Schramel, P., Eds.; Walter de Gruyter: Berlin, 1988; Vol. 5, pp 145 - 156.
78. Brätter, P.; Gercken, B.; Tomiak, A.; Roesick, U. In Trace Element Analytical Chemistry in Medicine and Biology; Brätter, P.; Schramel, P., Eds.; Walter de Gruyter: Berlin, 1988; Vol. 5, pp 120 - 135.
79. Gercken, B.; Barnes, R.M. Anal. Chem., submitted (1990).
80. Delaney, M.F.; Warren, Jr., F.V. Spectrochim. Acta 1983, 38B, 399-406.
81. Hagen, D.F.; Haddad, L.C.; Marhevka, J.S. Spectrochim. Acta 1987, 42B, 253 - 67.
82. Tsunoda, A.; Matsumoto, K.; Haraguchi, H.; Fuwa, K. Anal. Sci. 1986, 2, 99 - 100.
83. Hagen, D.F.; Belisele, J.; Marhevka, J.S. Spectrochim. Acta 1983, 38B, 377 - 85.
84. Firor, R.L. Am. Lab. 1989, 21(5), 40 - 48.
85. Quimby, B.D.; Sullivan, J.J. Anal. Chem. 1990, 62, 1027 - 34.
86. Sullivan, J.J.; Quimby, B.D. Anal. Chem. 1990, 62, 1034 - 43.
87. Hooker, D.B.; DeZwaan, J. Anal. Chem. 1989, 61, 2207 - 11.
88. Creed, J.T.; Mohamad, A.H.; Davidson, T.M.; Ataman, G.; Caruso, J.A. J. Anal. At. Spectrom. 1988, 3, 923 - 26.
89. Kawaguchi, H.; Auld, D.S. Clin. Chem. 1975, 21, 591 - 94.
90. Kawaguchi, H.; Vallee, B.L. Anal. Chem. 1975, 47, 1029 - 34.

91. McGregor, D.A.; Cull, K.B.; Gehlhausen, J.M.; Viscomi, A.S.; Wu, M.; Zhang, L.; Carnahan, J.W. Anal. Chem. 1988, 60, 1089A - 98A.
92. Luffer, D.R.; Galante, L.J.; David, P.A.; Novotny, M.; Hieftje, G.M. Anal. Chem. 1988, 60, 1365 - 69.
93. Galante, L.J.; Selby, M.; Luffer, D.R.; Hieftje, G.M.; Novotny, M. Anal. Chem. 1988, 60, 1370 - 76.
94. Motley, C.B.; Ashraf-Khorassani, M.; Long, G.L. Appl. Spectrosc. 1989, 43, 737 - 41.
95. Broekaert, J.A.C.; Toelg, G. In Inductively Coupled Plasma Emission Spectroscopy. Part II: Applications and Fundamentals; Boumans, P.W.J.M., Ed.; Wiley-Interscience: New York, 1987; Chapter 12.
96. Keliher, P.N. In Inductively Coupled Plasmas in Analytical Atomic Spectrometry; Montaser, A.; Golightly, D.W., Eds.; VCH Publishers: New York, 1987; Chapter 16.

RECEIVED July 16, 1990

Chapter 13

Analytical Method Aspects of Assessing Dietary Intake of Trace Elements

Stephen G. Capar

Division of Contaminants Chemistry, U.S. Food and Drug Administration, 200 C Street, S.W., Washington, DC 20204

The Food and Drug Administration (FDA) monitors toxic and nutrient elements in the US food supply through analysis of foods and calculation of dietary intakes in representative diets based on nationwide food consumption surveys. Toxic or nutrient elements routinely monitored by atomic absorption spectrometry include As, Se, Pb, Cd, and Hg. The nutrient elements Ca, Fe, K, Mg, Mn, P, Zn, Cu, and Na are monitored by inductively coupled plasma atomic emission spectrometry. Iodine is monitored by spectrophotometry. Neutron activation analysis provides background information on elements not routinely monitored. FDA's current method development program for monitoring elements in foods is designed (a) to reduce the number of different mineralization procedures, (b) to expand the number of elements determined, and (c) to use inductively coupled plasma atomic emission spectrometry for determining all elements of interest. These analytical methods are reviewed and the applicability of currently used methods for assessing dietary intakes is discussed.

The Food and Drug Administration's (FDA) Total Diet Study (TDS) has been a continuous program since 1961 (*1, 2*) for monitoring selected chemical contaminants in the United States food supply. Initially, the program monitored levels of radionuclide and pesticide residues; it was expanded in the 1970's to include nutrient elements (sodium, potassium, phosphorus, calcium, magnesium, iron, manganese, copper, zinc, selenium, iodine) and toxic elements (lead, cadmium, arsenic, mercury). Selection of foods to be analyzed by the current TDS was based on the 1977–78 United States Department of Agriculture Nationwide Food Consumption Survey and the Second National Health and Nutrition Examination Survey (1976–80)

(*3, 4*). The 234 individual foods selected compose the collection, which represents all foods eaten in the United States. The diets of eight age–sex groups (6– to 11–month–old infants, 2–year–old children, 14– to 16–year–old girls, 14– to 16–year–old boys, 25– to 30–year–old women, 25– to 30–year–old men, 60– to 65–year–old women, and 60– to 65–year–old men) are constructed from these 234 foods. From these diets and the analytical findings for the elements, the intakes of nutrients and toxic elements are calculated in order to assess the quality of the food supply. Approximately four collections are obtained and analyzed each year.

A collection is obtained from retail stores in one of four geographical areas of the United States (east, central, south, or west). Each food item is collected from three cities in the selected geographical area, and the collected foods are shipped for analysis to the FDA Total Diet Laboratory in

Table I. Levels of Elements in Total Diet Study Collections 26–29[a]

Element	Mean	Maximum	LOQ[b]	Percent of Samples Below LOQ (%)
		mg/kg		
Sodium	2,170	28,900	2	4
Potassium	1,970	14,900	5	1
Phosphorus	922	8,500	0.5	1
Calcium	455	7,700	2	1
Magnesium	220	1,920	1	1
Iron	16.0	560	0.1	2
Zinc	9.91	163	0.1	4
Manganese	2.84	49.0	0.02	11
Copper	1.41	229	0.05	9
Iodine	0.146	20.7	0.02	32
Selenium	0.065	0.800	0.05	71
Arsenic	0.023	3.22	0.02	94
Cadmium	0.011	0.171	0.01	67
Lead	0.008	0.730	0.02	85
Mercury	0.001	0.310	0.01	98

[a]936 samples.
[b]LOQ=limit of quantitation of analytical method.

Kansas City, MO. The three similar food items from the three cities are combined into a single composite for each of the 234 foods and prepared as if for consumption. All analyses are performed at the Total Diet Laboratory, except for radionuclides; after preparation, one collection per year is sent by the Total Diet Laboratory to FDA's Winchester Engineering and Analytical

Center in Winchester, MA, for radionuclide analysis (5). Analytical findings of the TDS are sent to FDA's Center for Food Safety and Applied Nutrition in Washington, DC, for calculation of daily intakes and evaluation. Daily intakes of TDS nutritional elements are compared to recommendations for nutrient intake (6); daily intakes of toxic elements are compared with recommended acceptable levels (7).

Currently, five different analytical methods are used to determine the concentration of 15 elements in the 234 food items of the TDS. Table I lists the minimum (*i.e.*, quantitation limit), mean, and maximum concentrations of the elements determined for four collections (designated collections 26–29) between August 1988 and June 1989 (936 samples). Most levels range over three or more orders of magnitude. New analytical methodology is being developed to expand FDA's analytical capabilities to include other elements of interest such as aluminum, chromium, cobalt, molybdenum, nickel, and tin. In addition, analytical methods are being developed which utilize resources more efficiently. The current analytical methodology employed by the Total Diet Laboratory, as well as research on new analytical methodology, is discussed in this paper.

Pre–Analysis Food Preparation Procedures

An institutional kitchen performs all required food preparation or cooking. Salt is added to the prepared food item if it is included in preparation instructions (4). Deionized water and stainless steel or aluminum utensils are used for food preparation and cooking. The foods collected from the three cities are chopped and blended into a single homogeneous composite. Water may be added to some canned foods to aid homogenization. The composites are stored in plastic containers, frozen until analyzed, and then thawed and mixed again before an analytical portion is taken. The homogeneity of the composites has been evaluated for some of the mineral elements (8). To avoid potentially violent reactions, the analytical portions of foods containing alcohol (*i.e.*, beer, wine, and whiskey) are dried on a steam bath before mixing with digestion acids. Prior to lead and cadmium determination, a special acid homogenization procedure is performed on selected canned foods to dissolve lead particles which may have migrated from the soldered seams of the cans (9, 10). The foods are not dried before analysis; results are reported on an "as is" basis.

Quality Assurance

For analysis, the 234 foods of a collection are divided into more manageable groups (series) of about 30 foods each. For each analytical method applied to a series, the following minimum quality assurance analyses are performed: two method reagent blanks, two duplicate test portions of food, two recovery spikes, and one reference material. Reference materials are obtained from the National Institute of Standards and Technology, Gaithersburg, MD, and are usually Rice Flour (SRM 1568) for arsenic and selenium; Oyster Tissue

184   BIOLOGICAL TRACE ELEMENT RESEARCH

(SRM 1566) for lead, cadmium, calcium, copper, iron, manganese, magnesium, phosphorus, potassium, sodium, zinc, and iodine; Spinach (SRM 1570) also for lead and cadmium; and Albacore Tuna (RM 50) for mercury. The analysis of the entire series is repeated if the average recovery falls outside the range 75–125%, or if any one recovery is below 50% or above 150%. For iodine the acceptable range is 70–110% because of the inherent low recoveries of the analytical method.

Control of contamination is an important aspect of trace element analysis of foods (11, 12). Stainless steel cutting blades are used on the blenders since chromium–plated brass blades were found to cause contamination with copper, iron, lead, and zinc (13). All laboratory ware is cleaned and rinsed with dilute nitric acid and deionized water in a specially designed automatic washer (14). High purity reagents are used where necessary and periodically checked for contaminants. All dilutions are made with deionized water. Clean air hoods are used when the test portion is being manipulated to reduce the potential of airborne contamination.

Analytical Methodology

Arsenic and Selenium. These elements are determined by acid digestion and hydride generation–atomic absorption spectrometry (15, 16). A test portion of approximately 5 g is digested with a 30–mL mixture of concentrated nitric, perchloric, and sulfuric acids (4:1:1). These three acids are required for complete destruction of refractory arsenic compounds for subsequent conversion to arsine in the determinative step (17, 18). Digestions are performed in 100–mL borosilicate Kjeldahl flasks on electrically heated racks with separate proportional input controls. Carbonization of the digest is avoided to prevent analyte loss. If carbonization begins, the flask is removed from the heat source and cooled in an ice water bath, 1 mL of nitric acid is added, and the digestion is continued. The acid mixture is boiled until only sulfuric acid remains (approximately 5 mL). Digests are transferred with the aid of water to 100–mL volumetric flasks containing 30 mL of concentrated hydrochloric acid and are diluted to volume. The heat generated by addition of the sulfuric acid digest to hydrochloric acid reduces Se(VI) to Se(IV), which is required for the subsequent generation of selenium hydride (19–21). Final acid concentrations are 5% sulfuric and 30% hydrochloric.

Hydrides and excess hydrogen are generated by a semiautomatic hydride generator which controls the addition of base–stabilized sodium tetrahydroborate solution to an aliquot of the test solution (15, 16). The generated gases are directed into the spray chamber of an atomic absorption spectrometer with a shielded, hydrogen (nitrogen–diluted) –entrained air flame for sequential determination of arsenic and selenium. Aliquots of test solution up to 20 mL (equivalent to 1 g of test portion) can be analyzed. For the determination of arsenic, 0.5 mL of sodium iodide solution (10% w/v) is added to the aliquot of test solution before hydride generation to pre–reduce As(V) to As(III). Absorbance is measured for arsenic at 193.7 nm and for selenium at 196.0 nm, and is recorded on a strip chart; peak heights are measured and compared to a standard calibration curve for quantitation.

Calcium, Copper, Iron, Manganese, Magnesium, Phosphorus, Potassium, Sodium, and Zinc. These elements are determined by acid digestion and inductively coupled argon plasma atomic emission spectrometry (Marts, R. W.; Meloan, C. E. Food and Drug Administration, Science Advisor Research Associate Program, **1982**, *61*, Report No. 105–79). A test portion of approximately 5 g is digested with a 30–mL mixture [91.3:8.7(v/v)] of concentrated nitric and sulfuric acids. After the solution is boiled until only boiling sulfuric acid remains, a maximum of 10 mL of 50%(w/w) hydrogen peroxide is cautiously added to remove carbonization, and the digest is boiled again until only boiling sulfuric acid remains. Digestions are performed in 250–mL borosilicate digestion tubes, using a temperature–programmable block digester with a capacity of 20 digestion tubes. Test solutions are prepared by quantitatively transferring the digest to 50–mL volumetric flasks with the aid of water and diluting to volume with water. The concentrations of the elements in the test solution are sequentially determined by inductively coupled argon plasma atomic emission spectrometry. To obtain an analytical response within the linear region of the instrument for foods which contain high levels of calcium, phosphorus, or sodium, additional dilutions of 250– or 1000–fold are required. The following wavelengths (nm) are used on the atomic emission spectrometer: Ca–393.4, Cu–324.8, Fe–259.9, Mn–257.6, Mg–383.2, P–214.9, K–769.9, Na–589.6, Zn–213.9.

Cadmium and Lead. These elements are determined by dry ash digestion and graphite furnace atomic absorption spectrometry (*22, 23*); some modifications to the original method are summarized in the following discussion. To minimize contamination of the test portions, all weighings and dilutions are performed under clean air hoods, and high purity reagents are used. A test portion of approximately 5 g in a 100–mL quartz beaker with 3 mL of 40% (v/v) sulfuric acid is initially dried at 100°C in an oven. To further minimize contamination, the quartz beakers are covered with larger borosilicate beakers while the test portions are drying. The test portion is ashed in a temperature–controlled muffle furnace through slow temperature increase to a maximum of 470°C. The ash is then treated with nitric acid, dried, and ashed again at 470°C. About 20 mL of water and 0.4 mL of nitric acid are added, and the test portion is heated on a hot plate to dissolve ash and quantitatively transferred to a 50–mL volumetric flask. Aliquots of test solution (10 μL for cadmium and 20 μL for lead) and 5 μL of matrix modifier (ammonium dihydrogen phosphate [4% (w/v)] and magnesium nitrate [0.2 % (w/v)]) are automatically introduced into the graphite furnace. Pyrolytic graphite platforms with pyrolytic graphite–coated grooved tubes are used. Individual temperature programs for graphite furnace atomization are used for cadmium and lead. The furnace is purged with argon except during atomization. Cadmium and lead are determined sequentially at 228.8 nm and 283.3 nm, respectively, with Zeeman background correction. Peak areas are measured and compared to a standard calibration curve for quantitation. Occasionally, test portions require dilution with nitric acid (0.8% v/v) to achieve a response within the analytical range.

Mercury. This element is determined by an acid digestion followed by flameless atomic absorption spectrometry (24). A test portion of approximately 5 g is partially digested with nitric and sulfuric acids with a vanadium pentoxide catalyst and hydrogen peroxide. A modified digestion procedure using a heated block system is used to digest about 20 test portions simultaneously (25). An additional 10 mL of nitric acid is used to digest foods high in carbohydrate or fat to reduce carbonization (26). The 250–mL digestion tubes used with the block digestion system act as air condensers and preclude the need for additional air– or water–cooled condensers to retain mercury during the digestion. Digested test portions are diluted to 100 mL with water. The entire test solution is mixed with stannous chloride to produce metallic mercury; then, with the aid of a special apparatus (24), the mercury is purged into an atomic absorption spectrometer and is measured at 253.7 nm. Absorbance is recorded on a strip chart, and peak heights measured and compared to a standard calibration curve for quantitation.

Iodine. This element is determined by a procedure using alkaline ashing with sodium hydroxide, iodide–catalyzed reduction of cerium (IV) by arsenic(III), and colorimetric measurement of the resulting loss of color (27, 28). A test portion of approximately 5 g is weighed into a 100–mL zirconium crucible and mixed with 3 mL of sodium hydroxide solution [50% (w/v)]. The mixtures are placed in an oven at 100°C, the temperature is slowly raised to 175°C, and the mixtures are allowed to dry overnight. The temperature is raised to 220°C for 2 hours; then the dried residues are placed in a muffle furnace at 375°C, the temperature is slowly raised to 500°C, and the dried residues are ashed for 45 minutes. The ash is allowed to cool, dissolved in 10–15 mL of water with warming, filtered, made acidic with 5 mL of 20 N sulfuric acid, and diluted to 100 mL with water. An automatic analyzer is used to control addition, mixing, and heating of the test solution with arsenious acid solution and ceric ammonium nitrate solution. The automatic analyzer also introduces the test solution into the spectrophotometer for measurement at 405 nm.

Limit of Quantitation and Intake Estimates

The daily intake of each element is calculated for the eight age–sex groups, based on the TDS collection findings. To calculate daily intakes, the level of each element in each food is multiplied by the corresponding food consumption figure for each age–sex group (3) to yield the intakes from each of the 234 foods. The sum of these individual food intakes for each group is the daily intake for the group, and this value is used to evaluate how much of an element is being consumed. Nutrient element intakes are compared with recommended daily levels or ranges (*6*); toxic element intakes are compared to acceptable or tolerable levels (*7*). If the concentration of an element for a particular food is below the limit of quantitation of the method but still above zero, this estimated level is reported but designated as "trace." Since the "trace" value (or zero, when the element is not found) is used, the intake calculated represents a minimum level. Ideally, an analytical method would allow quantitation of the elements at all levels found in all the foods. However, some elements are present at such a low level that quantitation is not always achievable.

Table II.   Average Daily Intakes[b] (µg/day) of Selected Elements Obtained by Substituting Zero and Limit of Quantitation (LOQ)[a] for Unquantifiable Findings from the Total Diet Study Collections 26–29

Element	Infants 6–11 mo. Zero	LOQ	Children 2 yr Zero	LOQ	Women 25–30 yr Zero	LOQ	Men 25–30 yr Zero	LOQ
Arsenic	3.2	28	9.3	39	25	67	36	96
Cadmium	2.7	14	5.6	19	9.1	28	12	38
Iodine	170	180	210	220	190	220	280	320
Lead	3.6	27	3.3	32	5.2	47	6.8	65
Mercury	0.14	13	0.52	15	1.5	23	1.9	32
Selenium	20	75	47	110	76	170	110	240

[a]LOQ values are given in Table I.
[b]Daily intakes are expressed to 2 significant figures.

Table I lists the percentage of foods from the collections discussed previously with findings below the limit of quantitation for each element. For the nutrient elements, very few foods contained concentrations below the limit of quantitation of the methods except in the case of selenium and iodine. For all the toxic elements, a much higher percentage of the foods had concentrations below the limit of quantitation of the methods. To evaluate the appropriateness of the limit of quantitation, average daily intakes for the four collections were calculated in which (a) zero and (b) the element's limit of quantitation were substituted for levels below the limit of quantitation. Higher daily intakes will result when the limit of quantitation is substituted instead of zero, but the conclusion on the adequacy or safety of the daily intakes may not change.

For calcium, copper, iron, magnesium, manganese, phosphorus, potassium, sodium, or zinc, there was no noteworthy difference between the average daily intakes whether zero or the limit of quantitation was used for unquantifiable findings. For arsenic, cadmium, lead, iodine, mercury, and selenium the average daily intakes were generally higher when the limit of quantitation was used. This result was expected, since many of the foods contained unquantifiable levels of these elements. Table II lists the daily intakes of these elements for four of the eight age–sex groups. The other age–sex group intakes behaved similarly.

The daily intakes for arsenic were also much greater when the limit of quantitation was substituted for unquantifiable values. Currently, there is no agreement on a maximum acceptable intake for total arsenic, but there is an adult Provisional Tolerable Daily Intake (PTDI) for ingested inorganic arsenic (29) of 2 µg As/kg body weight. A PTDI for arsenic was calculated for each age–sex group using mean age–sex weights (30) and was compared to the average daily intakes. Except for the infants and children groups, both ways of calculating intakes produced levels below the PTDI. For the infants and children groups the intakes calculated by substituting zero for unquantifiable findings were below the PTDI; intakes calculated by substituting the limit of quantitation were above the PTDI. Even though the PTDI may not be applicable to infants and children, these findings indicate that a lower limit of quantitation for arsenic may be necessary for assessing the exposure of infants and children to arsenic.

The average daily intakes for cadmium and mercury were compared to PTDIs (57–72 µg Cd/day; 43 µg Hg/day) calculated from Provisional Tolerable Weekly Intakes (PTWI) (31). The average intakes for cadmium and mercury did not exceed these PTDIs when either zero or the limit of quantitation of the element was substituted for unquantifiable findings in the calculation of daily intakes. Therefore, it appears that the limits of quantitation of the analytical methods for cadmium and mercury are adequate for assessing their dietary intake.

There is relatively little difference between the iodine intakes calculated by substituting the two values for unquantifiable findings. Comparison of either calculated intake to recommended dietary allowances (30) leads to the conclusion that allowances are exceeded. The limit of quantitation of the analytical method therefore appears adequate.

The average daily intakes for lead were compared to PTDIs (32, 46, and 429 µg Pb/day for infants, children, and adults, respectively) calculated from PTWIs (31, 32) and mean age–sex weights (30). The intakes calculated by substituting zero or the limit of quantitation did not exceed these PTDIs for any of the age–sex groups. However, the infants' and children's intakes calculated by substituting the quantitation limit were very close to the PTDIs.

Infants and young children are most susceptible to the effects of lead, and it is for these groups, and for women of child–bearing age, that the intake of lead is of most concern. As illustrated in Figure 1, the level of lead in foods has been declining in recent years, as have the guidelines for lead exposure. Based on the guidelines of various national and international health organizations, FDA currently has a provisional tolerable range for lead intake from food of 6–18 µg/day for a 10-kg child (33). The average daily intakes for infants and children exceed this range when the limit of

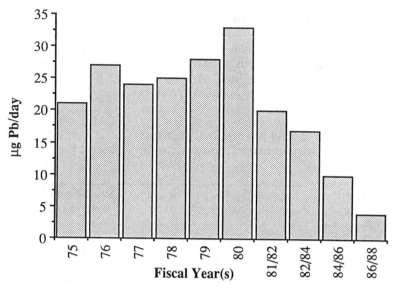

Figure 1. Average daily intakes of lead for infants
0–6 months old based on FDA Total Diet Study.

quantitation is substituted in the calculation, and they are close to the range
when zero is substituted.

A number of hypothetical daily intakes were calculated for 2–year–old
children to estimate the limit of quantitation necessary to ensure the ability to
estimate daily intakes of lead of about 10 μg/day. The calculations were
performed by substituting various concentrations of lead for the individual
foods with unquantifiable findings (<20 μg Pb/kg) as well as substituting
these concentrations for all the 234 foods. Table III lists the results of these
calculations. The calculated average daily intake of lead for collections
26–29 was ≥ 10 μg Pb/day when levels higher than about 5 μg Pb/kg were
substituted for unquantifiable lead findings for individual foods. When levels
≥ 7 μg Pb/kg were substituted for **all** 234 foods, the daily intake of lead was
≥ 10 μg Pb/day. Thus, to ascertain if the lead daily intake is on the order of
10 μg Pb/day, the limit of quantitation of the analytical method for lead
should be at least about 5 μg Pb/kg. To ascertain if the intake is **below** 10 μg
Pb/day, the limits would need to be somewhat lower, perhaps as low as 1 μg
Pb/kg.

There are differences between daily selenium intakes calculated by
substituting zero and the selenium limit of quantitation for unquantifiable
findings in the individual foods (Table II). Comparison of these intakes to
estimated safe and adequate daily dietary intakes (*30*) results in slightly
different conclusions. The intakes based on substituting zero for
unquantifiable findings are all within the acceptable range. However, the

range is exceeded for five of the age–sex group's intakes calculated by substituting the limit of quantitation for unquantifiable findings. Intakes of selenium were calculated by substituting various concentrations for unquantifiable levels of selenium to estimate an appropriate quantitation limit for selenium as previously described for lead. Based on these calculations, a method with a limit of quantitation of 0.01 mg Se/kg is needed to better assess the dietary intake of selenium.

Table III. Hypothetical Average Daily Intakes[a] of Lead Derived by Substituting Various Lead Concentrations for Individual Foods

Concentration Substituted (μg Pb/kg)	μg Pb/day	
	Substitution Only for Foods with Nonquantifiable Levels	Substitution for All Foods
0	3.3	0
0.5	4.0	0.75
1	4.7	1.5
3	7.6	4.5
5	10.4	7.5
7	13.3	10.5
10	17.6	15.0
20	31.9	30.1

[a]Total Diet Study collections 26–29 for 2–year–old children.

Analytical Methodology Research

Multielement Scheme. As described above, five different digestion procedures are used to determine the 15 elements in the TDS. The digestion of foods is the most time–consuming step of the analysis. Reducing the number of digestions required for the analysis of a collection will allow other analytes to be determined or more collections to be analyzed per year. An analytical scheme has been developed which uses a single digestion for the determination of 13 of these 15 elements and an additional 5 elements (18, 34–36). This scheme digests a test portion of approximately 5 g with a 30–mL mixture of nitric, perchloric, and sulfuric acids (4:1:1) in a 100–mL Kjeldahl flask. The digest is diluted with water to 50 or 100 mL in a volumetric flask. A small portion of this test solution (usually 5 mL) is analyzed by simultaneous inductively coupled argon plasma atomic emission spectrometry for calcium, iron, potassium, magnesium, manganese, phosphorus, sodium, and strontium. The remaining portion is concentrated and separated on an ion exchange resin. The initial test solution effluent from the resin is collected, diluted with hydrochloric acid and water, and analyzed for arsenic, selenium, and antimony by hydride generation atomic absorption spectrometry. The resin is then stripped with dilute nitric acid and the eluate is analyzed for aluminum, cadmium, cobalt, copper, lead, molybdenum, nickel, vanadium, and zinc by simultaneous inductively coupled argon plasma atomic emission spectrometry.

Analysis of a TDS collection using the multielement scheme produces results comparable to those from current methods and provides findings for aluminum (*37*) and for molybdenum, nickel, cobalt, vanadium, and strontium (*38*). Since the limits of quantitation of the inductively coupled argon plasma atomic emission spectrometer are usually insufficient for the determination of cadmium, lead, and nickel, the stripped eluate may be analyzed by using the more sensitive graphite furnace atomic absorption spectrometer. Analytical procedures using this technique are currently being evaluated.

Arsenic and Selenium. The hydride generation system currently used for determining arsenic and selenium requires much more glassware than continuous flow hydride generators that need only the flask containing the test solution (*39, 40*). A reduction in glassware cleaning and manipulation of the test solution would greatly reduce the time necessary for completing an analysis. Continuous flow generators are currently being investigated for determination of arsenic and selenium by atomic absorption spectrometry or simultaneous inductively coupled argon plasma atomic emission spectrometry. Our laboratory has successfully developed a method for arsenic and selenium that uses a common peristaltic pump and a specially designed gas–liquid separator (Alvarez, G. H., Food and Drug Administration, personal communication, 1989).

Iodine. The current analytical method for iodine discussed above occasionally produces erratic and frequently low recoveries for foods high in starch or lipids (*41*). Neutron activation analytical methods have been used to provide comparison information. A post–irradiation neutron activation analysis method was developed and used to analyze TDS collections for comparison of results with those from spectrophotometric findings (*42, 43*). This neutron activation procedure is very tedious and not amenable to routine analysis. A less tedious nuclear method using epithermal neutron activation analysis (*44*) has been applied to foods. This technique does not require any digestion or chemical separation of the food, but an appropriate nuclear reactor is required. Methods amenable to incorporating iodine determinations into the multielement scheme described above are being investigated. Areas of investigation include acid digestion of foods (*45*) and the use of inductively coupled plasma atomic emission spectrometry.

Other Elements. Analytical methods have been developed for providing background levels in foods for elements not routinely determined by the TDS. A method for tin in foods was developed and applied to a number of foods from a TDS collection (*46*). The method used hydride generation atomic absorption spectrometry to obtain a lower limit of quantitation than that from the commonly used flame atomic absorption spectrometry. Neutron activation has also been used to determine the levels of many elements in TDS foods. An instrumental neutron activation analysis method was developed and applied to a number of TDS collections (*47*). Besides providing a check on the TDS findings for nine elements, new elemental information was generated on silver, bromine, chlorine, cobalt, chromium, cesium, europium, rubidium, antimony, scandium, and vanadium. Radiochemical neutron activation analysis has been applied to the analysis of foods for aluminum (*48*) and for arsenic, chromium, molybdenum, antimony, and selenium (*49*). Preliminary investigations on the application to food of neutron capture prompt–gamma ray activation analysis has demonstrated its multielement capabilities (Anderson, D. L.; Cunningham, W. C.; Mackey, E. A. *Biol. Trace Elem. Res.* 1990, in press). The ability of this technique to determine boron in foods is of greatest interest, but it can also determine other elements such as sulfur, nitrogen, calcium, carbon, and sodium.

Future Analytical Approaches

Analytical improvements that enable many elements to be determined in more foods in less time are continually being sought. The simultaneous multielement inductively coupled argon plasma method has proven capable of rapidly determining many elements in foods, but lacks sufficient sensitivity for a number of the elements monitored in foods. Enhancement of these limits of quantitation would eliminate the need for multiple analytical methods and reduce analysis time. Potential means of reducing these limits of quantitation include instrumental and sample handling modifications. Instrumental modifications such as the use of a helium or mixed gas plasma (*50*) or alternative detectors (*51*) may prove beneficial. Modifications in handling test solutions such as flow injection (*52*), ultrasonic or frit nebulizers (*53*), and solvent extraction may also improve limits of quantitation. The relatively new inductively coupled argon plasma mass spectrometer has exceptionally good limits of quantitation for many of the elements of interest (*54*). A few investigators have applied this technique to food analysis (*55, 56*). The instrument's limits of quantitation for many of the toxic elements of interest, especially lead, have the potential of achieving the necessary limits of quantitation much lower than 1 μg/kg in a typical acid digest of foods.

Digestion of foods is the most time–consuming step for elemental analysis. Rapid microwave digestion has been demonstrated suitable for a few food matrices but is limited to smaller test portions than the approximately 5 g required by current methods. Any reduction in test portion mass would have to be compensated by a decrease in instrumental limits of quantitation. In addition, microwave digestion parameters need to be developed for all 234 foods of a TDS collection and their applicability demonstrated for elements of interest.

Chromium and fluorine are elements which should be included in the TDS. Our laboratory is investigating the modification of the analytical scheme described previously to include chromium with detection by graphite furnace atomic absorption spectrometry. An analytical method for fluorine (fluoride) by ion selective electrode has been developed (57), but its applicability to the 234 foods of a TDS collection still requires study.

Acknowledgments

R. Marts and B. Young of FDA, Total Diet Laboratory, Kansas City, MO, for their valuable discussions on the current analytical methodology used in the TDS.

Literature Cited

1. Lombardo, P. In *Environmental Epidemiology*; Kopler, F. C.; Craun, G. F., Ed.; Lewis: Chelsea, MI, 1986; Chapter 11.
2. Pennington, J. A. T. *J. Assoc. Off. Anal. Chem.* **1987**, *70*, 772–782.
3. Pennington, J. A. T. *J. Am. Diet. Assoc.* **1983**, *82*, 166–173.
4. Pennington, J. A. T. *Documentation for the Revised Total Diet Study: Food List and Diets*; National Technical Information Service: Springfield, VA, Accession Number PB 82–192154.
5. Cunningham, W. C.; Stroube, W. B.; Baratta, E. J. *J. Assoc. Off. Anal. Chem.* **1989**, *72*, 15–18.
6. Pennington, J. A. T.; Young, B. E.; Wilson, D. B. *J. Am. Diet. Assoc.* **1989**, *89*, 659–664.
7. Gunderson, E.L. *J. Assoc. Off. Anal. Chem.* **1988**, *71*, 1200–1209.
8. Marts, R. W. *FDA Laboratory Information Bulletin 2573* **1981**, Food and Drug Administration, Division of Field Science, Rockville, MD.
9. Jones, J. W.; Boyer, K. W. *J. Assoc. Off. Anal. Chem.* **1979**, *62*, 122–128.
10. Suddendorf, R. F.; Wright, S. K.; Boyer, K. W. *J. Assoc. Off. Anal. Chem.* **1981**, *64*, 657–660.
11. Boyer, K. W.; Horwitz, W. In *Environmental Carcinogens–Selected Methods of Analysis*; O'Neil, I. K.; Schuller, P.; Fishbein, L., Eds.; IARC Scientific Publications No. 71; Oxford: New York City, NY, 1986; Vol. 8, Chapter 12.
12. Jones, J. W. In *Quantitative Trace Analysis of Biological Materials*; McKenzie, H. A.; Smythe, L. E., Eds.; Elsevier Science: New York City, NY, 1988; Chapter 20.
13. Marts, R. W. *FDA Laboratory Information Bulletin 2141* **1977**, Food and Drug Administration, Division of Field Science, Rockville, MD.
14. Watson, H. E. *FDA Laboratory Information Bulletin 2963* **1985**, Food and Drug Administration, Division of Field Science, Rockville, MD.
15. Fiorino, J. A.; Jones, J. W.; Capar, S. G. *Anal. Chem.* **1976**, *48*, 120–125.
16. Capar, S. G.; Jones, J. W. *FDA Laboratory Information Bulletin 1900* **1976**, Food and Drug Administration, Division of Field Science, Rockville, MD.

17. Cassil, C. C. *J. Assoc. Off. Anal. Chem.* **1937**, *20*, 171–178.
18. Jones, J. W.; Capar, S. G.; O'Haver, T. C. *Analyst* **1982**, *107*, 353–377.
19. Verlinden, M.; Deelstra, H.; Adriaenssens, E. *Talanta* **1981**, *28*, 637–646.
20. Bye, R. *Talanta* **1983**, *30*, 993–996.
21. Piwonka, J.; Kaiser, G.; Tölg, G. *Fresenius Z. Anal. Chem.* **1985**, *321*, 225–234.
22. Young, B. E.; Faul, K. C. *FDA Laboratory Information Bulletin 2403* **1981**, Food and Drug Administration, Division of Field Science, Rockville, MD.
23. Faul, K. C.; Young, B. E. In *Environmental Carcinogens–Selected Methods of Analysis*; O'Neil, I. K.; Schuller, P.; Fishbein, L., Eds; IARC Scientific Publications No. 71; Oxford: New York City, NY, 1986; Vol. 8, pp. 401–408.
24. *Official Methods of Analysis*, Association of Official Analytical Chemists, Arlington, VA, 1984, 14th ed., secs 25.131–25.133.
25. Marts, R. W. *FDA Laboratory Information Bulletin 2602* **1982**, Food and Drug Administration, Division of Field Science, Rockville, MD.
26. Marts, R. W.; Blaha, J. J. *FDA Laboratory Information Bulletin 2708* **1983**, Food and Drug Administration, Division of Field Science, Rockville, MD.
27. Luchtefeld, R. G. *FDA Laboratory Information Bulletin 1678* **1974**, Food and Drug Administration, Division of Field Science, Rockville, MD.
28. Sandell, E. B.; Kolthoff, I. M. *J. Am. Chem. Soc.* **1934**, *56*, 1426.
29. *Evaluation of Certain Food Additives and Contaminants–27th Report of the Joint FAO/WHO Expert Committee on Food Additives*, World Health Organization, Geneva, Switzerland, 1983.
30. *Recommended Dietary Allowances*, National Academy of Sciences, 9th rev. ed., 1980, Washington, DC.
31. *Evaluation of Certain Food Additives and the Contaminants Mercury, Lead, and Cadmium–16th Report of the Joint FAO/WHO Expert Committee on Food Additives*, World Health Organization, Geneva, Switzerland, 1972.
32. *Evaluation of Certain Food Additives and Contaminants –30th Report of the Joint FAO/WHO Expert Committee on Food Additives*, World Health Organization, Geneva, Switzerland, 1987.
33. Food and Drug Administration *Fed. Regist.* June 1, **1989**, *54*, 23485–23489.
34. Jones, J. W.; O'Haver, T. C. *Spectrochim. Acta* **1985**, *40B*, 263–277.
35. Jones, J. W. *J. Res. Natl. Bur. Stand.* **1988**, *93*, 358–360.
36. Jones, J. W. In *Elements in Health and Disease*; Said, H. M.; Rahman, M. A.; D'Silva, L. A., Eds.; Hamdard: Karachi, Pakistan, 1989; pp 402–422.
37. Pennington, J. A. T.; Jones, J. W. In *Aluminum and Health–A Critical Review*; Gitelman, H. J., Ed.; Marcel Dekker: New York, 1989; pp 67–100.

38. Pennington, J. A. T.; Jones, J. W. *J. Am. Diet. Assoc.* **1987**, *87*, 1644–1650.
39. Panaro, K. W.; Krull, I. S. *Anal. Lett.* **1984**, *17*, 157–172.
40. Ek, P.; Huldén, S. *Talanta* **1987**, *34*, 495–502.
41. Blaha, J. J. *FDA Laboratory Information Bulletin 3045* **1986**, Food and Drug Administration, Division of Field Science, Rockville, MD.
42. Allegrini, M.; Boyer, K. W.; Tanner, J. T. *J. Assoc. Off. Anal. Chem.* **1981**, *64*, 1111–1115.
43. Allegrini, M.; Pennington, J. A. T.; Tanner, J. T. *J. Am. Diet. Assoc.* **1983**, *83*, 18–24.
44. Stroube, W. B.; Cunningham, W. B.; Lutz, G. J. *J. Radioanal. Nucl. Chem.* **1987**, *112*, 341–346.
45. Fischer, P. W. F.; L'Abbé, M. R.; Giroux, A. *J. Assoc. Off. Anal. Chem.* **1986**, *69*, 687–689.
46. Alvarez, G. H.; Capar, S. G. *Anal. Chem.* **1987**, *59*, 530–533.
47. Cunningham, W. C.; Stroube, W. B. *Sci. Total Environ.* **1987**, *63*, 29–43.
48. Cunningham, W. C.; Stroube, W. B. *Trans. Am. Nucl. Soc.* **1986**, *53*, 174–175.
49. Cunningham, W. C. *J. Radioanal. Nucl. Chem.* **1987**, *113*, 423–430.
50. Ohls, K. D.; Golightly, D. W.; Montaser, A. In *Inductively Coupled Plasmas in Analytical Atomic Spectrometry*; Montaser, A.; Golightly, D. W., Eds.; VCH: New York City, NY, 1987; Chapter 15.
51. Strasheim, A. In *Inductively Coupled Plasmas in Analytical Atomic Spectrometry*; Montaser, A.; Golightly, D. W., Eds.; VCH: New York City, NY, 1987; pp 106–112.
52. McLeod, C. W.; Zhang, Y.; Cook, I.; Cox, A.; Date, A. R.; Cheung, Y. Y. *J. Res. Natl. Bur. Stand.* **1988**, *93*, 462–464.
53. Clifford, R. H.; Montaser, A.; Sinex, S. A.; Capar, S. G. *Anal. Chem.* **1989**, *61*, 2777–2784.
54. Kawaguchi, H. *Anal. Sci.* **1988**, *4*, 339–345.
55. Munro, S.; Ebdon, L.; McWeeny, D. J. *J. Anal. At. Spectrom.* **1986**, *1*, 211–219.
56. Satzger, R. D. *Anal. Chem.* **1988**, *60*, 2500–2504.
57. Dabeka, R. W.; McKenzie, A. D. *J. Assoc. Off. Anal. Chem.* **1981**, *64*, 1021–1026.

RECEIVED July 16, 1990

Chapter 14

Direct Analysis of Biological Samples

Simultaneous Multielement Analysis—Atomic Absorption Spectrometry with Miniature Cup Solid Sampling

Ikuo Atsuya

Kitami Institute of Technology, 165 Koen-cho, 090 Kitami, Japan

Comprehensive studies on the direct simultaneous determination of trace elements in some biological samples using a simultaneous multielement analysis/atomic absorption spectrometer with a miniature cup-solid sampling technique were carried out. Optimum instrumental conditions based on a selection of analytical lines, heating programs and effects of ashing temperature, were established. For the direct determination of cadmium, effects of matrix modifier were also examined, and the matrix modification technique was subsequently applied for the direct simultaneous determination of Cd, Pb and Zn in certified reference samples of mussel and hair suplied by the National Institute for Environmental Studies, Japan.

The use of solid sampling technique with atomic absorption spectrometry (AAS) offers high sensitivity, facilitates rapid analysis, and minimizes contamination and loss of analytes. However, the direct AAS analysis of solid samples (especially powder samples) is still in its infancy, because of the difficulties encountered with background compensation; with the measurement of peak area absorbance; the introduction of powdered samples into an electrothermal atomizer without losing the element; and the removal of residues from the atomizer after the measurement. The problems of background compensation and of the measurement of peak area could be solved using Zeeman AAS, and AA data processer respectively (1-3). The introduction of powdered samples into the electrothermal atomizer was also facilitated through the use of small sophisticated containers such as the miniature cup by Atsuya(4-5), platform boat by Kurfurst(6), cup-in-tube by Vollkopf(7). These improvements have stimulated considerable interest on the development of direct AAS analysis of powder samples (8-12); and some new solid sampling techniques involving matrix modification

0097–6156/91/0445–0196$06.00/0

(e.g. Ni-H_2SO_4-HNO_3) for the direct determination of arsenic (13), and chelate-coprecipitation for the determination of µg/l level elements from natural waters (14-15).

On the other hand, it has become necessary to make the synthetic Standard Reference Materials (SRM), which correspond to the standard solutions in the solution analysis, for the direct analysis of biological samples by the solid sampling technique with AAS. We have reported that the use of the magnesium oxinate coprecipitates as the synthetic SRM in solid sampling with AAS was successful for the direct determination of several elements in biological samples of NBS-SRM(16).

As a further extension of our work with solid sampling-AAS, we wish to report here our investigations on the feasibility of direct simultaneous multielement determination in various biological samples.

EXPERIMENTAL METHODS

Apparatus. An Hitachi simultaneous multi-element Zeeman atomic absorption spectrophotometer, Model Z-9000, equipped with a cup-type, graphite furnace, and an Hitachi AA data processor to measure peak area and peak height, and also for recording absorption profiles was used.

A Mettler microbalance, Model M3, accurate to within 1 µg was used.

A Home-made miniature cup (o.d.,4.8mm; i.d.,4mm; depth,2.5mm; wall thickness,0.4mm; and bottam thickness,0.5mm; see Fig.1) was used in conjunction with the cup-type Hitachi furnace to overcome problems such as weighing small amounts of powdered samples, introduction of the sample into the furnace without any loss of material and removal of the residue. The incident light beam was not blocked by the mini-cup, which also facilitates the addition of reagents such as matrix modifiers. A given mini-cup could be used for at least 150 analysis. Furthermore, it was possible to weigh several by using several mini-cups at the same time and measure one after another successively because there were no individual differences among the cups.

Home-made forceps with tantalum tips were used for the insertion and removal of the mini-cup from the furnace. The tip was bent outwards to facilitate removal of the mini-cup.

PROCEDURE FOR THE SOLID SAMPLING TECHNIQUE

Powdered samples contained in the mini-cups were accurately weighed (0.1-1.5mg) by difference using microbalance. Each mini-cup was inserted into the furnace using the tantalum-tipped forceps and the analytical signals were measured using the furnace AAS. National Beureau of Standards (now National Institute of Standard and Technology or NIST) biological reference materials were used for calibration.

RESULTS AND DISCUSSION

Optimization of Instrumental Paramaters. For single-element determinations, it was possible to extend the dynamic range of calibration curves by changing the flow-rate of carrier gas or selecting a suitable analytical line. For multi-element determinations,
however, it was difficalt to extend the dynamic range by changing
the flow-rate of carrier gas, because, for example, when the
flow-rate of carrier gas for one element was increased in order
to extend the dynamic calibration range, it became impossible
to measure another trace level element. Therefore, the flow-rate
of carrier gas had to be kept constant; only the selection of
the analytical line was possible as shown in Table I for Cd, Cr,
Cu, Fe, Mn, Pb and Zn. Table II shows the relationship between
analytical line and dynamic calibration range for each of these
elements.

Table I. Instrumental conditions for Hitachi Multielement
Z-AAS

	Cd,Pb,Zn	Cr,Cu,Fe,Mn
Analytical line(nm)	Cd 228.8 Pb 283.3 Zn 307.6	Cr 359.4 Cu 324.8 Fe 373.7 Mn 403.1
Drying^a	80–120°C, 60s	
Ashig	400°C, 60s	600°C, 60s
Atomization^b	2600°C, 7s	2600°C, 15s

^{a} Drying step was necessary, when the matrix modifier was added.
^{b} Carrier gas flow-rate, 30ml/min

    For the simultaneous multielement determination, the concent-
ration ranges for analytes in samples, which depend on each analytical line, must balanced. Thus, the direct simultaneous determination of Cd, Pb and Zn in NBS pine needles, NIES mussel, and
hair was possible, since the dynamic ranges fell within the certified or provisional values.
    In the direct simultaneous analysis of biological samples
by a solid sampling technique with AAS, the ashing condition
was most important. Therefore, the effects of the ashing were
investigated in detail for each sample. Fig.1 shows typical ashing
curves for the simultaneous determination of the volatile element-
s, Cd, Pb and Zn in the NIES mussel reference sample. Note that
the Cd, Pb and Zn absorbances remained constant between 300 and
500°C, 300 and 500°C, and 400 and 700°C, respectvely (also see
Table III).

Table II. Relation Between the Analytical Line
and the Dynamic Range of Analytes

Element	Line (nm)	Relative Sensitivity	Dynamic Range (µg/g)
Pb	283.3	1	0.5 – 50
	261.4	0.003	200 – 2000
	368.3	0.002	250 – 2500
Zn	213.8	1	0.03 – 2.5
	307.6	0.001	30 – 350
Cd	228.8	1	0.02 – 0.80
	326.1	0.008	3.0 – 100
Fe	248.3	1	4 – 60
	373.7	0.08	50 – 500
Cr	359.4	1	0.5 – 10
Mn	279.5	1	0.2 – 10
	403.1	0.15	3 – 100
Cu	324.8	1	2 – 80
	327.4	0.34	6 – 120

Table III. Suitable Ashing Temperature for Cd,Pb and Zn

Sample	Element	Constant region of ashing temp. (°C)	Recommended ashing temp.(°C)
NBS SRM			
Bovine Liver and	Cd	300–500	300
Pine Needles	Pb	300–500	300
	Zn	400–700	500
NIES CRM			
Hair and	Cd	300–500	400
Mussel	Pb	300–600	500
	Zn	300–700	500

NBS SRM :National Bureau of Standards (now National Institute
of Standards and Technology) Standard Reference Material

NIES CRM :National Institute for Environmental Studies
Certified Reference Material

Fig.2 shows typical ashing curves for the not-so-volatile
elements, namely, Cu, Cr, Fe and Mn in the NIES hair reference
specimen. When non-volatile elements were measured, the condition of the ashing temperature was not significantly affected
by differences in the sample matrix, because of the conversion
of these elements to oxides during the ash-cycle. The ashing
temperature for these elements remained constant between 400
and 800°C.

Each of the points in Fig.1 and Fig.2 represent the average of five absorbance measurements based on five different sample
weights. The absorbance values plotted in these Figures were
normalized to correspond to 1mg of sample. Figures 1 and 2 also
suggest that it is possible to simultaneously determine volatile
and non-volatile elements. However, we did not investigate this
idea further.

Tables IV and V present the analytical results for Cd, Pb
and Zn in NIES mussel and hair, respectively. The samples were
analyzed with, and without matrix modifiers. The modifiers used
were $4NH_2SO_4$ and $10MHNO_3$. The selection of a suitable matrix
modifier was not easy. For example, only $4NH_2SO_4$ (5ml) proved
suitable for Cd in NIES mussel and hair ; even here, the results
agreed with the certified values only when NBS pine needles were
used for calibration. The results were poor with NBS bovine liver
as a calibration standard (Tables IV and V).

Table IV. Direct Simultaneous Determination of Cd,Pb and Zn
in NIES Mussel

Method	Standard for calibration	Proposed method ($\mu g.g^{-1}$) Cd	Pb	Zn
without matrix modifier	NBS Oyster Tissue	a	0.85±0.06	97.6± 3.7
	Pine Needles	0.66±0.10	0.87±0.06	90.6± 6.9
	Bovine Liver	0.75±0.11	0.76±0.05	91.0± 5.7
with $4N-H_2SO_4$	Pine Needles	0.85±0.09	0.71±0.09	257 ±26
	Bovine Liver	0.64±0.06	1.00±0.12	108 ±11
with $10N-HNO_4$	Tomato Leaves	a	0.84±0.07	81 ± 8
	Pine Needles	0.71±0.07	0.99±0.08	78 ± 8
	Bovine Liver	0.59±0.06	0.83±0.07	84 ± 8
Certified values:		0.82±0.03	0.91±0.04	106 ± 6

a not determined

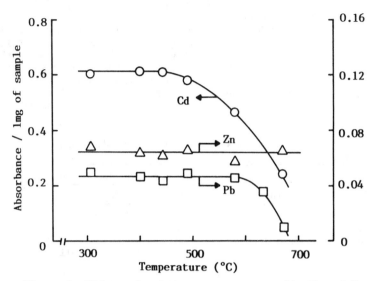

Figure 1. Effect of ashing temperature on Cd, Pb and Zn in NIES Mussel

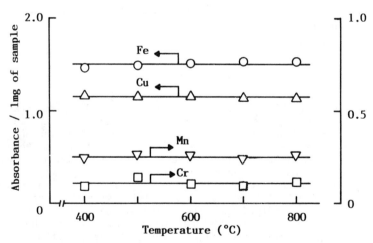

Figure 2. Effect of ashing temperature on Cr, Cu, Fe and Mn in NIES Hair
Cr 359.4nm, Cu 324.8nm, Fe 373.7nm, Mn 403.1nm

For Zn, good agreement was obtained with certified values when H_2SO_4 was used as the modifier and NBS bovine liver as the calibration standard (Tables IV and V). Use of HNO_3 (5μl) was also satisfactory for determining Cd in NIES hair, but was unsuitable for Pb and Zn in NIES hair (Table V) and Zn in NIES mussel (Table IV) due to the severe distortion of the absorption profile ; typical examples are shown in Fig.3 with NIES hair and NBS tomato leaves as matrices. It is clear that these modifiers are not suitable for the direct simultaneous determination of Cd, Pb and Zn in NBS tomato leaves and NIES hair.

Table V. Direct Simultaneous Determination of Cd,Pb and Zn in NIES Hair

Method	Standard for calibration	Proposed method ($\mu g.g^{-1}$) Cd	Pb	Zn
without matrix modifier	NBS Oyster Tissue	*a*	5.0±0.3	157±10
	Pine Needles	0.23±0.02	7.4±0.4	134±13
	Bovine Liver	0.17±0.02	5.6±0.3	151± 8
with 4N-H_2SO_4	Pine needles	0.21±0.03	6.7±0.7	372±25
	Bovine Liver	0.16±0.02	9.5±0.9	156±11
with 10N-HNO_4	Pine Needles	0.19±0.03	b	b
	Bovine Liver	0.16±0.02	b	b
Certified values:		0.20±0.03	(6.0)	169±10

a not determined

b Highly distorted absorption profiles

For this reason, NBS tomato leaves were not used as calibration standards in the present work. It was necessary to use matrix modifiers for Cd because of its potential for loss during the ash-cycle. Also the 307.6nm line had to be used for Zn because it was impossible to increase the carrier gas flow-rate when doing simultaneous multielement analyses.

Although the not-so-volatile elements (Cr, Cu, Fe, Mn) possess various analytical lines, the difficulties with the concentration ranges of the analytes and also in obtaining suitable standard materials for calibration remained. These analytes were, however, stable during the ash-cycle and therefore, it was felt that direct simultaneous determination with solid sampling-AAS might be feasible. The results are shown in Table VI for these elements in NIES mussel and hair. No matrix modifiers were required.

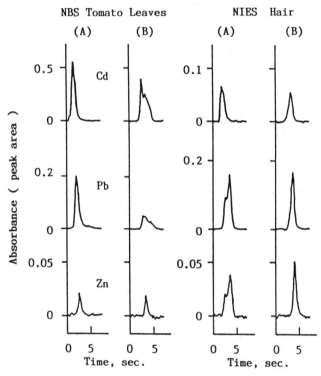

Figure 3. Absorption profiles for Cd, Pb and Zn in NBS Tomato
Leaves and NIES Hair with (A) 10N-HNO₃   (B) 4N-H₂SO₄
NBS Tomato Leaves A (0.393mg)   B (0.412mg)
NIES Hair   A (0.484mg)   B (0.502mg)

Table VI. Direct Simultaneous Determination of Cr,Cu,Fe and
Mn in NIES Mussel and Hair

NIES Sample	Standard for calibration	Proposed method ($\mu g\ g^{-1}$)			
		Cr	Cu	Fe	Mn
Mussel	NBS Oyster Tissue	0.59±0.16	5.1±0.3	158±12	16.4±1.6
	Orchard Leaves	0.74±0.20	6.2±0.3	160±12	16.3±1.6
	Citrus Leaves	0.80±0.22	6.5±0.4	157±12	15.9±1.6
	Certified values:	0.63±0.07	4.9±0.3	158± 8	16.3±1.2
Hair	NBS Oyster Tissue	1.1 ±0.3	16.2±0.7	242±14	5.1±0.8
	Orchard Leaves	1.4 ±0.3	19.6±0.8	245±14	5.0±0.7
	Citrus Leaves	1.5 ±0.4	20.8±0.9	241±14	4.9±0.7
	Pine Needles	1.5 ±0.4	17.8±1.2	250± 9	a
	Certified values:	1.4 ±0.2	16.3±1.2	225± 9	5.2±0.3

$a$not determined

Good agreement with certified values was obtained for all the
four elements in NIES mussel when NBS oyster tissue was used
as the calibration standard. Indeed, for Fe and Mn, NBS orchard
leaves and citrus leaves were equally good calibration matrices.
In the case of the NIES hair sample, excellent agreement with
certified values was obtained for all the four elements irrespective of the NBS calibration standard used.

CONCLUSION

Two conditions for the direct simultaneous determination of
multi-elements by AAS with a solid sampling technique had to
be satisfied : The first was in finding a suitable ashing temperature. In order to find the most suitable ashing temperature,
it was necessary to examine the effects of the ashing temperature
in detail for each sample. Since the volatility of analytes depen-
ded on the sample, the ashing condition had to be varied according to the difference of the chemical species even if the
analyte was the same. The second condition was the restriction
of the concentration range for the measurement of analytes, and
because of this, it was important to aim at one sample and to
find a suitable combination of elements in biological samples
for the direct simultaneous analysis. Also, when the matrix modification technique was applied for the determination of an element
such as cadmium, the effect of the matrix modifier on other elements such as lead, zinc, etc, had to be examined.
    We consider that single-element determination is preferab-
le to multielement determination for Cd because of the need
to use matrix modifiers.

ACKNOWLEDGMENTS

The author wishs to thank Mr. K.Abe, machinery engineer of Kitami Institute of Technology, for his great efforts to make our miniature cups. The author is grateful to the Ministry of Education, Japan, for a research grant (No.C-01540465).

LITERATURE CITED

(1) Koizumi,H.; Yasuda,K. Anal.Chem. 1975, 47, 1679-82.
(2) Koizumi,H.; Yasuda,K.; Katayama,M. Anal.Chem. 1977, 49, 1106-12.
(3) Koizumi,H. Anal.Chem. 1987, 50, 1101-05.
(4) Atsuya,I.; Itoh,K. Spectrochim.Acta 1983, 38B, 1259-64.
(5) Atsuya,I.; Itoh,K. Bunseki Kagaku 1982, 31, 708-12.
(6) Kurfurst,U. Fresenius'Z.Anal. 1983, 316, 1-5.
(7) Vollkopf,U.; Grobenski,Z.; Tamm.; Welz,B. Analyst(London), 1985, 110, 573-77.
(8) Atsuya,I.; Itoh,K.; Akatsuka,K. Fresenius'Z.Anal.Chem. 1987, 328, 338-841
(9) Grobecker,K.H.; Klussendorf,B. Fresenius'Z.Anal.Chem. 1985, 322, 673-77.
(10) Itoh,K.; Akatsuka,K.; Atsuya,I. Bunseki Kagaku 1984, 33, 301-05.
(11) Itoh,K.; Akatsuka,K.; Atsuya,I. Bunseki Kagaku 1986, 35, 122-27.
(12) Rosopulo,A.; Grobecker,K.H.; Kurfurst,U. Fresenius'Z.Anal-.Chem. 1984, 319, 540-44.
(13) Atsuya,I.; Itoh,K.; Akatsuka,K.; Jin,K. Fresenius'Z.Anal.-Chem. 1987, 326, 53-56.
(14) Akatsuka,K.; Atsuya,I. Fresenius'Z.Anal.Chem. 1987, 329, 453-57.
(15) Atsuya,I.; Itoh,K. Fresenius'Z.Anal.Chem. 1988, 329, 750-55.
(16) Akatsuka,K.; Atsuya,I. Anal.Chem. 1989, 61, 216-20.
(17) Ediger,R.D.; Peterson.G.E.; Kerber,J.D. At.Absorp.Newslett, 1974, 13, 61-65.

RECEIVED July 16, 1990

Chapter 15

Arsenic, Bismuth, Copper, Lead, Nickel, and Selenium in Some Biological Samples

Determination by Graphite Furnace Atomization—Atomic Absorption Spectrometry

Ni Zhe-ming[1], Shan Xiao-Quan[1], Jin Long-Zhu[1], Luan Shen[1], Zhang Li[1], and K. S. Subramanian[2]

[1]Research Center for Eco-Environmental Sciences, Academia Sinica, P.O. Box 934, Beijing, China
[2]Environmental Health Centre, Health and Welfare Canada, Tunney's Pasture, Ottawa, Ontario K1A 0L2, Canada

Graphite furnace atomic absorption spectrometric (GFAAS) methods have been developed for determining As, Bi, Cu, Ni, Pb and Se in urine; Al and Li in whole blood and serum; and Ni in some biological reference materials. Cu and Ni were separated from the urine matrix by coprecipitation with ammonium pyrrolidinedithiocarbamate (APDC); Bi and Pb were extracted with potassium iodide-methyl isobutylketone (KI-MIBK) and APDC-MIBK, respectively. The analytical sensitivities of these elements in the organic media were significantly improved when Pd was used as a matrix modifier. The spectral interference, which occurred in the determination of Se in urine, was eliminated by adding Rh. Improved sensitivities were obtained for As and Se in blood when the hydrides were deposited on Pd-coated graphite tubes prior to atomization. Aluminum was determined in blood using potassium dichromate as a matrix modifier. The use of Ta-coated graphite tube significantly minimized the memory effect and improved the sensitivity for Li determination.

Subramanian has comprehensively reviewed the determination of trace elements in human body fluids by GFAAS (1,2). GFAAS is undoubtedly one of the most sensitive and convenient analytica techniques. The use of modern furnace technology, especially stabilized temperature platform furnace atomization (STPF), is recommended for obtaining reliable results (see Subramanian, this issue). However, financial considerations often prevent us from using the STPF technology as it requires the use of expensive pyrocoated graphite tubes and platforms, and Zeeman background correction. Therefore, we chose alternate approaches that do not involve the use of these expensive components, and stil ensure the quality of our data. In this paper, we report our GFAAS methods for determining As, Bi, Cu, Ni, Pb and Se in human urine, and Al and Li in human whole blood and serum.

We developed our methods using a Perkin-Elmer Model 4000 atomic absorption spectrophotometer equipped with a HGA 400 graphite furnace, and a Model 056 strip chart recorder for measuring analyte absorbances under 'gas stop' and 'maximum power' conditions.

DETERMINATION OF Ni AND Cu IN URINE BY APDC COPRECIPITATION-GFAAS

The IUPAC Subcommittee on Environmental and Occupational Toxicology of Nickel has proposed a reference method for determining Ni in serum and urine (3). The reference method involved decomposition of the sample by acid-digestion; dilution of the digestate followed by pH adjustment to 7 with ammonium hydroxide; addition of APDC; extraction of the Ni-PDC chelate with MIBK; and measurement of the Ni concentration in MIBK by GFAAS. However, the chelate in the organic layer was not stable and tended to decompose gradually upon storage. Therefore, in our method, Ni in the digestate was preconcentrated and separated from the urine matrix by precipitation with APDC (4). Quantitative recovery (98-103%) was achieved in a single precipitation over a wide acidity range (0.08-2.0 M HNO_3, or 0.05-0.9 M $HCLO_4$); five different urine samples (endogenous Ni level = 0.5 ng/ml) were tested and each sample was spiked with 4 ng Ni/ml. The Ni-PDC precipitate was dissolved in a minimum amount of MIBK and Ni was measured at the resonance line of 232.0 nm. The Ni-PDC in MIBK was found to be stable for at least a week at ambient temperature. This feature is attractive for clinical and environmental laboratories with a large sample throughput.

The above method was also applied for determining Ni in three NIST standard reference materials: orchard leaves (SRM 1571), wheat flour (SRM 1567) and rice flour (SRM 1568). Our values of 1.2, 0.18, and 0.16 mg/Kg agreed well with the certified values of 1.3, 0.18, and 0.16 mg/Kg, respectively (4). The sample weights taken for analysis were 20, 50, and 50 mg, respectively.

The method for Ni described above was also adapted for determining Cu in urine (4). The Cu-PDC was quantitatively precipitated from 0.2-2.0 M perchloric acid. The recovery of Cu from five different samples spiked with 10 ng Cu/ml was 93-105%. The endogenous Cu levels in these samples were 6.7-16.3 ng/ml. We also determined Cu in the urine of a nephritic patient; the value was found to be 129 ng/ml. To detect such high levels, a 0.2-ml urine aliquot was sufficient.

DETERMINATION OF Bi IN URINE BY IODIDE-MIBK-GFAAS

The concentration of Bi in urine collected from local volunteers was found to be 0.02-0.07 ng/ml. The determination of such low levels requires some form of preconcentration. We used a solvent extraction procedure (5). Bi was extracted with MIBK from urine samples containing iodide and hydrochloric acid, and determined by GFAAS. As shown in Fig.1, the peak height absorbance signal of Bi in MIBK was very low even at a charring temperature of 200°C probably due to volatilization of Bi. However, in the presence of Pd, the sensitivity was greatly enhanced; also the charring temperature could be raised up to 1000°C without any loss of Bi probably due to the formation of thermally stable intermetallic compounds. The method was successfully applied for determining sub-ng/ml levels of Bi in six urine samples; the values ranged from 0.02-0.07 ng/ml. At supplemental levels of 0.05-0.20 ng/ml, the recovery of Bi was quantitative (90-98%) from these samples.

DETERMINATION OF Pb IN URINE BY APDC-MIBK-GFAAS

Unlike in the case of Ni and Cu, the Pb-PDC chelate was stable
in MIBK upon storage. Therefore, the less cumbersome solvent
extraction procedure was adopted (6). Fig. 2 shows the peak
height absorbance of Pb in MIBK as a function of charring
temperature. The absorbance initially increased with
increasing charring temperature and reached a maximum at 800°C;
beyond 800°C, however, there was a rapid decrease in
absorbance. This phenomenon suggested that at lower charring
temperatures the Pb chelate was partially decomposed to a
thermally stable compound. At the charring temperature of
800°C the decomposition was nearly complete, and therefore a
maximum occurred. With further increase in temperature, the Pb
atoms were lost from the graphite furnace and the peak height
absorbance decreased. However, when Pd was used as a modifier,
a charring temperature of 1200°C could be tolerated; also the
sensitivity was higher. Therefore Pd was employed as a matrix
modifier for successfully determining Pb in urine. The
advantage of the high charring temperature was that the urine
matrix could be burned off thereby minimizing interference.
The Pb values determined in nine urine samples were found to be
2, 4, 6, 8, 12, 19, 26, 40 and 60 ng/ml, respectively.

DETERMINATION OF Se IN URINE BY GFAAS USING RHODIUM MATRIX
MODIFICATION

The GFAAS determination of Se in human urine is complicated
because of its high volatility, interaction with the graphite
surface, and spectral and chemical interference (2). Se is
partially lost at temperatures above 300°C. Nickel (7) or
palladium (8) has been used as a modifier to stabilize Se up to
a charring temperature of 1200°C. However, these modifiers
could not overcome the spectral interference originating from
the decomposition products of iron and phosphorus salts (9,
10).
     The calcium phosphate present in urine caused a negative
interference at the 196.0 nm Se resonance line and this could
not be compensated for using the deuterium background corrector
(9, 10). The interference presumably originated from molecules
such as PO, PO$^+$, and P$_2$ within the narrow vicinity of the Se
line (10). The addition of Pt was found to remove this
interference by facilitating the formation of P atoms and
thereby decreasing the spectral interference. However, our
experimental results indicated that the amount of Pt or Pd
added to the sample should be >100 μg. Such high amounts
significantly suppressed the Se signal as shown in Fig.3 (b,
b'; c, c'). We found that the addition of Rh instead of Pt or
Pd not only stabilized Se up to 1200°C (Fig. 4), but also
helped overcome the spectral interference (Fig. 3 a, a'). Note
from Fig. 3 (a, a') that, unlike in the case of Pt and Pd,
there was no loss of sensitivity even at 100 μg Rh. Therefore,
we were able to develop a rapid and reliable method for
determining Se in human urine using deuterium background
correction in the presence of Rh.
     Table I summarizes the analytical results for Se in NIST
reference urine samples (SRM 2670) and also in five human urine
samples using Rh matrix modification. Good agreement was
obtained with the NIST certified values attesting to the
accuracy of our method. The Se values in the five urine
samples ranged from 49 to 77 ng/ml.

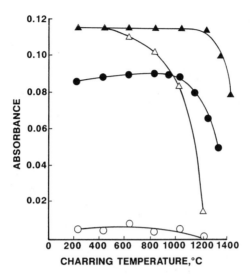

Fig. 1.   Effect of Palladium on the Charring Temperature
of Bismuth. (O) 0.4 ng Bi in MIBK; (▲) 0.5 ng
Bi in aqueous solution; (●) 0.4 ng Bi in MIBK +
1 μg Pd; (▲) 0.5 ng Bi in aqueous solution + 0.8
μg Pd. (Reproduced with permission from Ref. 5.
Copyright 1981 Polyscience Publication, Inc.)

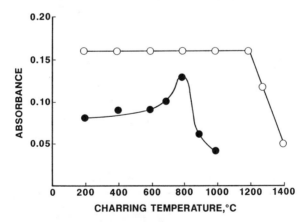

Fig. 2.   Effect of Charring Temperature on the Absorbance
of Lead in APDC-MIBK Extract. (●) 0.8 ng Pb +
321 μg La; (O) 0.8 ng Pb + 321 μg La + 2 μg Pd.
(Reproduced with permission from Ref. 6.
Copyright 1982 Polyscience Publication, Inc.)

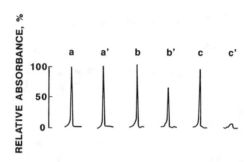

Fig. 3.   Effect of 100 µg of Rhodium, Palladium or
           Platinum on the Atomic Absorption Signal for
           Selenium. Charring: Rh, 1200°C; Pd, 1100°C;
           Pt, 1300°C. Atomization: 2400°C.
           (a)   0.5 ng Se + 1 µg Rh; (a') 0.5 ng Se + 100
                  µg Rh
           (b)   0.5 ng Se + 1 µg Pd; (b') 0.5 ng Se +
                  100µg Pd
           (c)   0.5 ng Se + 1 µg Pt; (c') 0.5 ng Se +
                  100µg Pt

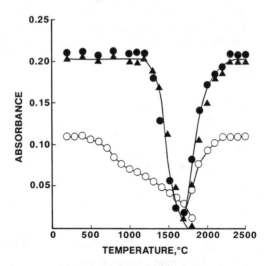

Fig. 4.   Effect of Charring and Atomization Temperature
           on the Atomic Absorption of Selenium with and
           without Rhodium. (O) 0.5 ng Se; (●) 0.5 ng Se
           1 µg Rh; (▲) 0.5 ng Se + 100 µg Rh.

Table I. Determination of Selenium in Human Urine

	Se Concentration (ng/ml)	
Sample	This Work	Certified
Lyophilized urine from NIST; SRM 2670:		
normal level	28 ± 2^b	30 ± 8
elevated level	440 ± 10	460 ± 30
	470 ± 10^c	460 ± 30
Human urine:		
1	58 ± 4	
2	77 ± 3	
3	73 ± 3	
4	60 ± 3	
5	49 ± 4	

[a] Samples were diluted fivefold; [b] Measure of precision is the standard deviation based on 8 determinations; [c] For this measurement a 10-fold dilution was used.

DETERMINATION OF As AND Se IN URINE BY HYDRIDE GENERATION-GFAAS

GFAAS has been used as a trapping medium and as an atomization cell for determining hydride-forming elements such as As and Se (11-13). However, radioanalytical studies showed that the efficiency of collecting arsenic by an uncoated graphite tube was only 24±2%; also, most of the arsenic was found to be deposited at the graphite-tube extremities which are inappropriate positions for the atomization of the analyte (14). We used Pd-coated graphite tubes for the sorption of the hydrides prior to atomization (15). Under optimal conditions, sensitivities were improved by factors of 2.5, 3.8 and 4.9 for As, Sb and Se, respectively. Also, the optimum sorption temperatures of the hydrides ($100-1000°C$ for As and Se; $100-1100°C$ for Sb) that we obtained on the Pd-coated graphite tubes were substantially higher than previously reported. A temperature of $300°C$ was selected to avoid prolonged heating of the graphite tube at high temperatures. The method was used for determining As and Se in the NIST urine standard reference material. Our values of 470 ± 10 and 450 ± 20 ng/ml for As and Se agreed favorably with the certified values of 480 ± 10 and 460 ± 30, respectively (14).

DETERMINATION OF Al IN HUMAN BLOOD BY POTASSIUM DICHROMATE MATRIX MODIFICATION

The 'normal value' for Al in human whole blood is believed to be 2 ng/ml (16, 17). The determination of such a low level is by no means an easy task. Although various modifiers have been used to eliminate matrix interferences and enhance the sensitivity of Al, magnesium nitrate has been recognized as the most suitable modifier (18, 19). However, we found that under optimal conditions potassium dichromate had definite advantages over magnesium nitrate (20). Thus, the characteristic mass values for Al based on peak height were

Table II. Effect of Matrix Modification on
Serum-Al Recovery

Matrix modifier	Sample	Al Concentration (ng/ml)			
		Al in serum	Al added	Total Al found	Recovery (%)
None	Serum 1	14.7	10.0	22.5	78
			15.0	26.2	77
			20.0	30.4	79
	Serum 2	9.9	10.0	17.7	78
			15.0	22.3	83
			20.0	26.2	82
$Mg(NO_3)_2$	Serum 1	14.7	10.0	23.1	84
			15.0	27.3	84
			20.0	31.4	84
$K_2Cr_2O_7$	Serum 1	14.7	10.0	24.5	98
			15.0	28.9	95
			20.0	33.9	96
	Serum 2	9.9	10.0	19.9	100
			15.0	24.4	97
			20.0	29.6	99

SOURCE: Reproduced with permission from ref.23. Copyright 1988 Royal Society of Chemistry

8.8, 5.8 and 4.4 pg/0.0044 A without matrix modification, with
magnesium nitrate modification and with potassium dichromate
modification, respectively. The corresponding peak area values
were 12.6, 11.0 and 8.8 pg/A.s. It is clear that the
sensitivity obtained with potassium dichromate is better than
that obtained without matrix modification and also with
magnesium nitrate as the modifier. Furthermore, as shown in
Table II, quantitative recovery of Al from serum samples was
obtained only with potassium dichromate as the modifier.
Potassium dichromate is available in high-purity, and this
helps to minimize or reduce reagent blank which is a problem in
determining Al due to its ubiquitous nature. The proposed
method was applied for determining Al in five human whole blood
samples and two serum samples. The Al values in the five blood
samples were 10.7, 5.7, 5.0, 3.9 and 3.6 ng/ml, respectively.
The serum values were 14.0 and 9.0 ng/ml, respectively.

DETERMINATION OF Li IN SERUM AND WHOLE BLOOD BY GFAAS

The published reference intervals for Li in whole blood and
serum vary widely probably indicating difficulties with its
determination (1, 21, 22). Li in the graphite tube tends to
form lithium carbide. The carbide is not readily atomized and
this may cause tailing of the signal, lowering of the
sensitivity, and memory effects. One way to overcome these
problems is to use metal-coated graphite tubes such as those of
boron, molybdenum, silicon, tantalum, tungsten and zirconium
(23).
Table III shows the different characteristics of the
various metal-coated tubes. It is apparent from Table III that

the analytical performances were distinctly improved with the
metal-coated tubes. Note that the Ta-coated tube gave the
highest permissible charring temperature, low characteristic
mass and relatively long tube lifetime. Therefore, we used the
Ta-coated graphite tube for determinig Li in human whole blood
and serum.
     The metal-coated tubes, however, could not prevent the
suppression of the Li signal by the chlorides present in the
blood matrix. This interference was caused probably due to the
formation of volatile LiCl and its subsequent loss from the
furnace as a gaseous molecular compound. We added ammonium
nitrate to the sample to overcome the interference.

Table III. Characteristics of Various Metal-coated
Graphite Tubes

Tube	Max. char temp (°C)	Atom. temp appearance (°C)	optimum (°C)	Characteristic mass (pg/0.0044 A)	Tube lifetime (No. firings)
P-E	1000	1450	2500	4.4	80
Locally Made	1050	1370	2100	1.5	220
Coated with					
B	1050	1250	2200	1.1	280
Mo	1000	1290	2200	1.7	200
Si	1100	1320	2000	1.1	320
Ta	1500	1620	2200	1.1	420
W	1000	1150	2100	1.0	460
Zr	1250	1440	2500	1.1	240

We applied the proposed method (23) for measuring Li in 14
whole blood, and 7 serum samples of apparently healthy adults.
The serum values in general were in the range of 1.2-2.7 ng/ml,
and the whole blood values ranged from 0.6 to 2.8 ng/ml. One
whole blood and one serum sample gave values of 6.1 and 7.3
ng/ml, respectively. We did not investigate the cause of these
high values.

CONCLUSION

GFAAS is a popular method for determining trace metals in human
whole blood, serum and urine due to its high sensitivity and
good selectivity. Procedures such as coprecipitation, solvent
extraction and hydride generation have been frequently used for
preconcentrating and separating trace elements. A convenient
approach to enhance sensitivity and minimize interference is to
use matrix modifiers. Pd, Pt or Rh proposed by us has now been
recognized as the universal modifier for easily volatile
elements such as arsenic, bismuth, lead and selenium. Metal-
coated graphite tubes may give better sensitivity, less memory
effect and longer tube life when determining carbide-forming
elements such as Li.

ACKNOWLEDGMENTS

Ni Zheming is grateful to the Organizing Committee of Pacifichem'89; the National Science Foundation of China; and the Research Center for Eco-Environmental Sciences for financing her attendance at the 1989 International Chemical Congress of the Pacific Basin Societies, December 17-22, 1989, Honolulu, Hawaii, U.S.A.

LITERATURE CITED

1. Subramanian, K.S. Prog. Anal. Spectrosc. 1986, 9, 237-334.
2. Subramanian, K.S. Prog. Anal. Spectrosc. 1988, 11, 511-608.
3. Brown, S.S.; Nomoto, S.; Stoeppler, M.; Sunderman Jr., F. Pure Appl. Chem. 1980, 53, 773-81.
4. Jin, L.; Ni, Z. Fresenius' Z. Anal. Chem. 1985, 321, 72-76.
5. Jin, L.; Ni, Z. Can. J. Spectrosc. 1981, 26, 219-23.
6. Shan, X.; Ni, Z. Can. J. Spectrosc. 1982, 27, 75-81.
7. Ediger, R. At. Absorpt. Newslett. 1975, 14, 127-30.
8. Shan, X.; Ni, Z. Acta. Chim. Sin. 1981, 39, 575-78.
9. Fernandez, F. J.; Myers, S. A.; Slavin, W. Anal. Chem. 1980, 52, 741-746.
10. Saeed, K.; Thomassen, Y. Anal. Chim. Acta 1981, 130, 281-87.
11. Sturgeon, R.E.; Willie, S.N.; Berman, S.S. Anal. Chem. 1985, 57, 2311-14.
12. Willie, S.N.; Sturgeon, R.E.; Berman, S.S. Anal. Chem. 1986, 58, 1140-43.
13. Sturgeon, R.E.; Willie, S.N.; Berman, S.S. J. Anal. Atom. Spectrom. 1986, 1, 115-18.
14. Whitley, J. E.; Hannah, R.; Littlejohn, D. Anal. Proc (London). 1988, 25, 246-48.
15. Zhang, L.; Ni, Z.; Shan, X. Spectro. Chim. Acta 1989, 44B, 339-46.
16. Andersen, J.R.; Reimert, S. Analyst (London) 1986, 111, 657-60.
17. Frech, W.; Cedergren, A.; Cederberg, C.; Vessman, J. Clin. Chem. 1982, 28, 2259-63.
18. Leung, F. Y.; Henderson, A. R. Clin. Chem. 1982, 28, 2139-43.
19. Bettinelli, M.; Baroni, U.; Fontana, F.; Poisetti, P. Analyst (London) 1985, 110, 19-22.
20. Shan, X.; Luan, S.; Ni, Z. J. Anal. Atom. Spectrom. 1988, 3, 99-103.
21. Bourret, E.; Moynier, I.; Bardet, L.; Fusellier, M. Anal. Chim. Acta 1985, 172, 157-66.
22. Stafford, D.T.; Saharovici, F. Spectrochim. Acta 1974, 29, 277-81.
23. Luan, S.; Shan, X.; Ni, Z. J. Anal. Atom. Spectrom. 1988, 3, 989-95.

RECEIVED July 16, 1990

Chapter 16

In Vivo and Biochemical Determination of Toxic Cadmium in Rats

Chien Chung[1] and Wen-Kang Chen[2]

[1]Institute of Nuclear Science, National Tsing Hua University, Hsinchu
30043, Taiwan
[2]Department of Biochemistry, Chung-Shan Medical and Dental College,
Taichung 40203, Taiwan

The feasibility of in vivo prompt gamma-ray activation of
toxic Cd concentrations in poisoned rats has been
demonstrated by partial body irradiation with minimal
radiation doses using a small mobile reactor neutron beam.
Rats chronically exposed to toxic Cd are diagnosed in vivo
and followed by biochemical tests. The results of
biochmical tests did not correlate to the Cd concentration
except when the rats were fatally poisoned. Features of
the in vivo determination of whole body Cd concentration
in rats are discussed.

Cadmium is one of the toxic heavy metals which is
used in various industries; it is absorbed occupationally
by workers via inhalation of polluted air or ingestion of
contaminated food and water. The initial Cd uptake is
transported from the lungs or gastrointestinal tract by
the blood to the liver and kidneys(1). Prolonged exposure to
Cd eventually leads to kidney problems, anemia, hypertension,
and liver damage(2,3).

In general, biological fluids such as urine and blood
are analyzed to deduce the toxic Cd concentration(4). Since
Cd exists only in trace amount in the urine and blood, in
vitro analysis may not show the quantity of Cd in the body or
in the organ directly(5). Thus, Cd level in blood and urine
cannot be used for predicting possible nephropathy. For
example, cadmium in the blood of exposed workers did not
correlate with any biochemical parameter that would indicate
potential renal dysfunction(6). Recently, in vivo prompt

0097–6156/91/0445–0215$06.00/0
© 1991 American Chemical Society

gamma neutron activation analysis(IVPGAA) became available to
directlmeasure the Cd concentration in human organs,
thereby making early therapy possible(7). The concept of
IVPGAA was first introduced by J. Anderson(8), followed by
other investigators(9,10). In Taiwan, IVPGAA study on living
beings by the use of nuclear reactor was begun in 1983 at
the National Tsing Hua University(11-15). An extensive
investigation in evaluating radiation safety and
phantom-simulated tests has been completed using reactor
neutron beams(16).

In this work, rats chronically poisoned by Cd as well
as a controlled group are diagnosed with both the IVPGAA
technique and biochemical tests. The results obtained may
serve as a basis for future clinical applications in human
beings poisoned by cadmium.

EXPERIMENTAL DETERMINATION OF CADMIUM IN RATS

To perform the in vivo medical diagnosis and biochemical
blood test, 19 Long-Evans rats were used. They were kept
in a controlled atmosphere where the temperature was
$22 \pm 1°C$ with $50 \pm 5\%$ humidity and the light/dark cycle was
12/12 hours a day. They were fed with water until fully
grown to 140 ± 20 g.

The Long-Evans rats were divided into chronic-poisoned
and control groups. The chronic Cd-poisoned group was fed
standard Cd solution(2μ g/ml) once a day until the injection
dose reached $200 \sim 875$ mg. In the control group, five rats
were fed normal food for two weeks.

The chronically Cd-poisoned and normal rats were
examined using the IVPGAA technique ascribed in our
previous investigation(17) at the modified critical assembly
of the Tsing Hua Mobile Educational Reactor(THMER)
facility(18). Each living rat destined for whole-body
irradiation was locked in an acrylic box lined with tissue
paper. When the neutron beam from THMER was directed at the
living rat, a 559 keV prompt gamma-ray was produced,
measured using a gamma-ray spectral system. The prompt
gamma-ray spectrum was stored, analyzed, and its absolute
count rate was eveluted. The whole body Cd concentration
in the rat was deduced using a conversion factor of
5.285 mg/g/cps obtained from the calibration using standard
$CdCl_2$ solution injected in the rat(17).

Cadmium poisoning is usually diagnosed by a combination
of body fluid analysis for Cd and evaluation of other
biochemical parameters. Therefore, we analyzed the rat
serum to obtain data in creatinine, uric acid, aggregated

protein, albumin, blood urea nitrogen(BUN), alkaline
phosphatase(AP), glutamic pyruvic transaminase(ALT), and
glutamic oxaloacctic transaminanse(AST) with a view to
correlate their variation with the level of Cd poisoning.

RESULT AND DISCUSSION

In Table I, the backgroud-subtracted net count rate of the
559 keV prompt gamma-ray from the Cd(n, γ)Cd reaction in
the Cd-poisoned and normal rats as well as the deduced

Table I.   In vivo Determination of Whole Body Cadmium
           Concentration in Rats.

Rat Number[a]	Count Rate (cps)[b]	Whole Body Cd Concentration(μ g/g)
1.	1.116(3)[c]	5898(16)
2.	1.084(3)	5729(16)
3.	1.002(3)	5296(16)
4.	0.948(3)	5010(16)
5.	0.977(3)	5163(16)
6.	0.799(3)	4223(16)
7.	0.762(3)	4027(16)
8.	0.791(3)	4180(16)
9.	0.785(3)	4149(16)
10.	0.744(3)	3932(16)
11.	0.652(3)	3446(16)
12.	0.546(2)	2886(11)
13.	0.374(2)	1977(11)
14.	0.336(2)	1776(11)
15.	0.0003(3)	1.6(16)
16.	0.0014(10)	7.4(53)
17.	0.004(3)	21(16)
18.	0.008(3)	42(16)
19.	0.0004(3)	2.1(16)

a. Rat numbers 1-14 were chronically exposed to Cd and
   numbers 15-19 were controls not exposed to Cd.
b. The 559 keV prompt gamma-ray in IVPGAA.
C. Number in parenthese represents the uncertainty in the
   last digit.

total body burden are listed. The chronically poisoned
rats (numbers 1-14) have whole body Cd concentration
ranging from 1800 to 5900 μ g/g, indicating nearly a full
absorption from the food dosed with Cd. On the other hand,
normal rats (numbers 15-19) contain only a trace amount of
Cd with a body burden range of 2-40 μ g/g, similar to the

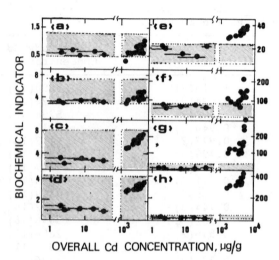

OVERALL Cd CONCENTRATION, μg/g

Fig.1 Relationship between whole body Cd concentration in
rat to biological indicator of (a) creatinine, iu/l,
(b) uric acid, iu/l, (c) aggregated protin, iu/l, (d)
albumin, iu/l, (e) blood urea nitrogen(BUN), mg/l,
(f) alkaline phosphatase(AP), iu/l, (g) glutamic
pyruvic transaminase(ALT), iu/l and (h) glutmaic
oxaloacctic transaminase(AST), iu/l. Shaded area
indicates the normal range in serum.

whole body Cd concentration in the Reference Man(19).
Results of biochemical test with their normal readings for
healthy rats as well as their correlation to the Cd body
burden are shown in Figure 1. The biochemical data for
rats in control group were all within the normal range of
reported values elsewhere(1). The cretinine, uric acid,
aggregated protein, and albumin values in fatally poisoned
rats were also within the normal range, implying such tests
were not correlated to the large intake of Cd. On the other
hand, increasing intake of Cd to a fatal level of 100 μ g/g
gave abnormnally high values of BUN, AP, ALT, and AST.

The correlation of whole body Cd concentration to in
vitro biochemical tests in rats is poor up to a Cd level of
50 μ g/g. Only at the fatally poisoned level of 100 μ g/g
for BUN, AP, ALT, and AST in serum tests gave abnormal
values. The renal tabular damage and organ dysfunction
begin to appear when the body burden of Cd for the renal
cortax in liver is > 300-400 μ g/g(20). Therefore, the
biochemical information of creatinine, uric acid,
aggregated protein, albumin, BUN, AP, ALT, and AST in
serum may not yield any clues to chronic Cd poisoning.

CONCLUSION

Rats chronically exposed to Cd were diagnosed in vivo
using the THMER facility. Biochemical tests were
performed on serum samples taken from the rats in order
to find possible correlation between the two results.
Biochemical data, however, did not show any correlation
to the whole body Cd concentration; only when the rats
were fatally poisoned by at least 100 μ g/g of Cd,
biochemical data for BUN, AP, ALT, and AST in serum
yielded abnormal readings. Thus for chronically poisoned
rats or patients, biochemical analysis may easily be
misdiagnosed as evidence of physical disease other than Cd
poisoning. On the contrary, the in vivo prompt gamma
activation analysis can diagnose the Cd concentration and,
accordingly, the patients can be treated properly.

ACKNOWLEDGMENTS

The authors wish to thank the staff of the Nuclear
Science and Technology Development Centre, National Tsing
Hua University, for their support of this work. Financial
aid for this work was supplied by the Atomic Energy
Council of the Republic of China under contract number
AEC-0979-025-K2.

Literature Cited

1. Lauwerys, R.R. Industrial Chemical Exposure:
    Guidelines for Biological Monitoring. (ed. Davis,
    C.A.) Biomedical Publication, NY, 1983; pp9-50.
2. Elinder, C.G.; Kjellstrom, T.; Lind, B.; Molander,
    M.L.; Silauder, T. Enviorn. Res. 1978, 17, 236-41.
3. Rowls, H.; Bernard, A.; Buchet, J.P.; Lauwery, R.;
    Masson, P. Pathol. Biol. 1978, 26, 329-36.
4. Buchet, J.P.; Rocis, H.; Bernard, A.; Lauwerys, R. J.
    Occup. Med. 1981, 23, 348-52.
5. Lerner, S.; Hong, C.D.; Bozian, R.C. J. Occup. Med.
    1979, 21, 409-12.
6. Murray, T.; Walker, B.R.; Spratt, D.M.; Chappelka, R.
    Arch. Environ. Health 1981, 36, 165-71.
7. Cohn, S.H. Nuclear Medicine and Biological Advances
    (ed. Rayaud, C.) Pergamon Press: New York, 1983; pp
    1049-53.
8. Anderson, J.; Osborn, S.B.; Smith, R.W. Lancet
    1964, 2, 1201-05.
9. Morgan, W.D.; Ellis K.J.; Varsky, D.; Yasumura, S.;
    Cohn, S.H. Phys. Med. Biol. 1981, 26, 577-90.
10. Jones, J.D.; Ludington, M.A.; Rigot, W.L. J.
    Radioanal. Chem. 1982, 72, 287-99.
11. Chung, C.; Yuan, L.J.; Chen, K.B.; Weng, P.S.; Ho.
    Y. H. Int. J. Appl. Radiat. Isot. 1985, 36,357-67.
12. Chang, P.S.; Chung, C.; Yuan, L.J.; Weng, P.S. J.
    Radioanal. Nucl. Chem. Articles 1985, 92, 343-56.
13. Chang, P.S.; Ho, Y.H.; Chung, C.; Yuan, L.J.;
    Weng, P.S. Nucl. Technol. 1987, 76, 241-47.
14. Chung, C. Int. J. Appl. Radiat. Isot. 1988, A39,
    93-96.
15. Chung, C.; Yuan, L.J. Int. J. Radiat. Appl. Instr.
    1988, A39, 977-79.
16. Chung, C.; Cheng, C.P.; Chang, P.S. Health Phys.
    1988, 55, 671-83.
17. Chen, W.K.; Chung, C. J. Radioanal. Nucl. Chem. Articles
    1989, 133, 349-58.
18. Chung, C.; Chen, C.Y. Nucl. Tech. 1990, 93, in press.
19. International Commission on Radiological Portection,
    Task Group On Reference Man, Publication ICRP-23,
    Elmsford, NY: Pergamon Press, New York, 1975.
20. Ellis, J. G.; Morgan, W.D.; Zanal, Z.; Yasumura, S.;
    Vartsky, D.; Cohn, S.H. Am. J. Indust. Med. 1980, 1,
    339-48.

RECEIVED July 16, 1990

Chapter 17

Trace Elements in National Institute for Environmental Studies Standard Reference Materials

Determination by Instrumental Neutron Activation Analysis

Shogo Suzuki and Shoji Hirai

Atomic Energy Research Laboratory, Musashi Institute of Technology, 971 Ohzenji, Asao-ku, Kawasaki 215, Japan

Concentrations of trace elements in standard reference materials prepared by the National Institute for Environmental Studies of Japan (NIES) were determined by instrumental neutron activation analysis (INAA). The reference materials analyzed were: human hair, mussel, tea leaves, vehicle exhaust particulates, Sargasso and unpolished rice flour. These specimens were irradiated under optimized conditions at the Musashi Institute of Technology Research Reactor. The activated samples were measured using three methods: conventional γ-ray spectrometry; anticoincidence γ-ray spectrometry; and coincidence γ-ray spectrometry. We could determine more than 50 elements in the majority of samples. The results obtained were in good agreement with the NIES certified and provisional values.

Trace elements play an important role in environmental, biological and medical fields. Therefore, environmental and biomedical samples should be analyzed with good accuracy and precision. The best approach to ensure the accuracy and precision of a measurement is to use appropriate standard reference materials. The National Institute for Environmental Studies of Japan (NIES) supplies various standard reference materials with certified values(1). In this paper, We report the concentrations of a number of trace elements in the NIES reference materials using instrumental neutron activation analysis (INAA).

In conventional INAA, the sample is irradiated with thermal neutrons, and subsequently counted using a coaxial Ge detector in conjunction with a multichannel pulse height analyzer. We used this system; in addition, we also used conventional thermal neutron irradiation without a cadmium filter; epithermal neutron irradiation with a cadmium filter; conventional γ-ray spectrometry with the coaxial Ge detector; and anticoincidence and coincidence γ-ray

0097–6156/91/0445–0221$06.00/0
© 1991 American Chemical Society

spectrometry with the coaxial Ge detector and a well-type NaI(Tl)
detector. The various approaches used permitted measurement of a
number of elements with high accuracy and precision(2).

EXPERIMENTAL METHODS

Preparation of Samples and Standards. Human hair, mussel, tea
leaves, vehicle exhaust particulates, Sargasso and unpolished rice
flour (all NIES reference materials) samples of each 8−20 portions
(150−1200 mg) were weighed into a clean polyethylene bag and then
double sealed in a polyethylene bag.
     Aqueous, acidic standard solutions generally recommended for
atomic absorption spectrometry were used as surrogate standards.
Aliquotes of 10−200 μ l were pipetted onto clean filter papers (Toyo
Roshi Co.) and the papers were doubly sealed in clean polyethylene
bag.

Irradiation and γ-ray Counting. The samples (reference materials)
and standards were inserted into polyethylene capsules, and were
irradiated either for 30 s, or for 5 h at the Musashi Institute of
Technology Research Reactor (MITRR). The two irradiations were
carried out with and without cadmium filter. The cadmium filter was
made of 1 mm thick cadmium cover.
     Irradiating and counting conditions, and the nuclides used for
measurement are shown in Table I. The samples were irradiated for
30 s; cooled for 1-4 min; and counted for 5 min using conventional
and anticoincidence γ-ray spectrometry. The same samples were then
cooled for 12-40 min and subsequently counted for 9−30 min. In the
second approach, the samples were irradiated for 5 h; cooled for
3-10 days; and counted for 3−14 h using conventional,
anticoincidence and coincidence γ-ray spectrometry. The same
samples were again cooled for 12-60 days and subsequently counted
for 7−40 h.
     An 85 cm^3 coaxial Ge detector (relative efficiency 16% and FWHM
1.8 keV at 1332 keV) was used for conventional γ-ray spectrometry.
For anticoincidence and coincidence γ-ray spectrometry, the coaxial
Ge detector and also a well-type NaI(Tl) detector (6in. ϕ ×6in.,
well 3in. ϕ ×4in.) were used. The γ-ray working ranges of the
NaI(Tl) detector were set to about 40−3500 keV and 90-170 keV in
anticoincidence and coincidence γ-ray spectrometry,
respectively(3).
     For monitoring the neutron flux, about 20 mg of Al wire
containing 0.15% Sb was irradiated together with the samples and the
standards. The neutron flux of samples and the standards was
corrected using the ^{122}Sb and ^{124}Sb activities of the irradiated Al
wires.

γ-ray Spectrometric Analysis System. The γ-ray spectrometric
analysis system called GAMA (Gamma-spectra Analysis of Musashi
Institute of Technology, Atomic Energy Research Laboratory) was
developed and constructed in-house(4-5). We built four models of
GAMA. The block diagram of the GAMA-IV system is shown in
Figure 1. It consists of an acquisition part and an analysis part.
The γ-ray spectrum is generated with the help of a mini-computer
(SORD computer Co.; model M343). The acquisition area of the

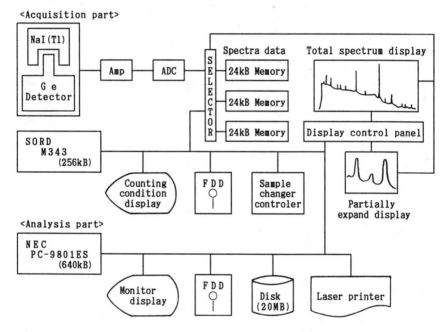

Figure 1.   Block Diagram of GAMA-Ⅳ System.

Table I. Irradiating and Counting Conditions

	Irradiation facility	Irradiation time	Cooling time	Counting time	Nuclide
I	Pneumatic transfer	30 s	1–4 min	5 min	^{27}Mg, ^{28}Al, ^{37}S, ^{49}Ca, ^{51}Ti, ^{52}V, ^{66}Cu,
			12–40 min	9–30 min	^{24}Na, ^{38}Cl, ^{56}Mn, ^{80}Br, ^{101}Mo, ^{116m}In, ^{125m}Sn, ^{128}I, ^{139}Ba, ^{165}Dy
	Central thimble	5 h	3–10 d	3–14 h	^{24}Na, ^{42}K, ^{72}Ga, ^{76}As, ^{82}Br, ^{99}Mo, ^{99m}Tc, ^{115}Cd, ^{115m}In, ^{122}Sb, ^{140}La, ^{142}Pr, ^{153}Sm, ^{166}Ho, ^{175}Yb, ^{187}W, ^{198}Au, ^{239}Np
			12–60 d	7–40 h	^{46}Sc, ^{51}Cr, ^{59}Fe, ^{54}Mn, ^{60}Co, ^{58}Co, ^{65}Zn, ^{75}Se, ^{86}Rb, ^{85}Sr, ^{95}Zr, ^{110m}Ag, ^{113}Sn, ^{124}Sb, ^{131}I, ^{134}Cs, ^{131}Ba, ^{144}Ce, ^{147}Nd, ^{152}Eu, ^{153}Gd, ^{160}Tb, ^{170}Tm, ^{169}Yb, ^{177}Lu, ^{181}Hf, ^{182}Ta, ^{203}Hg, ^{233}Pa
II	Pneumatic transfer	30 s	1–4 min	5 min	^{27}Mg, ^{28}Al, ^{49}Ca, ^{51}Ti, ^{52}V, ^{66}Cu,
			12–40 min	9–30 min	^{24}Na, ^{38}Cl, ^{56}Mn, ^{80}Br, ^{128}I, ^{139}Ba, ^{165}Dy
	Central thimble	5 h	3–10 d	3–14 h	^{24}Na, ^{42}K, ^{76}As, ^{82}Br, ^{99}Mo, ^{99m}Tc, ^{115}Cd, ^{115m}In, ^{122}Sb, ^{140}La, ^{153}Sm, ^{175}Yb, ^{187}W, ^{198}Au, ^{239}Np
			12–60 d	7–40 h	^{46}Sc, ^{51}Cr, ^{59}Fe, ^{54}Mn, ^{60}Co, ^{58}Co, ^{65}Zn, ^{86}Rb, ^{85}Sr, ^{95}Zr, ^{110m}Ag, ^{113}Sn, ^{124}Sb, ^{131}I, ^{134}Cs, ^{131}Ba, ^{144}Ce, ^{147}Nd, ^{152}Eu, ^{153}Gd, ^{160}Tb, ^{170}Tm, ^{169}Yb, ^{182}Ta, ^{203}Hg, ^{233}Pa
III	Central thimble	5 h	14–80 d	8–40 h	^{75}Se, ^{131}Ba, ^{181}Hf

				Nuclides
IV F-ring	5 h	3—9 d	4—15 h	^{24}Na, ^{42}K, ^{72}Ga, ^{76}As, ^{82}Br, ^{99}Mo, ^{99m}Tc, ^{115}Cd, ^{115m}In, ^{122}Sb, ^{140}La, ^{153}Sm, ^{166}Ho, ^{175}Yb, ^{187}W, ^{198}Au, ^{239}Np
		11—70 d	8—40 h	^{46}Sc, ^{51}Cr, ^{59}Fe, ^{54}Mn, ^{60}Co, ^{58}Co, ^{65}Zn, ^{75}Se, ^{86}Rb, ^{85}Sr, ^{95}Zr, ^{110m}Ag, ^{113}Sn, ^{124}Sb, ^{134}Cs, ^{131}Ba, ^{147}Nd, ^{152}Eu, ^{160}Tb, ^{170}Tm, ^{169}Yb, ^{177}Lu, ^{181}Hf, ^{182}Ta, ^{233}Pa
V F-ring	5 h	3—9 d	4—15 h	^{99}Mo, ^{99m}Tc, ^{153}Sm, ^{166}Ho
		11—70 d	8—40 h	^{46}Sc, ^{51}Cr, ^{59}Fe, ^{54}Mn, ^{60}Co, ^{58}Co, ^{65}Zn, ^{75}Se, ^{86}Rb, ^{85}Sr, ^{95}Zr, ^{110m}Ag, ^{113}Sn, ^{124}Sb, ^{134}Cs, ^{147}Nd, ^{160}Tb, ^{169}Yb, ^{181}Hf, ^{182}Ta, ^{233}Pa

I : Conventional γ-ray spectrometry without Cd filter
II : Anticoincidence γ-ray spectrometry without Cd filter
III : Coincidence γ-ray spectrometry without Cd filter
IV : Conventional γ-ray spectrometry with Cd filter
V : Anticoincidence γ-ray spectrometry with Cd filter

Thermal neutron flux : Pneumatic transfer 1.5×10^{12} n·cm^{-2}·s^{-1}
Cetral thimble 3.8×10^{12} n·cm^{-2}·s^{-1}
F-ring 1.8×10^{12} n·cm^{-2}·s^{-1} (Cd ratio 6)

spectrum (8192 channels) can be divided into 1/1, 1/2 or 1/4 units.
The total spectrum, the partially expanded spectrum of the region of
interest and the counting conditions are independently displayed at
the same time.
    The acquisition information includes the spectral data for 8192
channels and a label of 256 words. In this label, the start time,
end time, true time, and the live time for counting are
automatically designated. The name of the sample and the end time
of irradiation are designated by a keyboard letter. After the
counting, the acquired spectrum with the label is transferred to the
analysis part. The transfer time is about 0.5 s.
    Each γ-ray spectrum was analyzed by a 32-bit personal computer
(NEC PC-9801ES) using the analytical program GAMA98 written in
compiler basic language(6). The γ-ray peaks of the spectrum would
be searched automatically according to the analytical conditions
specified by the computer. The γ-ray peak areas were analyzed by
means of the non-linear least square method, and the energies and
intensities of the peaks were calculated. The intensities were
corrected for decay of the corresponding nuclide. The
concentrations of elements of interest in the sample were calculated
by comparison with the intensities of standards. Analytical results
and spectra were printed using a laser printer. The overlapping
multiple peaks could be analyzed precisely and easily by this
program.

RESULTS AND DISCUSSION

Human Hair Reference Material. Table II lists the results for 53
elements in the human hair, together with the NIES-certified and
provisional values. Errors show the standard deviation of the
analytical results for various sample weights. Note that the
observed values agree well with the NIES-certified and provisional
values except for Sr. The 514 keV peak of ^{85}Sr overlapped with the
511 keV annihilation peak. Concentration of As was determined using
anticoincidence γ-ray spectrometry without cadmium filter.
    Interferences accompanied by different nuclear reaction leading
to the same nuclide must be taken into account in determining some
elements. Determination of Al is achieved by using the
^{27}Al(n, γ)^{28}Al reaction. However Al is also produced through the
^{28}Si(n,p)^{28}Al reaction. The contribution from the interfering
reaction was evaluated by irradiating and counting a Si standard
under the same conditions as the samples, and the above interference
could then be corrected easily. Factors of interfering nuclear
reactions in our INAA are summarized in Table III.
    The lower limit of determination of this method for the 53
elements are also given in Table II. The lower limit of
determination of the individual elements were calculated using the
equation, $N = 3B^{1/2}$, where B is the number of counts in the
background and N is the minimum number of counts under each γ-ray
peak(7). Forty elements were detectable at levels below 1 μg/g.
Especially, the lower limit of determination of Sc and Au were
0.0003 μg/g.

Mussel Reference Material. Table IV lists the analytical results
for 51 elements in the mussel together with the NIES-certified and

Table II. Concentration of Elements in NIES No.5 Human Hair by Various Types of Irradiation and γ-Spectrometry (μg/g)

Element	Conventional γ-spectrometry without Cd filter		Anticoincidence γ-spectrometry without Cd filter	NIES[a]		Lower limit of determination
Na	26	±1		26	±1	0.3
Mg	220	±19		220	±19	40
Al	240	±16		(240)		1.4
Si	12500	±800				3000
S	49000	±1000				3000
Cl	250	±4		(250)		5
K	36	±8		34	±3	30
Ca	660	±40		728	±30	40
Sc	0.046	±0.001	0.047	(0.05)		0.0003
Ti	22	±7		(22)		10
V	0.67	±0.01				0.08
Cr	1.24	±0.07	1.49	1.4	±0.2	0.06
Mn	4.7	±0.1		5.2	±0.3	0.04
Fe	210	±6	195	225	±9	1.4
Co	0.099	±0.004	0.094	(0.10)		0.0015
Ni	1.73	±0.036	1.65	1.8	±0.1	0.3
Cu	16.1	±0.9		16.3	±1.2	5
Zn	164	±4	150	169	±10	0.13
As	<0.097		0.109			0.06
Se	1.29	±0.03	1.21[b]	(1.4)		0.03
Br	92	±2		(90)		0.06
Rb	0.157	±0.033	0.170	(0.19)		0.08
Sr	3.5	±0.8	2.7	2.3	±0.2	1.4
Zr	<2.3		<1.14			1.1
Mo	<0.12					0.12
Ag	0.095	±0.005				0.014
Cd	0.23	±0.10		0.20	±0.03	0.2
In	<0.0072					0.007
Sn	1.62	±0.37				1.0
Sb	0.073	±0.006	0.073	(0.07)		0.002
Te	<0.18					0.18
I	0.62	±0.08				0.3
Cs	0.0141	±0.0007				0.003
Ba	3.1	±0.6	3.0	(3.2)		1.2
La	0.184	±0.012				0.008
Ce	0.47	±0.03	0.49			0.05
Nd	0.30	±0.05				0.3
Sm	0.0197	±0.0014				0.0016
Eu	0.0037	±0.0006				0.003
Gd	<0.095					0.09
Tb	0.0034	±0.0017	0.0035			0.0015
Dy	<0.023					0.02
Ho	<0.34					0.3
Tm	<0.0094					0.009
Yb	0.025	±0.005				0.014
Lu	<0.0019					0.0019
Hf	0.0106	±0.0016	0.0147[b]			0.004
Ta	0.0064	±0.0005	0.0067			0.0009
W	<0.041					0.04
Au	0.0127	±0.0004	0.0120			0.0003
Hg	4.5	±0.3	4.9	4.4	±0.4	0.014
Th	0.022	±0.002	0.021			0.004
U	0.020	±0.004				0.008

a. NIES certified values. Data in parentheses are provisional values only.
b. Coincidence γ-ray spectrometry without Cd filter.

Table III. Correction for Interfering Nuclear Reaction

Element	Nuclear reaction	Interfering nuclear reaction	Factor of interfering nuclear reaction
Human hair			
Al	^{27}Al(n, γ) ^{28}Al	^{28}Si(n, p) ^{28}Al	35%
Mussel			
Al	^{27}Al(n, γ) ^{28}Al	^{31}P (n, α) ^{28}Al	10%
Mg	^{26}Mg(n, γ) ^{27}Mg	^{27}Al(n, p) ^{27}Mg	5%
Tea leaves			
Mg	^{26}Mg(n, γ) ^{27}Mg	^{27}Al(n, p) ^{27}Mg	15%
Na	^{23}Na(n, γ) ^{24}Na	^{27}Al(n, α) ^{24}Na	4%
		^{24}Mg(n, p) ^{24}Na	22%
Vehicl exhaust particulates			
Mg	^{26}Mg(n, γ) ^{27}Mg	^{27}Al(n, p) ^{27}Mg	10%
Unpolished rice flour			
Al	^{27}Al(n, γ) ^{28}Al	^{31}P (n, α) ^{28}Al	46~52%

Table IV. Concentration of Elements in NIES No. 6 Mussel by
Various Types of Irradiation and γ-Spectrometry (μg/g)

Element	Conventional γ-spectrometry without Cd filter		Conventional γ-spectrometry with Cd filter	NIES[a]	
Na	9900	± 300		10000	± 300
Mg	2100	± 130		2100	± 100
Al	260	± 20		(220)	
S	26000	± 3000			
Cl	17400	± 400			
K	5400	± 300	4900	5400	± 200
Ca	1370	± 270		1300	± 100
Sc	0.041	± 0.002			
Ti	<72				
V	0.82	± 0.19[b]			
Cr	0.66	± 0.17		0.63	± 0.07
Mn	15.3	± 0.6		16.3	± 1.2
Fe	161	± 10		158	± 8
Co	0.36	± 0.02		(0.37)	
Ni	0.92	± 0.22		0.93	± 0.06
Cu	<13			4.9	± 0.3
Zn	108	± 3		106	± 6
Ga	<1200				
As	9.7	± 0.5		9.2	± 0.5
Se	1.45	± 0.05	1.5	(1.5)	
Br	96	± 1			
Rb	2.5	± 0.1			
Sr	25	± 2		(17)	
Zr	<2.9				
Mo	1.10	± 0.24			
Ag	0.026	± 0.009		0.027	± 0.003
Cd	0.80	± 0.34			
In	<0.11				
Sn	1.28	± 0.46			
Sb	0.0116	± 0.0047			
Te	<0.18				
I	4.2	± 1.3			
Cs	0.032	± 0.006			
Ba	1.58	± 0.63			
La	0.176	± 0.012			
Ce	0.36	± 0.04			
Pr	<7.3				
Nd	<0.35				
Sm	0.045	± 0.002	0.039		
Eu	0.0073	± 0.0014			
Tb	0.0027	± 0.0005			
Dy	<0.20				
Yb	<0.023				
Lu	<0.0020				
Hf	0.0099	± 0.0029			
Ta	0.0024	± 0.0008			
W	<0.20				
Au	0.0048	± 0.0003	0.0054		
Hg	<0.058			(0.05)	
Th	0.041	± 0.003	0.049		
U	0.077	± 0.005			

a. NIES certified values. Data in parentheses are provisional
   values only.
b. Pile-up rejection method.

provisonal values. Again, the analyzed values were in good
agreement with the certified and provisional values except for Sr.
In the case of Al and Mg determination, corrections were made for
the interfering nuclear reactions (Table Ⅲ).
    Concentration of V was found to be $0.45 \pm 0.31 \mu g/g$ by
conventional γ-ray spectrometry without the cadmium filter. The
relative standard deviation of more than ±60% in this case was due
to the counting loss resulting from the high counting rate. So, a
pile-up rejector electric circuit was installed in counting ^{52}V by
conventional γ-ray spectrometry in order to correct for the
counting loss. Based on this correction, the concentration of V was
found to be $0.82 \pm 0.19 \mu g/g$, which we believe to be more acurate and
precise.

Tea Leaves Reference Material. Table V lists the analytical
results for 48 elements in the tea leaves, together with the
NIES-certified and provisional values. Se, As and W were determined
by conventional γ-ray spectrometry with cadmium filter, and the
pile-up rejection method was used for determining the concentration
of V. In the case of Mg and Na determination, corrections were made
for the interfering nuclear reactions (Table Ⅲ).

Vehicle Exhaust Particulates Reference Material. Table Ⅵ lists the
analytical results for 55 elements in the vehicle exhaust
particulates, together with the NIES-certified and provisional
values. Te, Nd and Hg could be determined only by anticoincidence
γ-ray spectrometry without cadmium filter; Ga only by conventional
γ-ray spectrometry with cadmium filter; and Ho only by
anticoincidence γ-ray spectrometry with cadmium filter. In the
case of Mg determination, a correction was made for the interfering
nuclear reaction (Table Ⅲ). Pr, Gd and Tm were not determined by
any combination of irradiation and counting methods.
    Typical γ-ray spectra are shown in Figures 2a and 2b. The
spectrum in Figure 2a was acquired by conventional γ-ray
spectrometry without cadmium filter while the spectrum in Figure 2b
was acquired by anticoincidence γ-ray spectrometry without cadmium
filter. The use of anticoincidence γ-ray spectrometry minimizes
the Compton background of ^{65}Zn 1116 keV. The ^{233}Pa, ^{58}Co and ^{54}Mn
peaks could be seen clearly in Figure 2b.

Sargasso Reference Material. Table Ⅶ lists the analytical results
for 52 elements in Sargasso, together with the NIES-certified and
provisional values. Zr, Mo, La and Th were determined by
anticoincidence γ-ray spectrometry without cadmium filter; Se and
Hf by coincidence γ-ray spectrometry without cadmium filter; and Ni
by anticoincidence γ-ray spectrometry both with and without cadmium
filter.
    The optimum irradiation and γ-ray spectrometric conditions for
determination of various elements in Sargasso is listed in Table Ⅷ.
Thus, many elements can be determined with better precision and
acuracy by optimizing the irradiation and γ-ray counting methods.

Unpolished Rice Flour Reference Material. Table Ⅸ lists the
analytical results for 31 elements in the unpolished rice flour,
together with the NIES-certified and provisional values. The rice

Table V. Concentration of Elements in NIES No.7 Tea Leaves by
          Various Types of Irradiation and γ-Spectrometry (μ g/g)

Element	Conventional γ-spectrometry without Cd filter		Conventional γ-spectrometry with Cd filter	NIES[a]	
Na	18.9	± 1.9		15.5	± 1.5
Mg	2300	± 400		1530	± 60
Al	760	± 70		755	± 20
Cl	790	± 50			
K	18600	± 600		18600	± 700
Ca	3200	± 200		3200	± 120
Sc	0.0101	± 0.0006	0.0117	(0.011)	
Ti	<57				
V	0.37	± 0.09[b]			
Cr	0.150	± 0.019		(0.15)	
Mn	700	± 20		700	± 25
Fe	98	± 7	90		
Co	0.119	± 0.004	0.120	(0.12)	
Ni	6.0	± 0.2		6.5	± 0.3
Cu	<21			7.0	± 0.3
Zn	33	± 1	32	33	± 3
Ga	<0.72				
As	<0.022		0.0160		
Se	<0.26		0.022		
Br	2.5	± 0.1			
Rb	5.8	± 0.1	6.1		
Sr	5.4	± 1.0		(3.7)	
Zr	<1.20				
Mo	<0.092				
Ag	<0.0084				
Cd	0.034	± 0.009		0.030	± 0.003
Sn	<0.79				
Sb	0.0129	± 0.0047	0.0135	(0.014)	
Te	<0.088				
I	<1.51				
Cs	0.022	± 0.001	0.022	(0.022)	
Ba	5.5	± 0.2	5.3	(5.7)	
La	0.049	± 0.004			
Ce	0.052	± 0.019			
Pr	<0.130				
Nd	<0.133				
Sm	0.0121	± 0.0014			
Eu	0.0027	± 0.0002	0.0026		
Tb	0.00120	± 0.00013	0.00139		
Yb	0.0052	± 0.0009			
Lu	<0.00128				
Hf	0.0033	± 0.0007			
Ta	0.00087	± 0.00029			
W	<0.058		0.050		
Au	<0.00049				
Hg	0.0126	± 0.0029			
Th	0.0048	± 0.0005			
U	<0.00135				

a. NIES certified values. Data in parentheses are provisional
   values only.
b. Pile-up rejection method.

Table VI. Concentration of Elements in NIES No.8 Vehicle Exhaust Particulates by Various Types of Irradiation and γ-spectrometry (μg/g)

Element	I	II	III	IV	V	NIES[a]
Na	1670 ± 80	1850 ± 50				1920 ± 80
Mg	1100 ± 150					1010 ± 50
Al	3000 ± 100					3300 ± 200
Si	18900 ± 1900					
S	50000 ± 9000					
Cl	840 ± 30					
K	1110 ± 90					1150 ± 80
Ca	4900 ± 200					5300 ± 200
Sc	0.51 ± 0.03	0.49 ± 0.02			0.52 ± 0.03	(0.55)
Ti	270 ± 30					
V	15.8 ± 1.0					17 ± 2
Cr	23 ± 2	25 ± 3			23 ± 3	25.5 ± 1.5
Mn	63 ± 7					
Fe	4600 ± 500	4500 ± 100			4200 ± 700	
Co	2.9 ± 0.2	2.9 ± 0.1			3.0 ± 0.3	3.3 ± 0.3
Ni	14.1 ± 0.9	17.9 ± 1.1		18.6 ± 0.9	18.5 ± 1.6	18.5 ± 1.5
Cu	70 ± 7					67 ± 3
Zn	900 ± 80	990 ± 90			1050 ± 120	1040 ± 50
Ga				0.47 ± 0.09		
As	2.6 ± 0.1	2.6 ± 0.1		2.6 ± 0.3		2.6 ± 0.2
Se	1.25 ± 0.12	1.34 ± 0.09	1.21 ± 0.07	1.37 ± 0.05		(1.3)
Br	57 ± 1	66 ± 4				(56)
Rb	4.1 ± 0.3	4.0 ± 0.1				(4.6)
Sr	115 ± 17	121 ± 7			107 ± 11	89 ± 3
Zr	12.4 ± 1.3	11.2 ± 2.0			<13.9	
Mo	6.0 ± 0.2					(6.4)
Ag	0.20 ± 0.01			0.20 ± 0.01	<0.46	(0.20)
Cd	1.34 ± 0.13	1.09 ± 0.21		0.90 ± 0.03		1.1 ± 0.1
In	0.078 ± 0.006					
Sn	9.5 ± 0.6	9.1 ± 0.3		9.0 ± 1.0	8.3 ± 1.1	

	I	II	III	IV	V	a
Sb	5.9 ±0.2	6.1 ±0.2		6.4 ±0.2	6.0 ±0.5	6.0 ±0.4
Te	<0.40	0.15 ±0.08		<1.60	0.21 ±0.03	(0.24)
I	3.0 ±0.3					
Cs	0.24 ±0.01	0.24 ±0.01		0.22 ±0.02		
Ba	128 ±5	117 ±9	108 ±5		129 ±13	
La	1.08 ±0.03					(1.2)
Ce	2.9 ±0.3	3.2 ±0.2		2.8 ±0.1	3.1 ±0.2	(3.1)
Pr	<1.06	<0.47		<2.5		
Nd		0.80 ±0.07				
Sm	0.193 ±0.010	0.186 ±0.008				(0.20)
Eu	0.049 ±0.008	0.051 ±0.005			<0.101	(0.05)
Gd	<0.37	<0.44				
Tb	0.031 ±0.006	0.030 ±0.001				
Dy	0.185 ±0.017					
Ho	<0.026	<0.030		<0.026	0.025 ±0.006	
Tm	<0.028	<0.019				
Yb	0.108 ±0.007	0.122 ±0.001				
Lu	0.020 ±0.002	0.0182 ±0.0006				
Hf	0.21 ±0.01	0.183 ±0.016	0.183 ±0.011		<0.23	(0.02)
Ta	0.116 ±0.007	0.135 ±0.008		0.125 ±0.014	0.136 ±0.009	
W	5.0 ±0.4	5.1 ±0.2				
Au	0.0029 ±0.0006	0.0029 ±0.0004		0.0031 ±0.0007		
Hg	<0.56	0.33 ±0.08				
Th	0.34 ±0.02	0.36 ±0.01			0.36 ±0.02	
U	0.104 ±0.015	0.098 ±0.012		0.101 ±0.007		(0.35)

I : Conventional γ-ray spectrometry without Cd filter
II : Anticoincidence γ-ray spectrometry without Cd filter
III : Coincidence γ-ray spectrometry without Cd filter
IV : Conventional γ-ray spectrometry with Cd filter
V : Anticoincidence γ-ray spectrometry with Cd filter
a. NIES certified values. Data in parentheses are provisional values only.

Table VII.   Concentration of Elements in NIES No. 9 Sargasso by Various Types of Irradiation and γ-spectrometry ($\mu g/g$)

Element	I	II	III	IV	V	NIES[a]
Na	15000 ±1000	17000 ±2000		12000 ±1000		17000 ±800
Mg	6900 ±600	6100 ±500				6500 ±300
Al	210 ±20	220 ±30				(215)
S	53000 ±6000					(12000)
Cl	43000 ±2000	43000 ±4000				(51000)
K	59000 ±3000	60000 ±5000				61000 ±2000
Ca	12000 ±700	11000 ±1000				13400 ±500
Sc	0.090 ±0.004	0.100 ±0.003		0.082 ±0.008	0.078 ±0.012	(0.09)
Ti	<75	<69				(9)
V	1.0 ±0.2	1.0 ±0.2				1.0 ±0.1
Cr	0.19 ±0.03	0.18 ±0.02		<2.0	<1.4	(0.2)
Mn	20 ±1	21 ±4				21.2 ±1.0
Fe	180 ±10	190 ±10		170 ±20	150 ±10	187 ±6
Co	0.11 ±0.01	0.11 ±0.01		0.12 ±0.01	0.12 ±0.01	0.12 ±0.01
Ni	<0.42	0.40 ±0.03		<0.38	0.43 ±0.03	
Cu	<51	<38		<28		4.9 ±0.2
Zn	13 ±1	14 ±1		15 ±1	16 ±2	15.6 ±1.2
Ga						
As	120 ±10	110 ±10		100 ±10		115 ±9
Se	<0.049		0.044 ±0.003	<0.13	<0.22	(0.05)
Br	270 ±10	290 ±10		210 ±10		(270)
Rb	22 ±1	23 ±1		21 ±1	21 ±2	24 ±2
Sr	990 ±40	1100 ±100		980 ±50	1000 ±100	1000 ±30
Zr	<4.7	4.3 ±0.5		<5.2	<4.8	
Mo	<0.58	0.56 ±0.20		<0.55	0.53 ±0.19	
Ag	0.30 ±0.01	0.30 ±0.02		0.28 ±0.02	0.29 ±0.04	0.31 ±0.02
Cd	<1.0	<0.89		<1.4		0.15 ±0.02
In	<0.11					

	I	II	III	IV	V
Sn	<1.3	<1.1	<2.4	<1.9	(0.04)
Sb	0.039 ± 0.004	0.039 ± 0.004	0.038 ± 0.004	0.034 ± 0.004	
Te	0.46 ± 0.07	0.43 ± 0.02			
I	520 ± 20	530 ± 30			(520)
Cs	0.042 ± 0.003	0.042 ± 0.004	0.043 ± 0.003	0.042 ± 0.003	(0.04)
Ba	9.4 ± 1.4	11 ± 1	8.7 ± 0.9		
La	0.089 ± 0.018	0.094 ± 0.010	<0.26		
Ce	0.18 ± 0.01	0.21 ± 0.01			
Nd	<0.52	<0.30	<2.0	<1.6	
Sm	0.083 ± 0.013	0.078 ± 0.014	0.083 ± 0.009		
Eu	0.0069 ± 0.0005	0.0074 ± 0.0012			
Gd	<0.12	<0.27			
Tb	0.0049 ± 0.0009	0.0046 ± 0.0008	0.0045 ± 0.0009	0.0046 ± 0.0006	
Dy	<0.21	<0.23			
Tm	<0.0098	<0.0093	<0.016		
Yb	0.024 ± 0.009	0.059	0.074	<0.14	
Lu	<0.0032		<0.056		
Hf	<0.011	0.011 ± 0.001	<0.029	<0.050	
Ta	<0.00085	<0.0030	<0.0027	<0.0034	
W	<0.41	<0.19	<0.40		
Au	0.0015 ± 0.0003	0.0010 ± 0.0001			
Hg	<0.061	<0.055	<0.017		
Th	<0.0079	0.0058 ± 0.0010		<0.010	(0.04)
U	0.41 ± 0.02	0.42 ± 0.03	0.40 ± 0.02		(0.4)

I : Conventional γ-ray spectrometry without Cd filter
II : Anticoincidence γ-ray spectrometry without Cd filter
III : Coincidence γ-ray spectrometry without Cd filter
IV : Conventional γ-ray spectrometry with Cd filter
V : Anticoincidence γ-ray spectrometry with Cd filter
a. NIES certified values. Data in parentheses are provisional values only.

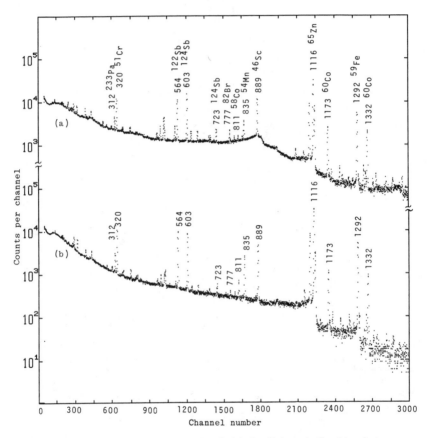

Figure 2.   γ-ray Specta of NIES Vehicle Exhaust Particulates.
            Irradiation time 5h, cooling time 18 day, counting
            time 2h, sample weight 451 mg.
            a. Conventional γ-ray spectrometry without cadmium
               filter.
            b. Anticoincidence γ-ray spectrometry without cadmium
               filter.

Table Ⅷ.   Optimum Methods for the INAA Determination of
Various Elements in Sargasso

γ-spectrometry	Irradiation	
	without Cd filter	with Cd filter
Conventional	Na, Al, S, Cl, Ca, Sc, Mn, Co, Br, In, La, Eu, Gd, Dy, Yb, Lu, Ta,	Ga, Ag, Sb, Cs, U
Anticoincidence	Mg, K, Ti, V, Cr, Fe, Cu, Zn, As, Rb, Sr, Zr, Cd, Sn, Te, I, Ce, Nd, Sm, Tm, W, Au, Hg, Th	Ni, Mo, Tb
Coincidence	Se, Ba, Hf	

Table IX. Concentration of Elements in NIES No.10 Unpolished Rice Flour ($\mu g/g$)

Element	Low level		Medium level		High level	
	This work	NIES[a]	This work	NIES[a]	This work	NIES[a]
Na	10.5 ±1.6	10.2 ±0.3	17.9 ±0.7	17.8 ±0.4	13.9 ±3.4	14.0 ±0.4
Mg	1500 ±90	1340 ±80	1490 ±80	1310 ±60	1410 ±100	1250 ±80
Al	4.6 ±1.6	(3)	2.2 ±1.5	(2)	1.6 ±1.4	(1.5)
S	1140 ±160		1310 ±130		1200 ±180	
Cl	250 ±10	(260)	300 ±20	(310)	230 ±10	(230)
K	2500 ±200	2800 ±80	2200 ±100	2450 ±100	2500 ±100	2750 ±100
Ca	91 ±8	93 ±3	79 ±13	78 ±3	95 ±7	95 ±2
Sc	0.00050±0.00006		0.00030±0.00007		0.00029±0.00004	
V	0.045 ±0.008		0.042 ±0.011		0.040 ±0.011	
Cr	0.062 ±0.008	(0.07)	0.186 ±0.015	(0.22)	0.059 ±0.008	(0.08)
Mn	35 ±2	34.7 ±1.8	32 ±2	31.5 ±1.6	41 ±3	40.1 ±2.0
Fe	12.5 ±0.7	12.7 ±0.7	12.3 ±0.7	13.4 ±0.9	10.9 ±0.9	11.4 ±0.8
Co	0.0172 ±0.0016	(0.02)	0.0181 ±0.0012	(0.02)	0.0058 ±0.0004	(0.007)
Ni	0.150 ±0.016	0.19 ±0.03	0.37 ±0.02	0.39 ±0.04	0.27 ±0.02	0.30 ±0.03
Cu	3.8 ±0.4	3.5 ±0.3	3.4 ±0.7	3.3 ±0.2	4.6 ±0.5	4.1 ±0.3
Zn	23 ±1	25.2 ±0.8	21 ±1	22.3 ±0.9	22 ±2	23.1 ±0.8
As	0.169 ±0.015	(0.17)	0.105 ±0.005	(0.11)	0.147 ±0.008	(0.15)
Se	0.061 ±0.005	(0.06)	0.020 ±0.002	(0.02)	0.068 ±0.005	(0.07)
Br	0.32 ±0.06	(0.3)	0.50 ±0.04	(0.5)	0.52 ±0.06	(0.5)
Rb	4.2 ±0.2	4.5 ±0.3	3.1 ±0.2	3.3 ±0.3	5.4 ±0.4	5.7 ±0.3
Sr	<0.35	(0.3)	0.36 ±0.09	(0.3)	<0.32	(0.2)
Mo	0.33 ±0.02	0.35 ±0.05	0.40 ±0.02	0.42 ±0.05	1.51 ±0.13	1.6 ±0.1
Ag	<0.0039		<0.0030		0.0097 ±0.0009	
Cd	0.026 ±0.009	0.023 ±0.003	0.28 ±0.02	0.32 ±0.02	1.72 ±0.09	1.82 ±0.06
Sb	<0.00085		<0.00075		0.00070±0.00007	
Cs	0.0082 ±0.0009		0.0083 ±0.0009		0.0073 ±0.0009	
Ba	0.70 ±0.10		0.40 ±0.06		0.61 ±0.06	
La	0.00083±0.00015		0.00044±0.00014		0.00085±0.00013	
Sm	0.00021±0.00004		<0.00010		0.00015±0.00003	
Hg	<0.0092	(0.004)	<0.0095	(0.003)	<0.0095	(0.005)
Th	0.00076±0.00009		<0.00062		<0.0010	

a. NIES certified values. Data in parentheses are provisional values only.

flour is certified for three different levels (low, medium, high) of the various elements. In the case of Al determination, a correction was made for the interfering nuclear reaction as per Table Ⅲ, and the precisions of Al results were in the vicinity of ± 30%. It was impossible to detect lower energy peaks, owing to the Bremsstrahlung reaction of β-rays mainly from ^{32}P (half life 14 days).

CONCLUSION

More than 50 elements were determined in various NIES reference materials by our INAA methods. The determined values were in good agreement with the NIES certified and provisional values. We conclude that the present method is effective in analyzing environmental and biomedical samples.

Literature Cited

1. Okamoto, K. Fresenius Z. Anal. Chem. 1988, 332, 524-27.
2. Suzuki, S.; Hirai, S. Anal. Sci. 1987, 3, 193-97.
3. Suzuki, S.; Hirai, S. Radioisotopes 1985, 29, 161-65.
4. Murata, Y.; Hirai, S.; Okamoto, M.; Kakihana, H.
   J. Radioanal. Chem. 1977, 36, 525-35.
5. Suzuki, S.; Hirai, S. Anal. Sci. 1987, 3, 91-2.
6. Suzuki, S.; Hirai, S. Radioisotopes 1989, 38, 315-18.
7. Yagi, M.; Masumoto, K. J. Radioanal. Nucl. Chem. 1987, 111, 359-69.

RECEIVED July 16, 1990

Chapter 18

Selenium in Human Serum

Determination by Rapid Wet-Ashing and High Performance Liquid Chromatographic–Fluorometric Method

Eiji Aoyama[1], Terumichi Nakagawa[1], Yoriko Suzuki[1], and Hisashi Tanaka[2]

[1]Faculty of Pharmaceutical Sciences, Kyoto University, Sakyo-ku, Kyoto 606, Japan
[2]Kyoto Pharmaceutical University, Yamashina-ku, Kyoto 607, Japan

A high performance liquid chromatography (HPLC)–fluorometric method has been developed for determining selenium in serum after its rapid acid decomposition in a sealed polytetrafluoroethylene (PTFE) vessel in a microwave oven. The wet-ashing was completed within 1.5 min. Total selenium was recovered as selenite and determined by a novel HPLC–fluorometric method based on the formation of selenotrisulfide. The recovery of selenium was 96% and the relative standard deviation for a standard reference serum was 4.5%. The limit of detection was 25 ng/mL. The serum volume needed was 0.4 mL.

Selenium is known to be both essential (1) and toxic (2) to man. The essentiality of selenium has been recognized by the fact that the element is involved at the active site of glutathione peroxidase. Recent studies (3–5) have suggested that selenium plays a preventive role in the etiology of cancer. Because serum selenium level is a good index of selenium status in mammals (6), its rapid and precise analysis is of practical demand.

As for the determination of selenium, the conventional fluorometry using diaminonaphthalene (DAN) needs intricate operations, though it is a sensitive method (7, 8). Atomic absorption spectrometry with hydride generation (9) and inductively coupled plasma atomic emission spectrometry (10) are also sensitive methods, but they require relatively large volumes of sample and pretreatment to obtain reliable results. Graphite furnace atomic absorption spectrometry (GF–AAS) (11) is a sensitive method and needs less than 0.1 mL of sample and no pretreatment, but severe spectral interferences were caused by iron and phosphorus salts (12–14).

Selenium(IV) reacts selectively with thiols to form selenotrisulfide (STS) involving an –S–Se–S– linkage (15):

$$4RSH + H_2SeO_3 \rightarrow RSSeSR + RSSR + 3H_2O$$

We found that the selenotrisulfide formed by the reaction of selenite with penicillamine was exceptionally stable (*16, 17*), and hence applied the reaction to the selective analysis of selenite. To permit sensitive detection, penicillamine selenotrisulfide was converted into a fluorophore by labeling the amino groups with 7-fluoro-4-nitrobenz-2,1,3-oxadiazole (NBD-F) (*18*). Selenium was determined by HPLC separation of the fluorophore followed by fluorometric detection. This method could easily determine selenium in small (0.2 mL) digested biological and environmental samples (*19*).

In this chapter, we extend the methods to determine selenium in serum samples. Because selenotrisulfide is formed by reaction of selenite with thiol, all of inorganic and organic Se compounds that may bind to proteins must be recovered as selenite by a suitable pretreatment procedure. Selenium is usually liberated from serum by a tedious digestion using a mixed acid solution with a high oxidation power (*3*). Wet-ashing in a microwave oven is a rapid and easy method (*20–23*), but it needs care to avoid loss of selenium as well as possible explosion. In this chapter, we describe improved digestion conditions to facilitate the rapid decomposition of serum. The pretreatment method thus developed was combined with a HPLC–fluorometric method for determining selenium in some standard serum samples.

Experimental Details

Materials. 7-Fluoro-4-nitrobenz-2,1,3-oxadiazole (NBD-F) was purchased from Dojin Laboratories Co., Kumamoto, Japan. D-Penicillamine, selenious acid, and other reagents used were of the highest purity available.

A 3-mg/mL ethanolic solution of NBD-F was freshly prepared and kept ice-cold to avoid degradation. A 0.3-mg/mL aqueous solution of D-penicillamine containing 0.2 M disodium EDTA was prepared daily. A standard solution of 10–100 ng/mL of selenium(IV) was prepared by diluting a 1000-mg/L Se(IV) stock solution (Wako Pure Chemical Industries, Ltd., Osaka, Japan) with 0.5 M HCl. Deionized distilled water was further purified with a Milli-Q system. Good's buffer solution (pH 7.5) was prepared by titration of 0.8 M 3-(*N*-morpholino)propanesulfonic acid (MOPS) solution with 5 M NaOH solution. Freeze-dried standard reference serum samples were obtained from the National Institute of Environmental Studies (NIES), Japan, and from the National Institute of Standard and Technology (NIST), USA. The freeze-dried serum samples were reconstituted by adding water prior to use.

A Hitachi MRO-M55 microwave oven was employed for sample decomposition. A 50-mL PTFE (polytetrafluoroethylene) digestion vessel was purchased from Sanplatec Co. (Osaka, Japan). A Twincle pump (Jasco, Tokyo, Japan) equipped with a Model 7125 sample injector (Rheodyne, Cotati, CA) was used. A column packed with Capcell Pak C18 (250 mm × 4.6 mm i.d., Shiseido Co., Tokyo, Japan) was used at 40 °C. A mixture of acetonitrile, water, and phosphoric acid (400/600/1, v/v/v) containing 10 mM lithium sulfate (pH 2.6) was used as a mobile phase at a flow rate of 1.2 mL/min. A Shimadzu RF-500 spetrofluorometer equipped with a 12-μL flow cell was employed for the detection at an excitation wavelength of 470 nm and an emission wavelength of 530 nm. The chromatogram and the peak area were recorded and processed on a Shimadzu C-R3A Chromatopak.

Recommended Procedure. A 0.4-mL portion of serum was mixed with 1.0 mL of 60% HNO_3 and 0.5 mL of 60% $HClO_4$ in a sealed PTFE vessel. After sonication for 1 min, the vessel was tightly capped and heated in a microwave oven at 500 W for 90 s. After cooling, the digested solution was transferred into a small conical beaker and heated at 130–150 °C to evaporate off the excess reagents and gases until white fumes of $HClO_4$ appeared. A 0.1-mL aliquot of 35% HCl was now added to a beaker and it was kept at 95 °C for 20 min to reduce the selenate to selenite. Water was added to make up the total volume to 2.0 mL.

A 0.2-mL portion of digested solution thus prepared was mixed in a capped glass tube with a 0.1-mL aqueous solution of D-penicillamine (0.3 mg/mL) containing 0.2 M disodium EDTA. The mixture was kept at 90 °C for 7 min and then cooled in ice–water. The solution was neutralized to pH 7.5 by adding an appropriate volume of 5 N NaOH and 0.3 of mL MOPS buffer solution. Then 0.1 mL of the solution was transferred to an aluminum-foil-wrapped tube containing 0.1 mL of NBD-F solution (3.0 mg/mL in ethanol). The tube was capped and kept at 70 °C for 1.5 min. After cooling in ice for 0.5 min, 0.05 mL of 2.4% hydrochloric acid was added to the reaction mixture to stop the reaction. A 25 μL portion of the solution was immediately injected into HPLC. Details of the determination procedure are described in a pervious paper (*19*).

Results and Discussion

Sample Decomposition. Since the present fluorometric–HPLC method can determine selenium in a smaller smaller sample size than DAN–fluorometry, atomic absorption spectrometry with hydride generation (HG–AAS) and inductively coupled plasma atomic emission spectrometry (ICP–AES), the digestion had to be done on a smaller scale (0.4 mL of serum and 1.5 mL of mixed acid). The use of small sample sizes permits safer digestion, because large samples increase the danger of explosion in wet-ashing. A closed vessel was used for microwave wetashing to avoid the loss of selenium by evaporation and damage to the oven by the acid fumes. Because microwave digestion proceeds fast in a mixed acid solution at high temperature and high pressure, the vessel must have mechanical and chemical stability. A durable PTFE vessel, rather than the glass or plastic vessels, was selected for this reason. To tightly seal the vessel, a double inner cap was used. To protect the oven from accidental leakage of acid fumes, the vessel was covered with a plastic cup, and an aspirated exhaust port was installed. The volume of the mixed acid and the digestion time were optimized to arrive at the procedure described above.

Standard Sample. The results for the selenium content of some standard reference serum samples determined by the present methods are shown in Table I, which also includes the results obtained by conventional acid digestion using a miniautoclave followed by the present fluorometric–HPLC method (*19*). Because certified values for selenium contents are not provided by NIST, the results reported in the previous papers (*24, 25*) are cited here. The results by the present method agreed well with these values.

Recovery and Precision. The recovery of selenium was studied by the standard addition method. Known amounts of selenite were added to standard serum solution after digestion. Selenite was also added to the sample before digestion to evaluate the loss during the digestion procedure. The results in Table II indicate quantitative recovery of selenium. This means that selenium can be determined

Table I. Selenium Contents (ng/mL) of Some Serum Standard Samples

Sample	Present Method	Previous Method[a]	Certified Value	Reported Value
NIES No. 4 serum	137 ± 6.3^{b}	150 ± 2.9^{c}	140^{d}	
NIST-SRM-909 human serum	105 ± 6.3^{c}	111 ± 0.4^{c}		106^{e}
				108^{f}
				106^{f}

[a]Digested in a miniautoclave at 130 ° C for 150 min.
[b]$n = 4$.
[c]$n = 3$.
[d]Provisional value.
[e]By AAS (*see* reference 24).
[f]By IDMS and AAS (*see* reference 25).

Table II. Recovery of Selenium from SRM-909[a]
Before and After Digestion

	Se Added[b] (ng/mL)	Se Recovery[c] (%)
Before digestion	25	96 ± 5
After digestion	50	103 ± 6

[a]NIST freeze-dried standard reference serum.
[b]Selenium was added to the NIST sample as H_2SeO_3.
[c]$n = 3$

Table III. Effect of Foreign Ions

Foreign ion	Concentration Added (mg/mL)	Se Recovery (%)
Cu(II)	1.1	101
Zn(II)	1.1	99
Fe(III)	1.1	97

without matrix interferences. Also, no significant loss occurred during the digestion procedure.

The within-run and between-run precisions, expressed as percent relative standard deviation, for selenium in the NIST standard serum were 6 and 4.5, respectively, and can be considered satisfactory.

Effect of Fortified Ions. The selenium recovery in the NIST standard serum was studied in the presence of zinc(II), copper(II), and iron(III), which are found at high concentrations in serum, and also form stable chelates with penicillamine. The results in Table III indicate no influences from added ions. Thus, the present method can reliably determine selenium in serum samples.

The present method could detect at least 25 ng/mL of serum selenium and precisely determine more than 50 ng/mL. A preconcentration method (26, 27), which we developed, permits application of the present method to samples with selenium content less than 50 ng/mL.

Literature Cited

1. Schwartz, K.; Foltz, C. M. *J. Am. Chem. Soc.* **1957**, *79*, 3292–3293.
2. Yang, G.; Wang, S.; Zhou, R.; Sun, S. *Am. J. Clin. Nutr.* **1983**, *37*, 872–881.
3. Das, N. P.; Ma, C. W.; Salman, Y. M. *Biol. Tr. Elem. Res.* **1986**, *10*, 215–222.
4. Schrauzer, G. N.; White, D. A.; Schneider, C. *J. Bioinorg. Chem.* **1977**, *7*, 23–34.
5. Greeder, G. A.; Milner, J. A. *Science* **1980**, *209*, 825–826.
6. Levander, O. A.; Alfthan, G.; Arvilommi, H.; Gref, C. G.; Huttunen, J. K.; Kataja, M.; Koivistoinen, P.; Pikkarainen J. *Am. J. Clin. Nutr.* **1983**, *37*, 887–897.
7. Hiraki, K.; Yoshii, O.; Hirayama, H.; Nishikawa, Y.; Shigematsu, T. *Bunseki Kagaku* **1973**, *22*, 712–718.
8. Shibata, Y.; Morita, M.; Fuwa, K. *Anal. Chem.* **1984**, *56*, 1527–1530.
9. Narasaki, H.; Ikeda, M. *Anal. Chem.* **1984**, *56*, 2059–2063.
10. Kimberly, M. M.; Paschal, D. C. *Anal. Chim. Acta* **1985**, *174*, 203–210.
11. Saeed, K.; Thomassen, Y.; Langmyhr, F. J. *Anal. Chim. Acta* **1979**, *110*, 285–289.
12. Saeed, K.; Thomassen, Y. *Anal. Chim. Acta* **1981**, *130*, 281–287.
13. Jowett, P. L. H.; Banton M. I. *Anal. Lett.* **1986**, *19*, 1243–1258.
14. Subramanian, K. S. *Prog. Anal. Spectrosc.* **1986**, *9*, 237–334.
15. Ganther, H. E. *Biochemistry* **1968**, *7*, 2899–2905.
16. Nakagawa, T.; Hasegawa, Y.; Yamaguchi, Y.; Chikuma, M.; Sakurai, H.; Nakayama, M.; Tanaka, H. *Biochem. Biophys. Res. Comm.* **1986**, *135*, 183–188.
17. Nakagawa, T.; Aoyama, E; Kobayashi, N; Tanaka, H.; Chikuma, M; Sakurai, H.; Nakayama, M. *Biochem. Biophys. Res. Comm.* **1988**, *150*, 1149–1154.
18. Imai, K.; Watanabe, Y. *Anal. Chim. Acta* **1981**, *130*, 377–383.
19. Nakagawa, T.; Aoyama, E.; Hasegawa, N.; Kobayashi, N; Tanaka, H. *Anal. Chem.* **1989**, *61*, 233–236.
20. Abu-Samra, A.; Morris, J. S.; Koirtyohann, S. R. *Anal. Chem.* **1975**, *47*, 1475–1477.
21. Fernando, L. A.; Heavner, W. D.; Gabrielli, C. C. *Anal. Chem.* **1986**, *58*, 511–512.

22. Lamothe, P. J.; Fries, T. L.; Consul, J. J. *Anal. Chem.* **1986**, *58*, 1881–1886.
23. Smith, F.; Cousins, B. *Anal. Chim. Acta* **1985**, *177*, 243–245.
24. Morisi, G.; Patriarca, M.; Menotti, A. *Clin. Chem.* **1988**, *34*, 127–130.
25. Lewis, S. A.; Hardison, N. W.; Veillon, C. *Anal. Chem.* **1986**, *58*, 1272–1273.
26. Aoyama, E.; Nakagawa, E.; Hasegawa, N.; Tanaka, T.; Chikuma, M.; Nakayama, M.; Tanaka, H. *Bunseki Kagaku* **1987**, *36*, 801–803.
27. Nakayama, M.; Tanaka, T.; Tanaka, M.; Chikuma, M.; Itoh, K.; Sakurai, H.; Tanaka, H.; Nakagawa, T. *Talanta* **1987**, *34*, 435–437.

RECEIVED August 6, 1990

SPECIATION   AND   BIOAVAILABILITY

Chapter 19

Speciation Chemistry of Foods
Problems and Prospects

D. J. McWeeny, H. M. Crews, and R. C. Massey

Food Science Laboratory, Food Science Directorate, Ministry of
Agriculture, Fisheries, and Food, Colney Lane, Norwich NR4 7UQ,
England

A 2-stage in vitro enzyme system has been used to simulate
gastric and intestinal digestion of food. Trace element
species in the digest can then be separated chromatographically
using ICP-MS as an on-line specific detector. This has
been used in examination of the cadmium species in kidney,
arsenic species in turkey and aluminium species in tea
infusions. An alternative approach developed in
collaboration with the University of Wales at Cardiff
involves computer modelling to predict the major chemical
forms of the method. The merits and limitations of
the two techniques are discussed.

Not all chemical forms of the inorganic elements are equally
biologically active. Thus methyl mercury is more toxic than
inorganic mercury salts, arseno-betaine is much less toxic than
inorganic arsenic, the biological value of iron in the form of
haem as compared to inorganic salts is well known and cobalt as
cyanocobalamin in liver has important vitamin activity. On the
other hand, in the conventional approaches to trace inorganic
analysis in food it is usual to take particular care to convert
all chemical compounds containing the analyte element into a single
convenient species – and to measure it in this form. In so doing
all information about the original chemical form of the analyte
is deliberately destroyed.
    Two fairly recent developments offer the prospect of being able
to investigate the metals in food in their original chemical form
on a general basis. One is based on an enzymolysis/chromatography/
instrumental approach (1), the other on a thermodynamic/computer
modelling approach (2). The present paper aims to outline both
of these, and to attempt to assess them in terms of their potential
and limitations in two particular respects ie their ability to:
    - work successfully at normal analyte concentrations
    - avoid altering the speciation significantly.

Enzymolysis/Chromatographic/instrumental Approach

This approach is based upon the view that in terms of biological
action the important information relates to the chemical species
which exist in the small intestine - the major point at which they
will be absorbed (if at all) by human beings. From the point of
view of the environmental scientist this may not be true - and
enzymolysis, in which the food is subjected to the digestive
enzymes normally encountered in the gastro intestinal tract - may
be irrelevant.
    The procedure involves a stepwise treatment with pepsin at
acid pH followed by trypsin at neutral pH.
    This technique was used initially to establish whether or not
there are significant differences between foods in terms of their
trace element speciation. This was done simply in terms of the
extent to which the metals are present in a water-soluble form
after treatment with digestive enzymes (1).
    Without going to any sophisticated separation techniques, this
simple treatment gives some indication of the effects which can be
encountered. Thus, the major differences are readily demonstrable
in the solubilities of Cu, Fe, Zn, Cd, Pb in the supernatant from
the enzymic digests of a range of foods(1, 3). In most cases,
only part of the trace element is soluble, but in others the
majority is still bound to the undigested residue - the relative
amounts vary from element to element, and from food to food. Some
further indication of the complexity of the situation involved
comes from a simple additional experiment in which the digest is
spiked with a small amount of the element of interest, in the form
of an inorganic salt. In the case of some food/metal combinations
the added salt and the naturally present element behave quite
differently. This indicates that in the digest the latter is not
present as a simple salt - nor in a form which is in rapid
equilibrium with such salts.
    This line of enquiry has been taken a stage further in
comparisons of the situations in individual foods with those
involving mixtures of foods(4) - the latter being what we normally
eat.
    This work demonstrated clearly that there are differences
when food items are considered in combination eg in terms of the
solubility of Zn and Cd in bread and from crab when digested
separately or in combination; there were also significant
differences between white and wholemeal bread in terms of their
effect upon the zinc from crab. On the basis of these relatively
simple experiments it has been concluded that there are real major
differences between foods in the speciation of the trace elements
they contain. Subsequent experiments have aimed to elucidate the
chemistry of these differences and to separate and identify the
substances involved.
    Where equilibria between species exist it is arguable that it
is not necessary to distinguish between species which equilibrate
in the timescale of a chromatographic separation - this being no
greater than the duration over which species are available for
absorption from the gut. Work based on this premise is described
below.

Chromatographic separation of soluble metallic species is, in
principle, straightforward – one can use gel permeation,
absorption or ion-exchange systems for the purpose – but until
recently detection systems were insufficiently sensitive for on-
line measurements on chromatographic eluates derived from foods
containing normal levels of trace elements, although it was possible
to measure aliquots from eluates by 'off-line' graphite furnace
atomic absorption spectrometry(5). The advent of ICP-MS instruments
has changed that; these instruments have an adequate sensitivity
for many trace elements in food – and moreover, can be used as an
'on-line' chromatography detector(6).
    The potential of this approach to metal speciation in food
digests can be illustrated by some recent studies.

Cadmium Speciation in Pig Kidney

The enzymolysis procedure was employed to examine the state of
cadmium in cooked pig kidney during digestion. Preliminary
experiments on simply buffer extracts of the uncooked pig kidney
revealed the presence of three cadmium binding species whose
SEC retention times corresponded to molecular weights of 1.2×10^6, 7.0×10^4 and $6-9 \times 10^3$ daltons. On cooking, the two higher
molecular weight species were not observed indicating that they
had been heat denatured. However, the lower molecular weight
component was still present and its molecular size and thermal
stability suggested that cadmium was associated with a
metallothionein (MT). On simulated gastric digestion of the
cooked pig kidney all of the cadmium present in solution eluted as
a single peak whose retention time corresponded to that of ionic
cadmium. This strongly suggests that the metal had dissociated
from the binding sites on MT under the acidic conditions. On
subsequent simulated intestinal digestion of the gastric digest
a single peak was again observed but in this case its retention
time corresponded to that of MT. These findings indicate that the
MT protein is resistant to gastrointestinal enzymolysis and is able
to rebind cadmium at the neutral pH at which the intestinal digest
was performed. This observation is consistent with the results
of animal feeding experiments in which orally administered Cd-MT
and ionic cadmium are transported differently following absorption,
the former being deposited in the kidney (the target organ for
cadmium toxicity) and the latter in the liver.
    In these studies cadmium concentrations in the extracts/
digests were in some cases around 10^{-9}g/ml but providing that
stringent precautions are taken with apparatus, reagents and
procedures to avoid contamination during the speciation
investigations, the instrumental sensitivity is adequate for studies
at the levels encountered in 'normal' tissue extracts.

Arsenic in Turkey Muscle

Arsenical drugs are widely employed in intensive poultry production
and this raises the question of the incidence of drug residues in
edible tissue from birds fed medicated rations – and the effect of
withdrawal periods prior to slaughter on these levels. 3-nitro
4-hydrophenyl arsenic acid is widely used in this way and an

analytical procedure broadly similar to that described above has
been used in the study of the arsenic species in muscle from birds
raised on feed containing this substance (Massey, R.C.; Baxter, M.;
Burrell, J.A.; Crews, H.M.; Ebdon, L.; Dean, J.R. unpublished
data). In this case a trypsin digest was used and the chromatographic separation was by reverse phase chromatography with ICP-MS
detection of arsenic at m/e 75. It was possible to demonstrate that
only very low drug residue levels existed in these birds ie.
up to 15 ng/g (as As) in muscle compared with levels <8 ng/g As
(limit of detection) in birds raised on a drug-free ration for a
week before slaughter. In both cases total As levels were around
180 ug/g, so drug residues made up no more than 0.01% of the total
As. Interestingly the chromatogram exhibited a fast-running
(ie., highly polar) species with m/e 75. Investigation showed this
to be due to an interfering molecular ion, $ArCl^+$, formed in the
plasma from chloride ion in the chromatographic eluate.

Aluminium in Tea

Tea leaf infusions typically contain 1-5 mg/l of aluminium. After
enzymolysis SEC/ICP-MS shows that at pH 5.5 the aluminium from a
tea infusion is present mainly as species with molecular weights
of 6000 and 12000 daltons (Massey, R.C.; Crews, H.M. unpublished
data). At higher pH five peaks corresponding to molecular weights
up to 110000 daltons were observed but at pH 2.5 there was only a
single low molecular weight (<2000 daltons) species. The
indications are that at these acidic conditions there may be
equilibrium between the aluminium from the tea leaf and that from
the water supply.

Computer Modelling Studies

The bioavailability of essential elements liberated from food
during digestion depends in part upon their chemical form in vivo.
The concentrations of these trace element species in food are often
so low that analytical determinations can only be carried out using
sophisticated techniques such as those outlined above and which are
not widely available. An alternative approach uses computerised
simulation techniques (2). For instance a protein food may be
considered to be completely broken down into its constituent amino
acids and metal ions which then take part in a series of complex
equilibrium reactions. The computer simulation of these equilibria
uses a thermodynamic approach to determine the concentrations of
up to 1000 different species which may be formed in vivo.
      In these computerised techniques each species present in the
simulation is assumed to be formed as a result of an equilibrium
reaction between metal ions and ligands. A thermodynamic equilibrium
constant can be calculated for this reaction, and once measured,
then the amount of complex which may be formed in aqueous solution
at any given concentrations of metal ions and ligands can be
calculated. Where more than one complex is formed, then a series
of equations can be obtained which, when solved, produce the
speciation profile for the metal under consideration. These
simulations can be carried out at any chosen concentrations and

are still valid at the very low concentrations which present such difficulties to experimental determinations even with modern instrumentation. There is also a more fundamental difference ie., the technique is based entirely on the equilibrium situation – it does not rely on assumptions about the equilibrium situation not being disturbed by the separation.

The nutritional composition of over 900 foods is given in the McCance and Widdowson food tables (7). This database can be interrogated by the computer program, MINCE, which is described by Robb et al(2).

The program MINCE is interactive and requires the user to first specify the number of items in the diet under consideration, up to a maximum of 25 foods.

Once details of all of the food items have been found, the user can direct the program to produce a file for use with the speciation program MINEQL.

Chemical Speciation Program

A number of computer programs exist which can be used to investigate chemical speciation in aqueous solution. In the work of Robb et al (2), a simplified approach was used which assumed that the protein in the food was completely dissociated into amino acids which then interacted with the metal ions found in the digested food; chloride and hydroxide ions were also included in the model. In this work, the program MINEQL was used after slight modification to simulate chemical equilibria in physiological solutions at $37^{o}C$. Up to 1000 different species and their appropriate thermodynamic constants can be considered in the calculations. Iron may be present as either Fe(II) or Fe(III) and simulations could be carried out considering both oxidation states. The modified program MINEQL can calculate the ionic strength of the final aqueous solution and repeat the simulations where necessary to achieve a situation where the ionic strength used in the computations is equal to that calculated on the basis of the predicted speciation profiles.

The use of this approach is illustrated by considering the speciation which may occur after the digestion of 100 g of raw tomato in a digest volume of 1^{-1} at pH 2 and at pH 7.

Each of the simulations carried out using the MINQL program produced a list of over 400 different species which may be formed in vivo. Species which may exist in excess of 1% of the total metal concentration are given in Table I.

The speciation profiles differ markedly in most cases when the pH of the simulated digest is changed from 2 to 7.

Calcium appears to be exceptional in that only relatively minor changes are predicted when the pH is varied. A slight reduction in the amount of Mg^{2+} and $(MgCl)^{+}$ present is predicted when the pH is raised along with a small increase in the amount of amino acid complexes present at the higher pH. Although very small amounts of copper- amino acid complexes are likely to be formed at pH 2, an increase in the amount of these complexes is predicted when the pH is raised. It is important to note that different complexes would be formed at the two pH values.

TABLE I.   COMPUTER PREDICTED SPECIATION PROFILE OF SOME ELEMENTS
IN ENZYMICALLY DIGESTED TOMATO($\underline{2}$)

At pH 2.0		At pH 7.0	
Species	Concentration (umol 1^{-1})	Species	Concentration (umol 1^{-1})
Total Calcium	320.0		
Ca^{2+}	269.0	Ca^{2+}	264.0
$(CaCl)^{+}$	49.3	$(CaCl)^{+}$	49.2
Total Magnesium	450.0		
Mg^{2+}	304.0	Mg^{2+}	294.0
$(MgCl)^{+}$	140.0	$(MgCl)^{+}$	138.0
Total Copper	1.60		
Cu^{2+}	1.20	$(Cu(histidinate)_2)$	1.36
$(CuCl)^{+}$	0.32	$(Cu(aspartate)_2)$	0.10
$(CuH(aspartate))^{2+}$	0.02	$(CuH(aspartate)_2)^{+}$	0.05
$(CuH(threoninate))^{2+}$	0.02	$(CuH(histidinate))^{+}$	0.03
Total Zinc	3.10		
Zn^{2+}	2.09	$Zn(OH)_2$	2.96
$(ZnCl)^{+}$	0.82	$(Zn(aspartate))^{+}$	0.06
$ZnCl_2$	0.14	$(Zn(histidinate))^{+}$	0.03
Total Iron	6.80		
Fe^{2+}	3.15	Fe^{2+}	2.74
$(Fe(II)Cl)^{+}$	3.65	$(Fe(II)Cl)^{+}$	3.23
or		$(Fe(II)(histidinate)^{+}$	0.51
Fe^{3+}	0.77	$(Fe(II)(aspartate)^{+}$	0.18
$(Fe(III)Cl)^{2+}$	3.12		
$(Fe(III)Cl_2)^{+}$	2.13	$Fe(OH)_3ppt$	6.80
$(Fe(III)OH)^{2+}$	0.59		
$(Fe(III)(aspartate)^{2+}$	0.09		

Some zinc amino acid complexes are formed at pH 7 although most of the zinc would be present as the dissolved hydroxide at this pH. The free zinc ion is expected to predominate under acidic conditions. Several species of iron (II) may be formed at pH 7 including small amounts of amino acid complexes. However, there are more iron (III) species predicted at pH 2 than at pH 7. This occurs since ferric hydroxide will readily precipitate in aqueous solution if present at a concentration in excess of 10^{-17} mol l^{-1} at the higher pH. Consequently, for iron (III) to be bioavailable, the iron must be complexed with a suitable ligand in the intestine since precipitated $Fe(OH)_3$ cannot be utilised by the body. In practice, iron (II) and iron (III) will both be present in the digested food. This problem can be solved in the future by measuring the redox potential of digested food and allowing the program MINEQL to determine the amount of each form of iron present in solution. This redox potential cannot be simply predicted since it depends upon the nature of the components in the digested food.

Although, only two pH values are considered in this example the program MINEQL is capable of calculating in speciation profiles of the metals in the digested food at any specified pH.

Future Method Development and Applications

Although this computer modelling approach is in the early stages of development, the results discussed here show the feasibility of using computerised techniques to study chemical speciation in digested foods. However, even though a considerable amount of data can be obtained concerning the nature of the chemical species which may be formed in vivo, a substantial amount of further development is clearly desirable. The food composition database is limited and can be expanded, not only to cover more foods, but also to include more components such as the amount of iron bound in haem compounds, toxic metal levels and other ligands such as inorganic phosphate. The assumption that all of the protein is degraded into simple amino acids will be modified as the chemistry of digestion becomes better understood. Models can then be developed to include polypeptides in the metal-ligand equilibria once the necessary thermodynamic constants have been measured. In addition, secretions in vivo can also contribute to the metal and amino acid concentrations in the digest and these factors should be considered in future studies. A number of other food components, for example, ascorbic acid, may be included in future models to enhance the accuracy of the simulations and the omission of citric acid from the work described above clearly needs to be rectified. It will also be necessary to be able to consider the behaviour of the metals simultaneously rather than separately.

At present, the models represent equilibrium situations and cannot take account of the dynamic situation during absorption from the intestine. This may be a fundamental limitation on the approach in its present form – particularly in attempting to predict bioavailability if the absorbed species is of relatively low abundance but is in rapid equibrium with a more abundant species which is not itself absorbed.

The computerised simulation method is rapid and can readily be used to obtain an assessment of the affect of changes in food composition or, for instance of dietary habits ie., the combination in which individual food items are consumed. For the moment computer models of trace element speciation in food rest on a number of assumptions and the conclusions reached can only be tentative. However, further software development, the experimental validation of proposed models and the use of an expanded food composition database will all enhance the value of this technique in research into the speciation of trace metals in digested foods.

The computerised simulation technique may also be valuable in indicating whether certain species are liable to be present in detectable quantities and may be of great value in targetting the design of analytical procedures.

Assessment

The enzymolysis approach has been valuable but is clearly inadequate in some respects – in particular it takes no account of the dynamic equilibria between species which exist in the digestive tract. Thus it may undervalue the role of any relatively minor species which is absorbed rapidly and is also in rapid equilibria with other species – overall uptake will be related to the sum of the species in rapid equilibria with the absorbed species.

The chromatographic systems when used in conjunction with ICP-MS are now capable of providing measurements at concentrations typical of those normally present in food – but may encounter difficulties when dealing with species which equilibrate quickly ie., within the timespan associated with chromatographic separations.

The computer models are potentially extremely powerful – not least in allowing an almost instant assessment of the changes which might follow from the inclusion of an additional food component in a digest. This ability to assess the implications of dietary composition may find many applications. However, like the other approaches, the present generation of computer models consider only equlibria situations – and do not allow consideration of the dynamics of the absorption process.

Imperfect though these tools may be – they now allow progress to be made where none was possible before ie., in terms of the development of a greatly increased information about the chemical forms of trace elements present in food digests. However, this is only one step towards having an assessment of bioavailability. The study of soluble species can be extended to that of species which are capable of crossing biological membranes – but it will still be necessary to distinguish between those which are utilised and those that are rapidly excreted unchanged. Ultimately it will also be essential to consider the biological control mechanisms which determine whether or not an intrinsically absorbable species is actually absorbed. The semantics surrounding the terms 'speciation' and 'bioavaiability' in relation to trace elements and in food look set for a long run (8) – but at least it is now possible to provide some experimental information for what has hitherto been a rather hypothetical discussion.

Conclusion

At present, work on metal speciation in foods in relation to
bioavailability relies upon imperfect tools – but in combination
they point to the clear conclusion that the chemical forms of trace
elements in food can vary widely and that, in terms of the chemical
situations encountered in the human digestive tract, it is important
to consider foods in the combinations in which they are consumed –
rather than as separate individual food items.

Literature Cited

1. Crews, H.M.; Burrell, J.A.; Mcweeny, D.J. J. Sci. Food Agric.
    1983, 34, 997–1004.
2. Robb, P.; Williams, D.R.; McWeeny, D.J. Inorg. Chim. Acta.
    1986, 125, 207–212.
3. Crews, H.M.; Burrell, J.A.; McWeeny, D.J. Z. Lebensm. u. Forsch.
    1985, 180, 221–226.
4. Crews, H.M.; Burrell, J.A.; McWeeny, D.J. Z. Lebensm. u. Forsch.
    1985, 180, 405–410.
5. Massey, R.C.; Burrell, J.A.; McWeeny, D.J.; Crews, H.M.
    Toxicol. Env. Chem. 1986, 13, 85–93.
6. Crews, H.M.; Dean, J.R.; Ebdon, L.; Massey, R.C. Analyst 1989,
    114, 895–899.
7. Paul, A.A.; Southgate, D.A.T. McCance and Widdowson's, The
    Composition of Foods; Her Majesty's Stationery Office,
    London, 1978.
8. Irgolic, K.J. Chem. Spec. Bioavail. 1989, 1, 127.

RECEIVED July 16, 1990

Chapter 20

Biological Reference Materials for Metal Speciation

National Institute for Environmental Studies Fish Tissue Reference Material for Organotin Compounds

K. Okamoto

National Institute for Environmental Studies, Environmental Agency of Japan, Onogawa 16–2, Tsukuba, Ibaraki, 305, Japan

A fish tissue reference material for use in the speciation of total tin, tributyltin (TBT) and triphenyltin (TPT) was prepared at the National Institute for Environmental Studies (NIES), Japan. Fish fillets of sea bass were minced, freeze-dried, ball-milled, homogenized and finally vacuum-packaged, together with an oxgen absorber, into polyethylene laminate bags. TBT and TPT in the reference material were determined by gas chromatography with flame photometric detection (GC–FPD) following methanolic HCl/ ethyl acetate extraction, cleanup through anion and cation exchange columns, and alkylation with a Grignard reagent. The GC–FPD method provides for very sensitive and selective determination of the organotin compounds. The certified value for TBT in the fish tissue reference material was determined to be 1.3 ± 0.1 µg/g as its chloride, based on the results of determinations provided by collaborative analyses. For the determination of TPT, however, there were analytical problems related to extraction, cleanup, measurement, etc. Development of an isotope dilution analytical method for TPT is under way.

Coastal water pollution with organotin compounds has recently become one of the most serious environmental problems in Japan. Organotin compounds such as tributyltin (TBT) and triphenyltin (TPT) have been used as antifouling paints for fish nets and ship's hulls. Organotin compounds have an adverse effect on the marine environment, particularly on the growth of bivalves and molluscs. The Environment Agency of Japan has been conducting surveys for four years on the concentrations of organotins in seawater, marine

0097–6156/91/0445–0257$06.00/0
© 1991 American Chemical Society

sediments and marine organisms. Figure 1 shows, for example, the
results of biomonitoring surveys conducted in 1986 and 1987 on TBT
concentrations in sea bass (1). Although the use of TBT on fish nets
in fish farms was prohibited about 3 years ago, it can be seen that
a relatively high level of TBT has still been detected in the fish
samples.
    A 1988-survey on organotin concentrations revealed that coastal
water pollution by TPT is a more serious problem than pollution by
TBT. Table I shows the results of the 1988-survey on the TPT concentrations in seawater, sediments and fish tissues (2). Relatively high
levels of TPT were detected from more than 50% of the samples of
seawater and sediments collected from the bays, and inland seas of
Japan. Such high levels can be attributed to the accumulation of
lipophilic TPT by fish (sea bass and several other species). These
TPT levels are usually about 2-3 times higher than TBT levels. It is
likely that the production and use of TPT, which has been used as an
antifoulant for ship's hulls, will shortly be banned by law in Japan;
however, the monitoring of TBT and TPT in the marine environment
must be continued.

Table I. Triphenyltin Concentrations in the Coastal Environment.
        Results of Triphenyltin (TPT) Monitoring in 1988
              by the Environment Agency of Japan (2).

Sample	Total Sample Number	TPT Detected (% Samples)	Concentration Range	Median	Arithmetic Mean
Seawater	119	61	0.005–0.088 µg/l	0.009 µg/l	0.015 µg/l
Sediment	129	77	0.001–1.1 µg/g^a	0.01 µg/g^a	0.048 µg/g^a
Fish tissue	144	82	0.02–2.6 µg/g^b	0.17 µg/g^b	0.39 µg/g^b

^{a} dry weight basis; ^{b} wet weight basis

    The two most common techniques for TBT determination are gas
chromatography with flame photometric detection (GC–FPD) and hydride
generation-atomic absorption spectrometry (HG–AAS). Both methods
require fairly extensive sample workup including extraction, cleanup
and derivatization. These analytical procedures invariably need skill
and experience, and therefore it is difficult to obtain accurate
analytical results for TBT in natural matrix materials. The Environment Agency of Japan has been using a TBT–ethanol solution to conduct
an interlaboratory comparison exercise by 80 participating environmental laboratories. Large discrepancies in the analytical results
were reported even for the "pure" TBT solution (3). High levels of
TPT in fish tissues were found by Takami, et al. (4) using the GC–FPD
technique. A careful cleanup procedure, using anion and cation
exchange columns (4) or a Florisil column (5), was required to
efficiently remove the lipids, salts and HCl remaining in the extract;
if not removed, these interferences would give erratic results in the
subsequent GC–FPD, or GC–ECD (electron capture detection) determination of TBT and TPT. The development and evaluation of analytical
methods for the simultaneous determination of TBT and TPT are
under way in our laboratory (6).

The most practical way to maintain the accuracy and precision of analytical results is through the use of appropriate certified reference materials (CRMs). With respect to organotin compounds, the marine sediment reference material PACS-1 issued by the National Research Council of Canada is the only currently available CRM certified for TBT, dibutyltin and monobutyltin (7). Metal speciation studies invariably require the development of so-called "hyphenated" methods, which combine a separation method and a detection technique in sequence, to determine the metal-containing compounds present usually at ppm (mg/kg) level or less in a complex matrix. Recent advances in GC, high performance liquid chromatography (HPLC), mass spectrometry (MS), and inductively coupled plasma mass spectrometry (ICP-MS) have made it possible to determine the chemical forms of elements in biological materials on a routine basis if standard materials are available. Preparation of biological reference materials for use in metal speciation studies is difficult because, in general, metal-binding compounds may not be stable during long-term storage. Depending on the type of samples and on the compound(s) to be measured, studies on sample homogenization, drying, packaging, preservation conditions, etc. should be performed.

Since 1975, NIES has been active in the preparation and certification of biological and environmental reference materials to serve the needs of scientists engaged in trace element analysis. The currently available CRMs from NIES include Pepperbush (8), Pond Sediment (9), Chlorella, Human Hair (10), Mussel (11). Tea Leaves(12) Vehicle Exhaust Particulates (13), Sargasso (14), and Rice Flour-Unpolished. Certified and reference values are given for "total" amount of elements in the NIES CRMs. NIES has recently initiated the preparation and certification of biological and environmental reference materials for use in metal speciation studies. A typical example is the preparation of a Fish Tissue reference material (NIES No.11) for the determination of total tin, TBT and TPT. This reference material is certified for total tin content with a value of 2.4 ± 0.1 μg/g. Details of analysis of total tin in this material by isotope dilution/ICP-MS will be published elsewhere (15). In this paper the preparation of Fish Tissue reference material and the determination of TBT and TPT by GC-FPD are described.

Preparation of Fish Tissue Reference Material

Sea bass (Leteolabrax japonicus) was selected because, along with mussel (Mytilus edulis), it is being routinely used in the biomonitoring surveys of organotins conducted by the Environment Agency of Japan. Sea bass is known to accumulate both inorganic and organic contaminants.

About 150 kg of sea bass collected in early July 1988 in Tokyo Bay was used. The fish samples (body length 50-60 cm, individual weight 1-2 kg) were stored in a liquid nitrogen jar until use. After thawing, the fish samples were dissected with a stainless-steel knife and only the fish fillets (about 50 kg) were used for further processing. These were cut into small pieces and a batch of about 2kg minced at 1,500 rpm for 3 min with a Tecator 1094 homogenizer. After repeating this mincing procedure for the remaining batches, the fish tissue homogenate was combined and freeze-dried in one lot. The

dry fish tissue (about 12 kg) was alumina ball–milled for 1 h and
then mixed in a V–blender for 2 h. Twenty gram samples of the homogenized fish tissue were packaged into polyethylene bags, which were
subsequently vacuum–packed doubly into polyethylene laminate bags
together with an oxgen absorber. The prepared NIES Fish Tissue reference material (600 samples, 20 g each) are currently stored at -20°C.

Determination of Tributyltin and Triphenyltin

TBT and TPT concentrations in the fish tissue reference material were
determined by GC–FPD, according to the method of Takami, et al.(4,16).
Sample preparation included extraction with methanolic HCl/ethyl
acetate, cleanup through anion and cation exchange cartridge columns,
and alkylation with a Grignard reagent.

Extraction. A sample (1 g) of the fish tissue reference material was
placed in a 200 ml separatory funnel and shaken with 50 ml of 1M HCl-
methanol/ethyl acetate (1:1, v/v) for 30 min. After filtration through
a Buchner funnel, the residue on a filter paper (Toyo No.5) was
extracted again with 20 ml of the 1M HCl–methanol/ethyl acetate.
One hundred ml of 10% sodium chloride was added to the combined
filtrates and then the organotin compounds were twice extracted with
two portions of 30 ml of ethyl acetate/hexane (3:2, v/v). Hexane
(100 ml) was added to the pooled extract, and the mixture was shaken
and allowed to stand for 30 min to allow the aqueous phase to separate.
After drying the organic layer with anhydrous sodium sulfate, the
solvent was concentrated to approximately 1 ml by a rotary evaporator
at 40°C. The solution was further concentrated to near dryness by
blowing nitrogen gas and then diluted to 10 ml with ethanol.

Cleanup. The ethanol solution was passed sequentially through anion
exchange (Bond Elut SAX, 500 mg) and cation exchange (Bond Elut SCX,
500 mg) cartridge columns, which were subsequently washed with 20 ml
of ethanol. The anion exchange cartridge, which adsorbed lipids and
salts remaining in the hexane/ethyl acetate layer, was removed. The
TBT and TPT adsorbed on the cation exchange cartridge was eluted with
15 ml of 1M HCl–methanol. After the addition of 15 ml of 10% sodium
chloride, the eluent was twice extracted with two portions of 2.5 ml
of hexane/cyclohexane (1:1, v/v). The extracts were combined and
concentrated to approximately 1 ml under a stream of nitrogen.
    This cleanup procedure is very efficient at removing lipids from
the fish tissue extracts and at eliminating salts and other extraneous
material. Utilization of this cleanup procedure is necessary when
analyzing fish which can have a high lipid content.

Alkylation. Two ml of propylmagnesium bromide (2M in tetrahydrofuran)
were added dropwise to the concentrate. The mixture was allowed to
stand for 30 min at 40°C and the excess of reagent was destroyed by
careful addition of 10 ml of 1M sulfuric acid. After the addition of
10 ml of methanol and a suitable amount of hexyltributyltin as an
internal standard, the TBT and TPT were extracted with two portions
of 2 ml of hexane. The extracts were combined and made up to 7 ml
with hexane.

GC–FPD. A Shimadzu GC–7A gas chromatograph equipped with a flame photometric detector (FPD) was used for the determination of TBT and TPT. The FPD was equipped with a 600 nm filter to monitor the SnH molecular emission. Chromatographic separations were carried out on a glass capillary column Ultra–1 (crosslinked methylsilicon gum, 25 m x 0.32 mm i.d. x 0.52 µm). Column temperature was programmed at 80°C for 4 min and then heated to 250°C at 8°C/min. The detector temperature was maintained at 270°C and the injection port was at 240°C. Sample aliquots of 2 ul were injected at room temperature in the split mode.

Figure 2 shows a typical GC–FPD chromatogram of the propyl derivatives of TBT and TPT extracted from the fish tissue reference material, together with hexyltributyltin added as an internal standard. As can be seen, the FPD used provides for very selective tin detection with no interferences of any significance. This method showed excellent sensitivity for the organotin compounds. The determination limits of the method were 0.01 µg/g for TBT and 0.02 µg/g for TPT, when 1 g of the fish tissue reference material was analyzed. Analytical values of 1.26 ± 0.06 µg/g as tributyltin chloride and 6.2 ± 0.4 µg/g as triphenyltin chloride in the reference material were obtained in our laboratory.

Collaborative Analysis

A collaborative study on the analysis of the fish tissue reference material for TBT and TPT concentrations was carried out by 9 participating laboratories. These laboratories have been engaged in the determination of organotin compounds in biological and environmental samples. Table II summarizes the results of the collaborative analyses of TBT in the fish tissue reference material. The TBT was determined by GC–FPD, GC–ECD and GC–MS as its chloride, hydride or propyl derivative. Analytical values for TBT showed good agreement (Table II). The certified value for TBT in the fish tissue reference material was determined to be 1.3 ± 0.1 µg/g as tributyltin chloride, based on the results of determinations by three different analytical methods.

Table II. Determination of Tributyltin in NIES Fish Tissue Reference Material: Results for 9 Laboratories

| Lab. | Tributyltin chloride (µg/g) | | |
	Content	Method	Derivative
1	1.30 ± 0.06	GC–FPD	propyl TBT
2	1.31 ± 0.04	GC–ECD	TBT hydride
3	1.30	GC–ECD	TBT chloride
4	1.30 ± 0.09	GC–FPD	propyl TBT
5	1.4	GC–FPD	propyl TBT
6	1.07 ± 0.03	GC–FPD	propyl TBT
7	1.18 ± 0.02	GC–FPD	TBT hydride
8	1.12 ± 0.07	GC–ECD	TBT chloride
9	1.26 ± 0.06	GC–FPD	propyl TBT
9	1.24 ± 0.07	GC–MS	TBT hydride

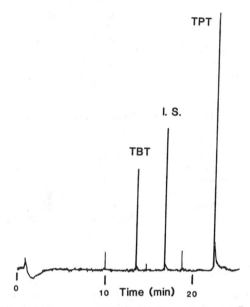

Figure 1. Tributyltin concentrations in sea bass. Results of
biomonitoring survey in 1986 and 1987 by the Environment Agency
of Japan.

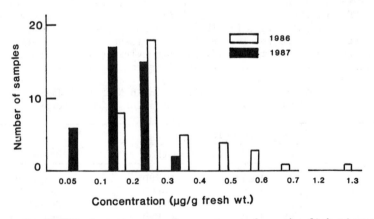

Figure 2. GC–FPD chromatogram of an extract from the fish tissue
reference material. TBT: propyl tributyltin, TPT: propyl triphenyltin, I. S.: hexyl tributyltin (0.6 ng) added as the internal
standard.

Analytical values for TPT by GC–FPD and GC–ECD, however, showed considerable disagreement among 9 laboratories. Values ranging from 4.2 to 6.4 µg/g for TPT as its chloride were reported. A number of factors can account for this variability. For example, the recovery of TPT from fish homogenates greatly depends on extraction conditions, particularly on a combination of extraction solvents. Use of methanolic HCl/ethyl acetate, however, gave quantitative recovery of TPT from fish tissues (4, 5). Also, TPT easily decomposes during the sample preparation procedure if excess mineral acids are present. Therefore, rapid and efficient removal of the excess of HCl and H_2SO_4 from the extracts are required. We hope to overcome these problems by the development of an isotope dilution technique using GC–MS and HPLC–MS in conjunction with a deuterated triphenyltin (TPT–d_{15}) compound.

Conclusion

The NIES Fish Tissue is the only currently available "biological" reference material certified for TBT content. Certification of TPT in this reference material is urgently required for analytical quality assurance in biomonitoring surveys of TPT in the marine environment. In addition to the analytical results obtained by GC–FPD and GC–ECD, the use of an isotope dilution technique in conjunction with GC–MS and HPLC–MS would hopefully provide accurate results, and may lead to the certification of TPT. NIES Fish Tissue reference material will be of practical use for quality assurance, development of analytical methods and calibration of instruments in marine and environmental laboratories engaged in TBT and TPT analysis.

Acknowledgments

The author is grateful to the following collaborating laboratories: K. Imaeda (Hoshi University), M. Takeuchi (Tokyo Metropolitan Research Laboratory of Public Health), K. Takami (Environmental Pollution Control Center of Osaka), K. Shinohara (Kitakyushu Environmental Health Center), Y. Shirane (Hiroshima Prefectural Research Center for Environmental Science), S. Hori (Osaka Prefectural Institute of Public Health), K. Ozaki (Niigata Prefectural Research Laboratory for Health and Environment), Y. Tsuda (Shiga Prefectural Environmental Research Center), T. Yamamoto (Osaka City Institute of Public Health and Environmental Sciences), and K. Baba (Nagasaki Prefectural Institute of Public Health and Environment Science).

Literature Cited

1. Environment Agency of Japan, Sangyo Kougai 1988, 24, 56–59.
2. Environment Agency of Japan, Sangyo Kougai 1989, 25, 715–717.
3. Environment Agency of Japan, Report on Interlaboratory Comparison Exercise 1989, pp 89–104.
4. Takami, K.; Okumura, T.; Yamazaki, H.; Nakamoto, M. Bunseki Kagaku 1988, 37, 449–455.
5. Takeuchi, M.; Mizuishi, K.; Yamanobe, H.; Watanabe, Y.; Michiguchi M. Bunseki Kagaku 1989, 38, 522–528.
6. Kadokami, K.; Uehiro, T.; Morita, M; Fuwa, K. J. Anal. Atom. Spectromet. 1988, 3, 187–191

7. National Research Council Canada, MACSP Marine Sediment
    Reference Material PACS-1 Certified for Butyltins 1989.
8. Okamoto, K. Res. Rep. Natl. Inst. Environmental Studies; 1980,
    No. 18.
9. Okamoto, K. Res. Rep. Natl. Inst. Environmental Studies; 1982,
    No. 38
10. Okamoto, K.; Morita, M.; Quan, H.; Uehiro, T.; Fuwa, K. Clin.
    Chem. 1985, 31, 1592–1597.
11. Okamoto, K.; Fuwa, K. Analyst 1985, 110, 785–789.
12. Okamoto, K.; Fuwa, K. Fresenius Z. Anal. Chem. 1987, 326,
    622–626.
13. Okamoto, K. Anal. Sci. 1987, 3, 191–192.
14. Okamoto, K. Marine Environ. Res. 1988, 26, 199–207.
15. Okamoto, K. Spectrochim. Acta (in press)
16. Takami, K.; Okumura, T.; Yamazaki, H.; Nakamoto, M. Bunseki
    Kagaku 1988, 37, 117–122.

RECEIVED July 16, 1990

Chapter 21

Trace Elements Associated with Proteins

Neutron Activation Analysis Combined with Biological Isolation Techniques

Susan F. Stone,[1] Rolf Zeisler,[1] Glen E. Gordon,[2] Raphael P. Viscidi,[3] and Erich H. Cerny[4]

[1]National Institute of Standards and Technology, Center for Analytical Chemistry, Gaithersburg, MD 20899
[3]Department of Chemistry and Biochemistry, University of Maryland, College Park, MD 20742
[3]The Eudowood Division of Infectious Diseases, Department of Pediatrics, Johns Hopkins University School of Medicine, Baltimore, MD 21205
[4]Fondation pour la Recherche Medicale, University of Geneva, 64 Avenue de la Roseraie, 1211 Geneva 4, Switzerland

Combinations of biological techniques with neutron activation analysis (NAA) can be employed for protein quantification in biological samples. A protein of interest can be isolated physically by a high resolution separation technique such as polyacrylamide gel electrophoresis (PAGE) or by an immunochemical technique, such as a specific antibody-antigen reaction. NAA is then used for quantification of the isolated protein by determining either an element that is structurally intrinsic, or one that has been introduced as a label. An example of the first method is the determination of phosphoproteins by PAGE-NAA. This method is discussed and results for two phosphoproteins, α-casein and phosvitin, are presented for several materials, including an α-casein concentration of 26 mg/mL in low-fat milk. The second method presented is an immunoassay with a colloidal gold tag and detection by NAA.

Neutron activation analysis (NAA) has been used to determine trace element concentrations in many types of biological samples. We have employed NAA and related nuclear methods for the analysis of a wide variety of biological samples, ranging from the determination of more than twenty trace elements in human livers (1) and 45 elements in bivalve samples (2), to the determination of ultratrace concentrations of Pt in human livers (3) and Cr in blood (4). It is also possible to quantify specific macromolecules in biological samples by combining NAA with protein "isolation" techniques. We describe here the application of NAA combined with two such techniques: polyacrylamide gel electrophoresis

0097–6156/91/0445–0265$06.00/0

(PAGE), which gives spatial protein separation, and an enhanced immunoassay with colloidal gold, which utilizes the specificity of an antibody-antigen reaction to isolate the desired analyte.

PAGE-NAA For Determination of Intrinsic Elements

In contrast to previous work, where the emphasis was on the determination of trace elements associated with isolated proteins on an individual basis, we have developed a method to simultaneously analyze elements in proteins that have been isolated from complex protein mixtures. The method was used to determine concentrations of elements associated with specific proteins and thus determine the concentrations of the proteins. The sharp resolution of PAGE, combined with the sensitive detection of NAA, was used to detect and quantify proteins in complex samples. The feasibility of the PAGE-NAA method was shown by our analysis of covalently bound P in phosphoproteins (5).

The advantages of PAGE-NAA can be summarized as follows:

(1)   A protein sample is separated into its components by electrophoresis on a single polyacrylamide gel matrix. This takes advantage of the high resolution and simultaneous multi-component analysis that is possible with this system.

(2)   The whole gel is subjected to neutron activation; hence, each activatable element contained in the gel matrix, in principle, can be detected, so the method is not limited to selectively introduced labels. Since the analytical separation is done prior to neutron bombardment, and the positions of the proteins are fixed in the gel, the Szilard-Chalmers effects induced by recoil from prompt γ-ray emission do not affect this technique. The separated proteins are visualized via the activated element (in this case, ^{32}P) after exposure to x-ray film and subsequent development, yielding an autoradiograph of the activated gel with its separated proteins.

(3)   An additional dimension for identification is added to the usual molecular weight parameter of PAGE, as the intrinsic elements are specific for certain proteins. This relationship makes the technique superior to staining techniques.

(4)   Quantitative assay of proteins is possible from determination of the associated trace elements if the protein-element stoichiometry is known. The areal densities, measured as absorbance readings of the visualized areas,

are proportional to the mass of the element
present in the separated band and/or standard. By comparing the areal densities of
standards, which contain known amounts of the
element, with the areal densities of the
separated proteins, the mass of the element
in the protein band is determined. If the
element-protein stoichiometry of the particular protein is well-characterized, the mass
of the protein can be assayed.

<u>Materials</u>. Acrylamide, N,N',N'-methylenebisacrylamide
(bis), (both analytical grade reagents) and TEMED were
obtained from SERVA Fine Biochemica (Westbury, NY).
Glycine, tris hydroxymethyl-aminomethane (Tris), ammonium persulfate, phosvitin, α-casein and sodium
caseinate were purchased from Sigma Chemical Company
(St. Louis, MO), and sodium dodecyl sulfate (SDS),
from Pierce Company (Rockford, IL). The two reference materials used were the National Institute of Standards and Technology Standard Reference Material (SRM)
1845, Cholesterol in Egg Powder, and the International
Atomic Energy Agency (IAEA) Certified Reference
Material (CRM) A-11, Milk Powder. Ammonium dihydrogen
phosphate (purity 99.0%) was from Baker. All water
used to prepare the gels and buffers was purified
through a Nanopure triple exchange-column system. To
minimize contamination, all preparation and sample handling were done under a Class 10 (< 10 particles per
cubic foot) laminar flow clean hood (Environmental Air
Co.). The Hoefer series 600 unit for slab gels was
used for the electrophoresis. The filter papers used as
gel backings were either Whatman 41 or Bio-Rad filter
paper. The X-ray film used was Kodak X-OMAT AR ™
film, 20- x 25-cm (8 x 10- in.), and the developing and
fixing chemicals were the Kodak GBX brand. An LKB 2202
UltroScan laser densitometer (LKB-Produckter, Bromma,
Sweden) equipped with a helium-neon laser source (632.8
nm) was used to scan the areas on the developed
autoradiographs.

<u>Method</u>. Samples were separated by gradient polyacryl-
amide gel electrophoresis in a discontinuous buffer
system (<u>6</u>). Polyacrylamide slab gels, 15 x 15 x 0.15
cm, were prepared in Tris/glycine buffers with a stacking gel of 5% T and 2.6% C (where T = wt/vol % of
monomers, and C = g bis/100 g (acrylamide + bis)) and
a resolving gel of 10 - 17% T and 2.6 % C. The
gradient gels were prepared using a gradient maker
with a peristaltic pump. The final molar concentrations of Tris were 0.375 M (pH 8.8) in the separating gel and 0.125 M (pH 6.8) in the stacking gel. The
concentration of SDS in both gels was 0.1%. The gels
were polymerized chemically by the addition of 0.025%
TEMED (by volume) and 0.02% (wt/vol %) ammonium persul-

fate.   Duplicate gels, one for irradiation and one for
non-specific protein staining were cast simultaneously.
     Protein samples, generally obtained as lyophilized pow-
ders, needed to be dissolved in sample buffer prior to application to the gels.   The Tris/glycine sample buffers
also contained the detergent SDS (10 %, wt/vol)   and a
reducing agent, such as 2-mercaptoethanol or dithio-
threitol, to   achieve uniform, rod shaped detergent-
polypeptide complexes.   Ethylenediamine tetraacetic acid
(EDTA, sodium salt) was also included in the sample buffer to chelate Ca^{2+},   which promotes formation of casein
aggregates or micelles.   The prepared sample solutions
contained either approximately 1 mg/mL of the isolated
proteins or 2 mg/mL of the lyophilized composites.   The
dairy products were sampled directly as liquids and
diluted 1:1 in a sample buffer containing 50 mM Tris, 0.3
mM EDTA, 3% SDS and 12.5% mercaptoethanol ([7]).   Sample
loadings ranged from 2 µL to 50 µL, the maximum possible
loading in the electrophoresis system employed.
     Electrophoresis was run at constant amperage, between
15 and 20 mA per gel, for 6-8 hours, until the leading
buffer line was close to the bottom of the gels.   The
gels were cooled to 12-15 °C during the electrophoresis
run   by   circulating   cold   water   from   a   refrigerated
circulator   through   the   heat   exchanger   on   the
electrophoresis unit.   Two identical gels were run simultaneously so that the PAGE-NAA results could be compared
with standard protein detection methods such as Coomassie
Brilliant   Blue™ staining.
     Following   electrophoresis,   the   separated   proteins
were fixed in the gels by   soaking in 50% ethanol / 0.1%
formaldehyde, usually overnight.   One gel was stored in
20%   ethanol prior to drying for irradiation,and the
other gel was stained with Coomassie Brilliant   Blue™.
If detergent (SDS) removal was needed, the procedure of
Olden and Yamada, as modified by Schibeci and Martonosi
([8]), was followed.
     Prior to drying and irradiation, all gels were condi-
tioned with 10% acetic acid/ 1% glycerol.   They were
then vacuum-dried on a low background filter paper at
80 °C for two hours.   Each gel was individually packaged
in linear polyethylene (a   high pressure extruded film of
polyethylene made from virgin materials, not treated with
catalysts), and then all gels were wrapped together in
another polyethylene bag for irradiation.
     The phosphoproteins phosvitin (10.0 % P) and α-
casein (1.10% P) were selected because of their high P
content (relative to molecular weight) and because they
were commercial available in purified form (>90%).   A
commercially available phosphoprotein composite, sodium
caseinate, as well as two reference materials, SRM 1845,
Cholesterol in Egg Powder, and   IAEA CRM A-11, Milk Powder, were also studied.   Finally, to establish the
method for analysis of natural   samples, several dairy
products were analyzed for their casein content: low-fat

(2%) milk, homogenized milk, and "half-and half" (50% regular milk, 50% cream).

Irradiations were performed in a specially designed irradiation cassette at the University of Virginia reactor in Charlottesville, VA (5). The container consists of an aluminum cassette, which is placed underwater in the pool, adjacent to the reactor core. A continuously circulated dry gas (N_2) surrounds the gel package in the sealed cassette and provides positive pressure to reduce the possibility of water leakage into the cassette. The cassette is 20 cm square and holds a sample stack up to 3 cm in thickness. The reactor was operated at half power (1 MW) at a fluence rate of $1.7 \cdot 10^{13}$ $n \cdot cm^{-2} \cdot s^{-1}$, since at full power, γ-ray heating damaged the gels so much that they were unusable. Irradiation times ranged between 4 and 6 hours.

Following a 9-10 day decay period, autoradiographs were made by exposing the gels to X-ray film. The autoradiographs were usually done by direct exposure in cardboard cassettes. Exposure times varied from one day to three weeks, depending on the type of protein and the amount of protein that was applied.

A standard process for film development was followed. The film was developed for 5 min, with gentle agitation every minute for approximately 15 seconds, rinsed in distilled water for 30 sec, and fixed for 3 min with constant agitation. The film was rinsed in running distilled water for 5 min, then dried in air for at least 1 hr. Film was stored on flat surfaces in protective plastic covers.

A laser densitometer was used to scan the areas on the developed autoradiographs. Each band was scanned across the mid-section and through the length of the band at a scanning speed of 20-40 cm/min. Peak areas were integrated on an Apple IIe computer that was interfaced to the densitometer using the LKB 2190 Gel Scan software. The peak areas are given in absorbance units (au), where a full scale peak with a full width at half maximum of 1.0 has an intensity of 1000 units. These absorbance units have been related to standard calibrated glass filters (9).

The peaks were integrated over a selected background with a fitted Gaussian curve and the area under the curve was then calculated. If it was not possible to describe the peak with a single curve, several Gaussian curves were fitted and the integrated areas were summed. The quoted accuracy level of the Gaussian fit method is 5% from the GelScan program documentation (10). The accuracy of the software fit is visually demonstrated by reconstructing the calculated curve and comparing it to the actual scanned absorbance data as an overlay. Peak fits can also be done manually, with results virtually identical to the software fit.

Ammonium dihydrogen phosphate solutions were prepared

as P standards. Dilutions were prepared from a primary
stock solution and ranged from 20 ng P/mL to 1000 ng
P/mL. Two µL of each solution was pipetted onto the dried
gels containing the separated proteins to be irradiated.

Results. Phosphorus standards were applied to the gels
before irradiation and autoradiographs were developed
from the irradiated gels following the described procedures. Since the proteins and the P standards were on
the same gel, there was no concern about variations
between standards and samples in terms of exposure times
to the x-ray film or times required for film development. Absorbance measurements were made of both the
resultant spots from the ^{32}P of the activated standards
and the protein bands, and calibration curves were
plotted for each set of standards. In the autoradiograph, the absorbance of the total area is proportional
to the mass of P. The absorbance of the total area,
which can be called the areal density, must be obtained
in order to compare the standards and samples because of
the different shapes of the bands and spots. The spots
are cylindrically symmetric with an intensity distribution that is gaussian about the center of the circle.
The bands are symmetric across the major axis with a
Gaussian distribution about that axis.

A representative standard calibration curve of total
areal densities versus mass of P in ng is given in
Figure 1. The mass of P is obtained by comparing the
areal density of an unknown to these areal densities of
standards with known P masses, on the same
autoradiograph. The mass of protein corresponding to the
mass of P can then be calculated from the stoichiometry
of the particular phosphoprotein being analyzed. For instance, there is 1.10% P in the phosphoprotein α-casein,
so one would divide the mass of P by 0.0110 to obtain
the mass of α-casein.

The method was first tested with commercially
available, isolated phosphoproteins. The results of the
PAGE-NAA method applied to previously isolated phosphoproteins also yielded absorbance values proportional
to the mass of sample applied (5). Replicate aliquots of
an α-casein solution were separated on a number of gels
and the gels were then analyzed to assess the reproducibility of the method. The test was sensitive to errors
due to sample loading, fluence corrections which were
based on activities of Fe foils placed at various locations in the cassette, and estimates of areal density
for a band (9). A comparison of results from two gels is
shown in Table I.

The mass of α-casein in the original protein
solution was used to calculate mass of P that was
applied. This calculated mass was compared with the experimental results from the PAGE-NAA method. A 0.96 mg
solid/mL buffer solution was prepared, with the α-casein

material certified by the manufacturer to have at least a 90% purity, so the expected mass of P was 47.5 ng P (assuming 90% protein), at a 5 mL application, and a P content of 1.10%. Due to the presence of some of the reagents in the buffer solution used to dissolve the protein, such as Tris, mercaptoethanol, SDS, and EDTA, several different protein determinations were not successful because of the interference caused by these reagents. From Table I, the overall mean from the replicate applications, 48.7 ng, is in good agreement with the expected mass of P. A similar calculation for the second α-casein solution yielded a value of 225 ng P, close to the expected value of 195 ng P.

Table I.   Reproducibility assessment of PAGE-NAA method using replicate aliquots of α-casein
Means are quoted as ± the standard error, where

$$SE = \frac{1\sigma}{\sqrt{n}}$$

Gel #	α-casein (replicates)	Measured P (ng)
1	1	47.1
	2	47.2
	3	55.7
	4	40.8
	Mean =	47.7 ± 3.0
2	1	53.7
	2	42.2
	3	54.2
	Mean =	50.0 ± 3.8
	Overall Mean =	48.7 ± 2.3

A sample of sodium caseinate was separated on a discontinuous gradient gel with sample aliquots of 10 - 50 µL from a protein solution of 2 mg/mL. An autoradiograph from a 9 day direct exposure was obtained from the irradiated gel and is shown in Figure 2.

The results from the determination of phosphoproteins in several samples are given in Table II. The uncertainties in the table represent an observed standard deviation relative to the mean of four measurements, except for IAEA A-11, which is the uncertainty for a single measurement. At the present detection limits, only the α-casein was quantified in the milk samples. The identification of the protein on the autoradiograph was made on the basis of molecular weight determinations, and its known high P content.

These measurements represent a direct quantitation of the constituent in milk. In studies reported in the

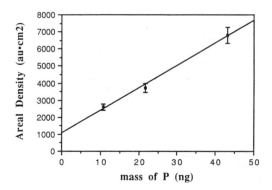

Figure 1. Standard Calibration Curve for P standards; areal density (au • cm²), obtained from absorbance measurements of an autoradiograph, versus mass of P (ng) applied to a gel prior to irradiation.

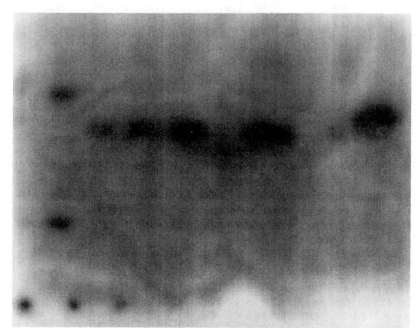

Figure 2. Autoradiograph of an irradiated gel containing aliquots of sodium caseinate samples separated by SDS-PAGE (track 3: 10 µL; 4: 20 µL; 5: 30 µL: 7: 40 µL; 10: 50 µL).

literature, the α-casein composition is often given as percentages of total casein, using a non-specific protein staining method following casein isolation and electrophoresis (11). The percentages are obtained from densitometric scanning of the stained, separated proteins. Casein results are given relative to each other or, if the total protein content of isolated casein is determined, then concentrations of the caseins can be calculated from the ratio. The accuracy of the staining method is dependent on the constancy of staining for each protein, and the protein upon which standardization is based is not always presented.

Another procedure for determining caseins involves quantifying each of their unique terminal amino acids (12), and requires an amino acid analyzer or sequencer. Precise sample preparation is vital in this method; prior isolation of the casein proteins is necessary, usually by isoelectric precipitation. Contamination by additional proteins greatly jeopardizes the results because the instruments involved in the sequencing cannot handle extraneous materials, other than the polypeptides that are being sequenced.

Table II. Concentrations of phosphoproteins in several materials.
(Concentration units are given relative to the original form of the material sampled; mg/mL if a liquid sample, %(g/g) if a solid sample)

Sample	α- casein	phosvitin
Sodium caseinate	6.3 ± 1.1 %	
IAEA A-11, Milk Powder	0.309 ± 0.036 %	
Low-fat (2%) milk	25.7 ± 5.2 mg/mL	
"half and half"	2.0 ± 0.2 mg/mL	
(50% milk and 50% cream)		
SRM 1845, Egg Powder		1.28 ± 0.12 %

The PAGE-NAA method can be extended for the determination of other types of proteins. However, the proteins need to be selected carefully. This is because some types of trace elements associated with proteins, such as those with weak ionic bonds, would not remain associated under electrophoretic separation. Other covalently bound protein-associated elements, such as selenium, would be obvious candidates for this research. Also, for most other protein-associated elements, autoradiography of the whole gel would not be appropriate. Other detection methods would be needed, e.g. γ-ray spectrometry of gel sections or scanning gels

with a well-collimated γ-ray detector. The feasibility
of determining separated selenoproteins is currently
being investigated, using both autoradiography and γ-ray
spectrometry.

The advantage of the PAGE-NAA method is that it is a
direct determination, not a relative one and requires
little treatment of the sample prior to electrophoresis.
It can be employed to determine individual proteins in
complex samples by using the resolution power of
electrophoresis and the sensitivity of neutron activation analysis.

Enhanced Immunoassays Using Colloidal Gold Labels

In contrast to the physical separation of proteins prior
to quantification by NAA, a protein such as an antibody
can be "isolated" by its immunological properties. In
this procedure, the NAA determined element is not
intrinsic to the structure of the protein (antibody) as
were the phosphoproteins in the PAGE-NAA analysis, but
it is introduced as a tag on the immunoreagent that
reacts specifically with the antibody. This type of
protein analysis is based on the specificity of antibody-
antigen reactions. If the immunoreagent (antigen) is
tagged with colloidal gold, the antibodies can be determined, since the amount of colloidal gold tag determined
by NAA is directly proportional to the amount of immunoreagent bound to the antibody (the analyte). This
method is similar to the enzyme immunoassays, but has
the possibility of enhanced sensitivity due to the high
neutron cross section of ^{197}Au. Colloidal gold with a
particle size of 20 nm contains approximately $2 \cdot 10^5$
gold atoms, which provides a considerable factor of
amplification compared to the binding of a single atom.

The details of the application of this method to quantify human IgG antibodies are described elsewhere (Zeisler, R.; Stone, S.F.; Viscidi, R.P.; Cerny, E.H., Clin.
Chim. Acta, submitted.). Briefly, individual wells of
microtiter strips were coated with human IgG samples,
and affinity purified goat anti-human IgG labeled with
colloidal gold were then added and allowed to incubate.
The excess unbound antigen was washed off and the gold
labeled conjugate bound to the adsorbed IgG was measured
via NAA using a γ-ray spectrometer. Since the amount of
colloidal gold tag is constant per antigen, and only a
single antigen binds to an antibody, the amount of gold
determined is proportional to the antibody analyte.

Figure 3 depicts the results of the immunoassay as applied to a series of IgG dilutions, from 10 to 0.0003
μg/mL. There is a definite linear working range where
quantification is possible; above this range, the lack
of surface area decreases the probability of antigen-
antibody reaction, and saturation occurs. The limit of

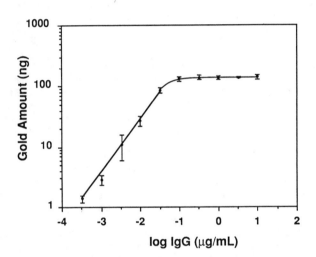

Figure 3. Immunoassay of human IgG with gold labeled anti-human IgG and subsequent NAA determination of gold. Uncertainties represent observed standard deviation of the mean of four determinations.

detection is currently determined by the gold blank, not so much by the reagent blank, but rather by non-specific binding to the solid phase. The measurements shown in Figure 3 were well above the mean blank level of 0.3 ng Au. Another IgG dilution series with a more dilute gold reagent was closer to this blank level, and gave a detection limit of 2 • 10^{-16} mole of IgG. Further investigations with this method will include testing a variety of buffers, and minimizing the incubation time with the immunoreagent.

In addition to the quantification of various antibodies, the method could also be applied to a wide variety of compounds for which antibodies are designed to react with great specificity. By eliminating some of the blank problems, the theoretical limit of detection could well extend below the attomole range.

Conclusion   Through the two methods described here, NAA has been shown to be an important technique in the development of quantitative protein analysis for biochemistry and clinical chemistry. By combining NAA with biochemical techniques used for protein isolation, such as PAGE and immunoassays, these macromolecules taken from complex natural matrices can be quantified down to ultratrace levels; in the case of antibodies, down to below 10^{-16} mole.

Acknowledgments   The authors acknowledge the assistance of the personnel at the University of Virginia Nuclear Reactor Facility, especially B. Hosticka, in the gel irradiations, and B. Clayman for technical assistance with the enzyme immunoassays. The work using PAGE-NAA is taken from a dissertation submitted to the Graduate School, Univeristy of Maryland, by S. F. Stone in partial fulfillment of the requirements for the Ph.D. degree in Chemistry. Certain commercial equipment, instruments, or materials are identified in this paper in order to adequately specify the experimental procedures. Such identification does not imply recommendation or endorsement by the National Institute of Standards and Technology, nor does it imply that the materials or equipment identified are necessarily the best available for the purpose.

Literature Cited:

1. Zeisler, R.; Greenberg, R.R.; Stone, S.F., J. Radioanal. Nucl. Chem. (1988) 124, 47-63.
2   Zeisler, R.; Stone, S.F.; Sanders, R., Anal. Chem. 1988, 60, 2760-2765.
3. Zeisler, R.; Greenberg, R.R.; In Trace Element Analytical Chemistry in Medicine and Biology; Brätter, P. and Schramel, P., Eds.; Walter de Gruyter, Berlin/New York, 1988; pp 296-303.

4. Greenberg, R.; Zeisler, R.R., J. Radioanal. Nucl. Chem. (1988) 124, 5-20.
5. Stone, S.F.; Zeisler, R.; Gordon, G.E., In Trace Element Analytical Chemistry in Medicine and Biology; Brätter, P. and Schramel, P., Eds.; Walter de Gruyter, Berlin/New , York, 1988; pp 157-166.
6. Laemmli, U.K., Nature 1970, 227, 680-681.
7. Basch, J.J.; Douglas, F.W.; Procino, L.G.; Holsinger, V.H.; Farrell, H.M.Jr., J. Dairy Sci., 1985, 68, 23-31.
8. Schibeci, A.; Martonosi, A., Anal. Biochem. 1980, 104, 335-342.
9. Stone, S.F., "Determination of Trace Elements Associated With Proteins by Polyacrylamide Gel Electrophoresis and Neutron Activation Analysis", Ph.D. Thesis, University of Maryland, College Park, Maryland, 1989
10. Heilmann, P., GelScan Program Documentation, LKB-Produckter, 1984.
11. Ramos, S.; Sanchez, R.M.; Olano, A., Chem. Mikrobiol. Technol. Lebens. 1985, 9, 24-27.
12. Ribadeau-Dumas, R., Biochim. Biophys. Acta , 1968, 168, 274-281.

RECEIVED August 6, 1990

Chapter 22

Speciation and Determination of Selenium and Mercury Accumulated in a Dolphin Liver

Kazuko Matsumoto

Department of Chemistry, Waseda University, Okubo, Shinjuku-ku, Tokyo
169, Japan

Selenium and mercury species accumulated
in a dolphin liver were separated and
purified using ultrafiltration and
various chromatographic methods. One of
the compounds identified is water-soluble
; has a molecular weight of < 1000;
contains amino groups; and has a Se to Hg
molar ratio of 1:1. Selenum and mercury
in the various fractions were determined
by carbon-furnance atomic absorption
spectrometry with palladium matrix-
modification. The addition of palladium
to the sample solution eliminated the
volatile loss of Hg and Se, and also
minimized the negative interference due
to the biological matrix.

Selenium is an essential element amd is known to be
present in some enzymes or proteins of vertebrates
mainly bound to an amino acid, viz., selenocysteine
(1-3). The toxicity of mercury compounds consistently
decreases in the presence of selenium (4-7). It is
also known that many sea animals such as seals and
dolphins sometimes accumulate Se and Hg at a 1:1 molar
ratio in their livers; the concentrations can be > 100
μ g/g (8-11). Burk et al. (12) and Chen et al. (13)
reported the formation of a high molecular weight
(mol wt) complex containing Se and Hg at 1:1 molar
ratio in rabbit livers when the animals were
simultaneously administered with mercuric chloride and
sodium selenite. This accumulation is important from a
bioinorganic viewpoint, because of the obvious
environmental and toxicological significance. However,
the chemical forms and the mechanism of accumulation of
Se and Hg are not known. In the present work, we wish

0097–6156/91/0445–0278$06.00/0

to report our studies on the separation, purification
and identification of the Se and Hg species in a
dolphin liver. We also report a sensitive carbon-
furnace atomic absorption spectrophotometric method
with Pd matrix-modification for the determination of Se
and Hg.

Palladium Addition Method for Selenium and Mercury
Determinations

The main obstacle that discourages speciation studies
of Se and Hg compounds in dolphin liver is the lack of
sensitive analytical methods for these elements.
Although carbon-furnace atomic absorption spectrometry
usually provides rapid and sensitive methods for these
elements, it is not applicable to sample solutions
containing organic materials (14). Organic matrices
suppress the atomic absorption signal of Se and Hg,
and reduce their sensitivities considerably. We found
that these severe negative interferences could be
removed by addition of 100 μg/ml Pd to the sample
solution (14-16). The interferences due to organic
materials and the effect of Pd addition on Se and Hg
absorbance are easily observed in Fig. 1. Although
detection limits (a concentration corresponding to
twice the relative standard deviation of the background)
of Se and Hg in carbon-furnace atomic absorption
spectrometry were 10 ng/ml and 2 ng/ml, respectively,
for simple inorganic solutions, no absorption signal
was observed when extract solution spiked with Se or
Hg were used (Fig. 1). These signals were, however,
restored to those of the original inorganic levels with
a 100 μg/ml Pd addition (Fig. 1). With the Pd
addition method, the detection limits for Se and Hg
were 17 ng/ml and 10 ng/ml, respectively, even in the
presence of 1.5% of albumin, a typical organic matrix.
Furthermore, we found that the addition of albumin or
any other organic material to the sample solution
enhanced the Se sensitivity by approximately a factor
of two, provided the drying and ashing conditions were
carefully selected (17). This is in contrast to the
common observation that organic materials accelerate
the volatilization of the analyte during the ashing
stage and thereby reduce sensitivity (18). The
optimized temperature program we employed using our
instrument, a Hitachi H-180-50 atomic absorption
spectrometer with a Hitachi GA-3 graphite furnace
atomizer, was as follows: dry, 30-150 °C , 40s; ash,
150-400°C , 60s; 400-800°C , 60s; 800-1200 °C , 60s; and
atomize, 2400 °C , 6s.
The mechanism of the enhancement effect was
studied with electron microscopy (17). The microscopy
picture of a vertical section of a carbon furnace,
which had been applied with palladium and albumin
solution and heated to 1200 °C , showed that new

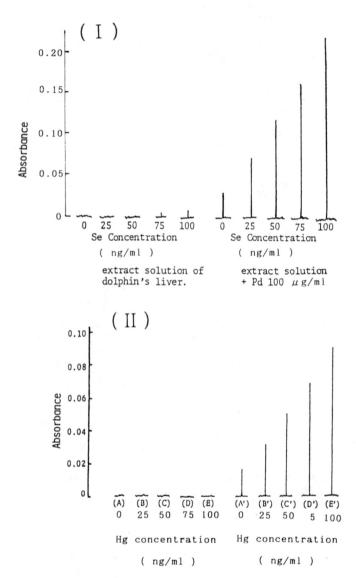

Fig. 1. Atomic Absorption Profiles of Selenium and Mercury in Dolphin's Liver Extract with and without Palladium Addition. (I) selenium calibration curves. (II) mercury calibration curves. (A) to (E), respectively are 0, 0.25, 0.5, 0.75 and 1.0 g/ml mercury additions without Pd modification. (A') to (E') are the same as (A) to (E), but with 100 μ g/ml Pd added.

graphite layers were formes from albumin on the surface of the tube. A combined study using electron microscopy and energy-dispersive X-ray spectrometry of the tube surface clearly indicated the existence of both Se and Pd as fine particles dispersed on the tube surface; also, the Se to Pd molar ratio was apporximately 1:1. This fact proved that Se was retained on the tube surface at higher temperatures as palladium selenide. In addition, thermal stability of the palladium selenide eliminated volatilization loss during the ashing stage and increased the sensitivity of Se determination. The formation of graphite layer possibly contributed to the retention of palladium selenide in or between the layer structure.

Extraction and Ultrafiltration of Se- and Hg-Containing Compounds from a Dolphin Liver

A dolphin liver was homogenized with the addition of acetone (10 ml/g liver) and was dried. The Se and Hg concentrations were 56.9 and 138 μg/g, respectively, and the Hg to Se molar ratio was 0.95. The liver powder was tested for extraction efficiency with various solvents (see below) and the extract-solution was subjected to ultrafiltration with an Amicon YM-2 membrane for obtaining a low-mol-wt (1000) fraction (19).

Distribution of High- and Low-mol-wt Hg- and Se-containing Compounds

Six solvents were tested for their suitability to extract the Hg- and Se- containing compounds from dolphin liver as shown in Table I. When inorganic solutions without surface-active agents were used for extraction, the extraction efficiency was less than 15%; among the inorganic solutions, 0.2M ammonium acetate gave the highest yield of 14% for both Hg and Se. The yields of Hg and Se were always nearly equal in any extraction solution and in any retentate or filtrate after ultrafiltration. This impiled that Hg and Se were present in more than one compound, all of which contained Hg and Se at a 1:1 molar ratio. On the other hand, sodium dodecyl sulphate (SDS) or Triton X-100 enhanced the extraction efficiency, but mainly of the high-mol-wt(>1000) fractions. Also these results showed that the compounds containing Hg and Se at a 1:1 molar ratio exist in the low-mol-wt(<1000) fractions. It is thus conceivable that the low-mol-wt molecule is the basic structural unit existing also in the high-mol-wt fractions. Therefore, further purification procedures were carried out only for the low-mol-wt compounds.

Table I. Extraction Ratios of Hg and Se in Various Extraction Solutions

Extraction solution	Extraction ratio, %					
	Filtrate of ultrafiltration (A)		Retentate of ultrafiltration (B)		Total (A)+(B)	
	Hg	Se	Hg	Se	Hg	Se
1% NaCl	5.6	4.9	0.59	0.60	6.2	5.5
0.20M phosphate buffer (pH 7)	6.3	6.0	1.0	1.0	7.3	7.0
0.20M phosphate buffer containing 1% SDS^a	16	17	15	18	31	35
0.20M phosphate buffer containing 2% Triton X-100	--	--	--	--	--	--
0.20M CH$_3$COONH$_4$	8.2	8.1	5.4	5.5	14	14
0.20M CH$_3$COONH$_4$ containing 4% SDS^a	16	17	43	41	59	58

^{a}Sodium dodecyl sulfate.

Chromatographic Behavior of the Hg- and Se-Containing Species

Both Hg and Se were detected mainly in unadsorbed fractions and numbers 85-95 in Dowex 50W-X4 chromatography, as shown in Fig. 2. The mercury-to-selenium molar ratios were not 1:1 in the unadsorbed fractions. Selenium in these fractions could be simple compounds such as inorganic selenite or selenocysteinic acid. On the other hand, Hg and Se exist at a 1:1 molar ratio in fractions 85-95. A separate experiment using standard amino acids showed that this elution position was just after the position at which leucine was eluted. Fractions 90-95 were pooled and applied to Sephadex G-15 gel filtration. Mercury and selenium appeared in fractions 50-59 at a 1:1 molar ratio (Fig. 3). From the elution pattern, the fractions were divided into three groups, 48-53, 54-56, and 57-59, and each pooled fraction was further purified by paper chromatography.
    In the paper chromatograms shown in Fig. 4, Hg and Se were found at R_f 0.3-0.6. Mercury and selenium in these cases were present at a 1:1 molar ratio. The materials from R_f 0.3-0.6 and 0.3-0.5 in Fig. 4 were combined and further purified using anion exchange chromatography.
    Both Hg and Se appeared in fractions 4-7 at a 1:1 molar ratio in Dowex 1-X2 anion exchange chromatography (Fig. 5). Most of the histidine present as the main impurity could be removed by this chromatographic procedure. Fractions 4-7 were pooled and further purified by Chelex-100 ion-exchange chromatography.

Behavior of the Hg- and Se-Containing Compounds on Reversed-Phase HPLC and Chelex-100 Chromatography

The Hg and Se species were not adsorbed on reversed-phase columns (Senshu Pak SC-8-1251 or Sedox-DE-613). The hydrophilicity of the compounds was more than that of the amino acids usually found in proteins, which were adsorbed on the column under the same consitions. Mercury and selenium were detected in fractions 18-29 at a 1:1 molar ratio in Chelex-100 chromatography. The elution positions for Hg and Se were almost identical with those of Ca^{2+} and Mg^{2+}. Therefore, the Hg- and Se-containing compound was weakly adsorbed on the chelate resin. The adsorption behavior was distinctly different from, and weaker than that of simple Hg^{2+} ion, showing that Hg in the Hg-, and Se-containing compound was tightly bound to the molecule and could not be easily removed.
    In all chromatographic studies, except for the unadsorbed fractions in Dowex 50W-X4 chromaotgraphy, Hg and Se appeared at the same positions, with a 1:1 molar ratio. The most purified material contained 0.22 mol of both Hg and Se. Inductively coupled plasma atomic

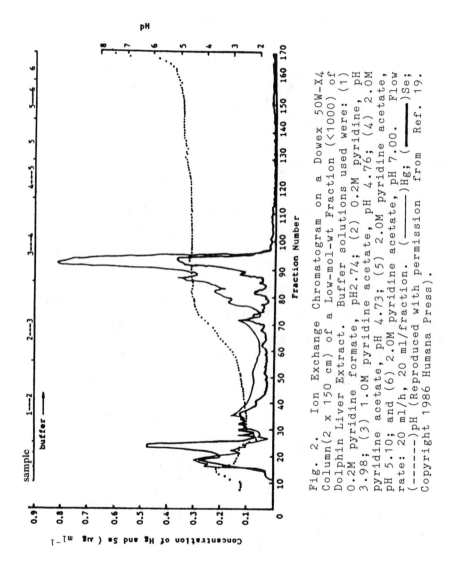

Fig. 2. Ion Exchange Chromatogram on a Dowex 50W-X4 Column(2 x 150 cm) of a Low-mol-wt Fraction (<1000) of Dolphin Liver Extract. Buffer solutions used were: (1) 0.2M pyridine formate, pH2.74; (2) 0.2M pyridine, pH 3.98; (3) 1.0M pyridine acetate, pH 4.76; (4) 2.0M pyridine acetate, pH 4.73; (5) 2.0M pyridine acetate, pH 5.10; and (6) 2.0M pyridine acetate, pH 7.00. Flow rate: 20 ml/h, 20 ml/fraction. (———)Hg; (———)Se; (------)pH (Reproduced with permission from Ref. 19. Copyright 1986 Humana Press).

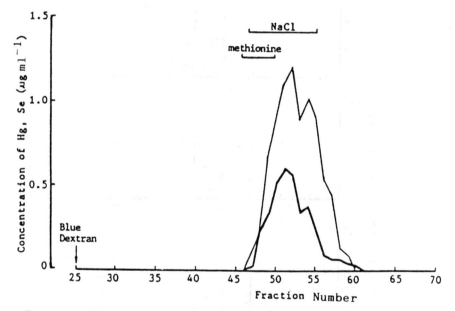

Fig. 3.   Elution Patterns of Hg and Se on Sephadex G-
15.   Sample   fractions 90-95 on ion exchange
chromotography (Dowex 50W-X4) of dolphin liver low-mol-
wt fraction (<1000); column: Sephadex G-15 (2 x 80 cm);
buffer: 0.1M CH$_3$COONH$_4$ (pH7); flow rate: 60 ml/h,
3ml/fraction. (————)Hg; (▬▬▬)Se (Reproduced with
permission from Ref. 19. Copyright 1986 Humana Press).

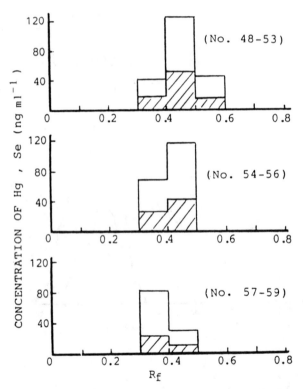

Fig. 4. Distribution Patterns of Hg and Se in Paper Chromatography. Sample nos. 48-53, 54-56, 57-59 on Sephadex G-15 column(2 x 80 cm) of dolphin liver low-mol-wt fraction (<1000). solvent: *n*-butanol-acetic acid-water (4:1:2 v/v/v). (☐)Hg; (▨)Se(Reproduced with permission from Ref. 19. Copyright 1986 Humana Press).

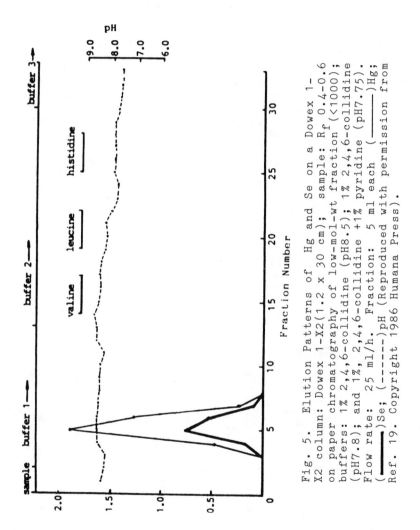

Fig. 5. Elution Patterns of Hg and Se on a Dowex 1-
X2 column: Dowex 1-X2(1.2 x 30 cm); sample: Rf 0.4-0.6
on paper chromatography of low-mol-wt fraction(<1000);
buffers: 1% 2,4,6-collidine (pH8.5); 1% 2,4,6-collidine
(pH7.8); and 1%, 2,4,6-collidine +1% pyridine (pH7.75).
Flow rate: 25 ml/h. Fraction: 5 ml each (————)Hg;
(————)Se; (-----)pH (Reproduced with permission from
Ref. 19. Copyright 1986 Humana Press).

emission analysis of the final purified solution showed
that there were no other metals in the solution.
The recovery at each chromatographic step was more
than 80% for both Hg and Se, except for the cation
exchange chromatography (50%).

Chemical Nature of the New Compound Containing Hg and
Se

The study described above clearly demonstrated the
existence of a new compound containing Hg and Se at a
1:1 molar ratio in the dolphin liver. The chemical
nature of the compound was found to be as follows: it
is water-soluble; has a mol wt of less than 1000;
contains amino groups, since it is ninhydrin positive;
and it has tightly bound Hg. It is highly probable
that Hg is coordinated by the selenol anion RSe⁻, since
the Hg-Se bond is very stable, and the mercury could
not be removed as a cation by Chelex-100 resin.   The
compound is probably a complex of methylmercury.
However, the yield of the final product was
insufficient to obtain direct evidence for the
existance of a methul group. Alternately, the compound
may exist as a structural unit in the larger molecules
found in dolphin livers.
The development of highly sensitive analytical
method for Hg and Se, i.e., palladium addition method
(14-16), for the first time enables us to perform such
purification studies. The results obtained here would
render an important clue to the consideration of the
mechanisms of Hg and Se accumulations and detoxication
of Hg by Se.

Literature Cited

1. Mosenbocker, M. A.; Tappel, A. L. Biochim. Biophys.
    Acta 1982, 709, 160-65.

2. Mosenbocker, M. A.; Tappel, A. L. Biochim. Biophys.
    Acta 1982, 704, 253-60.

3. Sliwkowski, M. X.; Stadtman, T. C. J. Biol. Chem.
    1985, 260, 3140-44.

4. Parizek, J.;   Ostadalva, I. Experientia 1967, 23,
    142-46.

5. Ganther, H. E.; Grudie, C.; Sunde, M. L.; Kopecky,
    M. J.; Wanger, P.; Oh, S. H.; Hoekstra, W. G.
    Sience 1972, 175, 1122-24.

6. Stillings, B. R.; Lagally, H.; Bauersfeld, P.;
    Soares, J. Toxicol. Appl. Pharmacol., 1974, 30,
    243-46.

7. Stoewstand, G. S.; Bache, C. A.; Lisk, D. J. Bull. Environ. Contam. Toxicol., 1974, 11, 152-57.

8. Koeman, J. H.; Peeters, W. H. M.; Koudstaal-Hol, C. H. M.; Tjioe, P. S,; De Goeij, J. J. M. Nature 1973, 245, 385-86.

9. Kosta, L.; Byrne, A. R.; Zelenko, V. Nature 1975, 254 238-39.

10. Smith, T. G.; Armstrong, F. A. J. Fish. Res. Board Can. 1975, 32, 795-99.

11. Koeman, J. H.;   van de Ven, W. S. M.; Goeij, J. J. M.; Tjioe, P. S.; Van Haaften, J. L. Sci. Total Environ. 1975, 3, 279-304.

12. Burk, R. F.;   Foster, K. A.; Greenfield, P. M.; Kiker, K. W. Proc. Soc. Exp. Biol. Med. 1974, 145, 782-84.

13. Chen, R. W.; Whanger, P. D.; Fang, S. C. Pharmacol. Res. Commun. 1974, 6, 571-72.

14. Ping, L.; Wei-Lei; Matsumoto, K.; Fuwa, K. Anal. Sci. 1985, 1, 257-61.

15. Ping, L.; Fuwa, K.; Matsumoto, K. Anal. Chim. Acta 1985, 171 , 279-84.

16. Ping, L.;   Lei, W.; Matsumoto, K., Fuwa, K. Bull. Chem. Soc. Jpn. 1985, 58, 3259-63.

17. Nishimura, J. E. T.;   Tominaga, T.; Katsura, T.; Matsumoto, K. Anal. Chem. 1987, 59, 1647-51.

18. Saeed, K.;   Thomassen, Y.; Langmyhr, F. J. Anal. Chim. Acta 1979, 110, 285-89.

19. Ping, L.; Nagasawa, H.; Matsumoto, K.; Suzuki, A.; Fuwa, K. Biol. Tr. Elem. Res. 1986, 11, 185-99.

RECEIVED July 17, 1990

Chapter 23

Synchrotron Radiation and Its Application to Chemical Speciation

B. M. Gordon and K. W. Jones

Department of Applied Science, Brookhaven National Laboratory, Upton, NY 11973

Synchrotron radiation can be used in Extended X-ray Absorption Fine Structure (EXAFS) and X-ray Absorption Near Edge Structure (XANES) experimental modes to extract information concerning chemical speciation of elements at trace concentration levels. The structure and relative position of an absorption spectrum at high energy resolution in the absorption edge region provide information regarding the oxidation state of the element and symmetry of the molecule in the immediate vicinity. Speciation is elucidated by comparison of spectra with those of model compounds. Examples of such studies in the literature are presented. The technique, which generally requires no chemical preparation, is sensitive in the mg/kg concentration range in biomedical tissue samples. The recent proliferation of intense and dedicated synchrotron radiation sources provides wide access to the technique.

Trace elements have long been recognized as playing an important role in the functioning of living organisms. This realization has been fostered by the rapid development of analytical techniques capable of quantitation at ever improving sensitivities and decreasing spatial resolutions. Trace elements have been shown to be both essential and harmful to the well-being of organisms ($\underline{1}$). For some elements the range between deficiency and toxicity is indeed narrow.

The advanced capabilities for trace element determination has also brought the realization that knowledge of the chemical speciation of the trace elements is possibly the most important component of data that can be collected. For example, in the field of nutrition, the bioavailability of an essential element requires that it be in a reactive form rather than one that is incapable of being metabolized. Many elements, particularly first row transition metals, are components of biologically important compounds such as proteins, enzymes, etc. Each of these compounds has specific

0097–6156/91/0445–0290$06.00/0
© 1991 American Chemical Society

functions such as metabolism, detoxification, oxidation-reduction catalysis, transfer reactions, etc.

Specific techniques for speciation include NMR spectroscopy, dialysis, gel filtration, electrophoresis, anodic stripping voltammetry, and radioactive tracers. Many of these methodologies require sample processing for separation of and identification of the desired elemental species from matrices such as food, serum, tissue, etc. In some cases species identity may be sacrificed and lead to incorrect results. Another method which is less susceptible to species alteration involves the excitation of K- and L-shell x-ray fluorescence where the energy of the fluorescence is determined using high resolution crystal spectrometers (2). The resultant spectra are compared with those of model compounds for possible speciation.

Recently, the x-ray fluorescence technique has been improved using a high-intensity x-ray source which has been made possible by the recent development of synchrotron radiation sources (3). The method is carried out by scanning the energy of a highly collimated x-ray beam through the absorption edge of the element in question with a narrow energy interval ($\Delta E/E \sim 10^{-4}$). The spectra are taken by measuring the transmittance of a major or minor constituent and by fluorescence x ray detection for trace constituents. Again, the spectra obtained are compared with those of model compounds. These comparisons can provide, as illustrated below, information concerning the oxidation state, symmetry, molecular bonding and surrounding structure of the absorbing atom. Under the most favorable circumstances, these measurements can be made *in situ* at 1 mm^2 spatial resolution and 5 mg/kg concentration level (4).

THEORY

The type of x-ray spectroscopy (5) used to elucidate structural and speciation information in an atomic environment have been termed EXAFS (Extended X-ray Absorption Fine Structure) and XANES (X-ray Absorption Near Edge Structure). Fig. 1 illustrates schematically a scattered radiation spectrum from a fictitious sample containing a trace amount of zinc and shows the major features of the photoelectric effect (PE) resulting in K-shell fluorescence, Compton inelastic scattering and Rayleigh elastic scattering. The line widths shown for the fluorescence and Rayleigh lines are actually much narrower, but are widened for ease of viewing. The width of the Compton scattering reflects the large acceptance solid angle of a typical ion chamber since the energies of the scattered photons are angle dependent. In the case shown the excitation energy is above the Zn-edge. If the excitation energy were below the Zn-edge, the Zn fluorescence would not exist. In addition, but not shown, the spectrum includes K-shell fluorescence lines of elements present with $Z \leq 29$ and L-shell fluorescence lines of elements with $Z \leq 71$. It should be noted that the intensities of all radiations except the Zn fluorescence line are essentially constant over the narrow energy range being scanned and constitute a slowly varying background in the measurement.

The abrupt change in the photoelectric absorption represents the ionization energy where the photon has sufficient energy to excite electrons of a particular shell to unoccupied states and the continuum. The absorption spectrum is rather unremarkable until one observes the edge region with sufficient fine energy resolution. The K-edge region for argon gas is shown in Fig. 2 (6). Unbound argon, with an electron configuration $1s^22s^22p^63s^23p^6$, shows a simple spectrum with allowed dipole

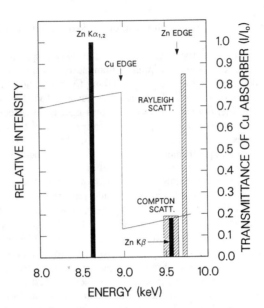

Figure 1. Schematic scattered energy spectrum with excitation energy above Zn edge for 1000 mg/kg sample in carbon matrix. The right-hand scale for the line graph is the transmittance for Cu absorber between sample and detector.

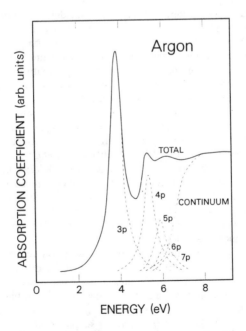

Figure 2. Absorption spectrum for gaseous argon resolved into excitations to allowed Rydberg states. (Reproduced with permission from Ref. 6. Copyright 1959 American Institute of Physics.)

transitions to the Rydberg states 1s → 4p,5p,6p... according to the selection rule $\Delta l = \pm 1$. At energies above the ionization potential, the spectrum shows a smooth fall with increasing energy.

The situation is a great deal more complex in the usual case where the absorbing atom is bound in a condensed phase. An absorbing atom A is surrounded by nearest neighbors and next nearest neighbors. The photoelectron waves emanating from A, which in the case of gaseous argon would travel out without hindrance, are backscattered by the atoms in the near vicinity, causing interference with the waves from A. The interference manifests itself as modulation in the absorption spectrum above the absorption edge. Moreover, at the edge, molecular transitions take place and provide significant information concerning speciation. A typical absorption spectrum is shown in Fig. 3 identifying the separate XANES and EXAFS regions (7).

EXAFS provides structural information for the molecule in the immediate vicinity around absorber A. EXAFS data can contribute to speciation as illustrated below, but it is the XANES region that is most important. In general it is difficult to calculate XANES spectra from first principles, although there have been significant advances in recent years. This procedure is applicable in cases where fine points of structure in a known compound are being investigated. For chemical speciation of an unknown sample, the procedure generally involves comparison of the sample absorption spectrum with those of model compounds containing the absorbing atom. The model compounds would cover the range of oxidation state, types of ligands, and types of molecular symmetries exhibited by the element.

EXPERIMENTAL METHOD

SYNCHROTRON RADIATION.   Synchrotron radiation as used here is electromagnetic radiation emitted by relativistic electrons accelerated in a constant orbit (3). The properties of synchrotron radiation most useful for speciation are: a wide and continuous spectral energy range of very high intensity x rays extending to 30 keV (at the National Synchrotron Light Source, NSLS); a vertical divergence angle of ~ 0.2 mrad permitting focusing to small beam spots; a high degree of polarization to minimize scattering background; and an ability to work on living cells in air or helium atmosphere. The intensities are five orders of magnitude greater than the continuum spectra of rotating anode x-ray tubes. Storage rings now under construction will provide intensities greater by three to four orders of magnitude.

EXCITATION. A typical beamline for EXAFS studies is illustrated in Fig. 4. The beam originating at a tangent point to the stored electron beam is collimated for a particular horizontal acceptance angle and a vertical divergence consistent with the optical elements of the beamline. The first optical element may be a collimating mirror to reduce the angular divergence before monochromation. This serves to improve the energy resolution in that differentiation of Bragg's Law results in the relationship $\Delta E/E = \cot\theta \Delta\theta$. The continuous energy ("white") beam then passes through the monochromator where an energy interval is selected. The monochromatic beam then may be focused at the sample position. The exit height of the beam will change during a scan in accordance with Bragg's Law. In reality the beam height can be kept constant by using a four-crystal arrangement or a two-crystal arrangement as shown in which the second crystal translates while being rotated. The four-crystal

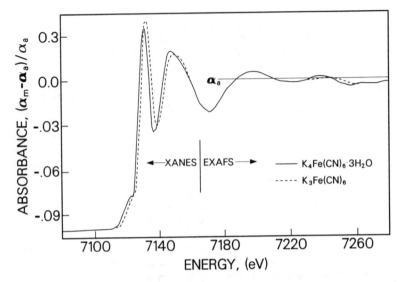

Figure 3. Absorption spectra of $K_3Fe(CN)_6$ and $K_4Fe(CN)_6$ illustrating EXAFS and XANES regions. The absorbance is plotted relative to the continuum absorption coefficient, α_a, at higher energies. (Reproduced with permission from Ref. 7. Copyright 1982 American Institute of Physics.)

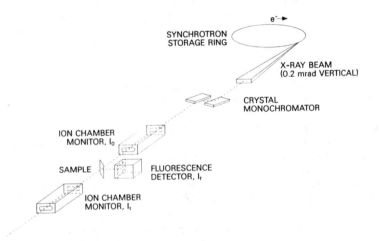

Figure 4. Schematic diagram of typical EXAFS beamline. The ring to sample distance at NSLS X-19A is 20 m.

arrangement is made up of a pair of two-crystal arrangements in sequence to bring the exit beam to the same height as the entrance beam.

The range of available excitation energies essentially covers the entire periodic table. Elements of $Z \geq 10$ can be studied outside the storage ring vacuum by using diamond film windows and helium atmospheres. The elements of $Z \geq 40$ can be investigated by L-shell excitation.

DETECTION. Upon passage through a target, photons undergo interactions with the target which remove them from the transmitted beam. They may cause K- and L-shell photo-excitation resulting in fluorescence of target atoms or they may undergo Rayleigh and Compton scattering. Rayleigh scattering does not change the photon energy while Compton scattering reduces the excitation photon energy by 1 to 2% in the energy range of interest at a 90° viewing angle. The PE process is predominant and is illustrated in Fig. 1 where the peak heights represent Zn at 1000 mg/kg in a carbon matrix (8). All cross sections except the PE cross section of the element being studied are slowly varying in the energy range of an EXAFS scan. The observed signal is a product of the cross section and the elemental concentration. Because of the differences in concentrations of the trace and matrix elements, a comparison should be made among the PE cross sections of the trace element at the edge, PE cross sections of lower Z trace elements and the sum of the scattering and PE cross sections of the matrix elements. At concentrations of 1% and greater, the effect of the PE cross sections of the investigated element is large and a determinable change in transmission can be seen at the edge. When the concentration of the element is at a trace level, the inner-shell excitation is small compared to scattering events and inner-shell excitation of matrix atoms and therefore, there is no easily discernible change at the edge. In such a case the fluorescence signal must be extracted.

The schematic in Fig. 4 shows the principal methods of detection. For samples where the absorbing element is a major or minor constituent, the transmission (I_T/I_o) is determined as a function of photon energy. The detectors are ionization chambers with filling gases whose mass absorption coefficients are appropriate to the energy range. For study of elements at trace levels one uses a fluorescence detector (9), which is usually a specially constructed ionization chamber. The main features of the x-ray spectrum are the fluorescence peaks of the element being investigated, the Compton and Rayleigh scattering peaks at energies slightly above the fluorescence peaks, and the fluorescence peaks of matrix elements capable of penetrating the detector window. The PE of matrix elements in biomedical samples produce low energy x rays incapable of entering the detector. An ionization chamber for fluorescence detection has provision for insertion of a critical absorber between the sample and detector in order to preferentially absorb the higher energy scattered radiation. The detector also has Soller slits to limit the efficiency of detection for the characteristic x rays of the absorber.

Cramer, et al. (10) have recently developed an intrinsic germanium array detector system with thirteen separate solid-state detectors on a common cryostat. The detectors have their own dedicated electronics and each can be set to detect a selected energy range, usually the fluorescence peak of the desired element. Separate detectors greatly increase the efficiency of the detector system by alleviating dead-time problems. The detector system has an active area of 1300 mm². Fig. 5 shows a XANES spectrum for iron in the NIST reference material, Orchard Leaves (SRM 1571) at 300

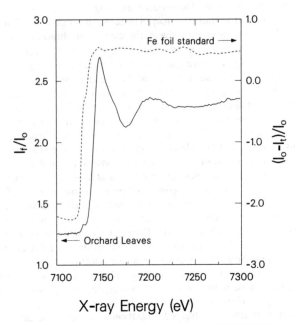

Figure 5. Absorption spectrum for Fe (300 mg/kg) in NIST SRM 1571, Orchard Leaves standard, taken in the fluorescence mode using Ge-array detector.

mg/kg Fe, obtained with the Ge-array detector (11). The edge jump ratio is increased four-fold compared to using a fluorescence ionization chamber.

ENERGY DISPERSIVE OPTICS. There are numerous potential applications for a system capable of time-dependent studies. In such a system as performed at the Stanford Synchrotron Radiation Laboratory (12), a continuous "white" beam illuminates a bent diffraction crystal whereby the continuously changing angle of reflection provides a dispersion of the photon energies. The dispersed x rays then fall on a position-sensitive detector providing a measurement of the energy spectrum. The curvature of the crystal can be so arranged that the dispersed energies intersect at a point between the crystal and detector. The sample is positioned at this polychromatic focal point. Thus a full EXAFS spectrum can be obtained on a millisecond time scale.

Hastings, et al. (13) have used a wavelength dispersive detector with a multi-crystal pyrolytic graphite surface in a "barrel" geometry tuned to Fe $K\alpha$ radiation. They obtained EXAFS spectra of 75 ppm Fe in a Cu matrix at 1 mm^2 spatial resolution.

X-RAY MICROPROBE. The x-ray microprobe at the NSLS X-26 beamline (14) is capable of quantitation of femtogram (10^{-15} g) quantities of trace elements at spatial resolutions of a few micrometers and in the mg/kg range of concentrations. The sensitivity will be improved with the introduction of a monochromator, a focusing mirror, and a crystal spectrometer. The microprobe will be of considerable use to speciation by identification of areas of large concentrations. For example, the average concentration of chromium in human tissue and food stuffs is less than the sensitivity for speciation by XANES (24). The microprobe is capable of scanning a tissue sample to identify those areas with chromium concentrations amenable to speciation. Site selection in this way is more likely to focus on areas with single species and higher concentrations rather than large analysis areas with multiple species and smaller average concentrations. There are plans to adapt the microprobe beamline to XANES studies and to reduce the resolution below that now available on focused beamlines.

DISCUSSION

In using XANES and EXAFS spectra for chemical speciation, an attempt is made to relate spectral features to those of model compounds. The EXAFS region represents electrons of higher kinetic energy that undergo single scattering processes resulting in information concerning atomic distances and Z values of neighboring atoms. Low energy electrons near the edge undergo multiple scattering interactions and can provide a wealth of information on characteristics such as spatial arrangements, bonding, and charge densities. The theory of XANES and the calculation of spectra are beyond the scope of this paper (25). The process of speciation is simpler in that comparisons are made with model compounds, but is complicated by the possible occurrence of more than one species of the same element. This is no doubt true for bulk samples. The use of a microprobe with high spatial resolution is more likely to provide information on specific species.

It is useful to consider the various of features that characterize XANES spectra. One is the position of the edge. The outer valence electrons provide some shielding of the nucleus from the core electrons. This shielding is reduced in atoms of higher oxidation states, thus requiring more energy for excitation. Cramer, et al. (15)

reported on the linear relationship between edge position and coordination charge for molybdenum compounds ranging in formal oxidation state from +2 to +6. The coordination charge is basically the oxidation number corrected for ligand electronegativities. The best correlation of position with oxidation state is observed in transitions to core states. However, in many of these cases the transitions result in weak dipole forbidden peaks. In the absence of such peaks, one can observe shifts of the edge and the large multiple scattering structures which do not usually show a clear linear dependence (16).

The transition elements form a variety of coordination compounds because of interactions with d orbitals. The elements vanadium through zinc have been established as being vital to the functioning of living organisms. The XANES spectra of these elements exhibit many common characteristics which are useful to keep in mind before discussing specific elements and their spectra. Many of the complexes have octahedral symmetry in which the center of symmetry forbids the $1s \rightarrow T_2$ dipole transition. However, such quadrupole transitions with greatly reduced oscillator strengths, or mixing in of p character can result in a weak pre-edge peak. This small peak is observed for all the octahedral coordination compounds of transition elements except for Zn(II) and Cu(I) compounds with a filled $3d$ shell. The tetrahedrally coordinated compounds, and others lacking a center of symmetry, have an intense and sharp pre-edge peak as shown for the chromate ion in Fig. 6 (17). The pre-edge peak position varies with oxidation state as predicted for a series of tetrahedral structures.

There is an extensive literature on the XANES studies of compounds of interest to the biomedical community (20). This literature has been reviewed recently (20). In almost all cases these studies are structural studies on known preparations and are not examples of chemical speciation. However, these studies do provide a data base for the model compounds to be compared with XANES spectra in speciation studies. Examples of XANES studies for elements of biomedical interest are provided below.

SPECIATION STUDIES. The actual speciation studies to date have been few in number and are predominantly in the environmental and geochemical fields. However, they can serve as an illustration of the technique to be applied to biomedical studies. Jaklevic, et al. (18) reported on the speciation of Zn, Fe, and Cu in air particulate samples as a function of particulate size. The particulates were separated by automatic sampling into fine and coarse particles with a division criterion of 2.4 μm. The concentration of the elements on the substrate filter was generally 200 to 1000 mg/kg and therefore the fluorescence mode was used. The EXAFS spectra for the two fractions are shown in Fig. 7 and the corresponding spectra for relevant model compounds are shown in Fig. 8. It is clear from comparison of the spectra that the Zn in the "fine" fraction is predominantly $ZnSO_4$ or $Zn(NH_4)_2(SO_4)_2$ and the "coarse" fraction is mostly ZnO. The method can be made quantitative. The spectrum of a 1:1 mixture of ZnO and $ZnSO_4$ was fit linearly to the model compound spectra, resulting in a $ZnSO_4$ coefficient of 0.46 ± 0.03 and a ZnO coefficient of 0.54 ± 0.03. The spectra of the two sulfate compounds were too much alike to measure relative abundances. Likewise, the Fe and Cu spectra indicated the Fe to be distributed between $FeNH_4(SO_4)_2$ and Fe_2O_3 and the Cu to be present as $CuSO_4$ at mole fractions of 0.80 to 1.0.

Maylotte, et al. (19) reported on the speciation of vanadium in coal in an effort to determine the fractionation of vanadium compounds upon pre-combustion processing. The processing consists of pulverization and density separation by flotation

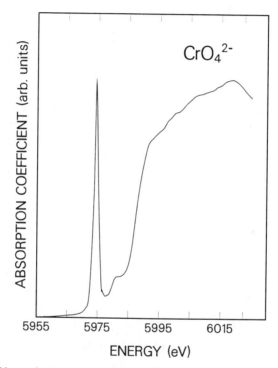

Figure 6. Absorption spectrum of tetrahedral CrO_4^{2-} ion showing the intense allowed pre-edge transition. (Reproduced with permission from Ref. 17. Copyright 1980 American Institute of Physics.)

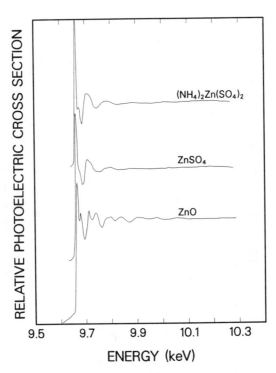

Figure 7.   Absorption spectra of Zn model compounds in aerosol fractions. (Reproduced with permission from Ref. 18. Copyright 1980 American Chemical Society.)

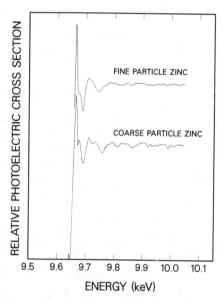

Figure 8. Absorption spectra of Zn model compounds for analysis of aerosols. (Reproduced with permission from Ref. 18. Copyright 1980 American Chemical Society.)

resulting in float and sink fractions. The concentration of vanadium was the order of 1000 ppm. The XANES spectra of model compounds are shown in Fig. 9. The spectra b and c represent octahedral coordination, d represents tetrahedral coordination, and e and f represent square pyramidal coordination. Note the strong pre-edge peaks for the structures without a center of symmetry. The pre-edge peak in d, as well as other V(V) compounds occurs at 5 to 6 eV. Pre-edge peaks of V(IV) compounds with nitrogen-bound ligands, such as vanadium phthalocyanine and porphyrin occur at 3.5 to 4.0 eV. Fig. 10 shows the XANES spectra for unseparated coal, the float and sink fractions, and a liquefaction residue, the result of treatment with stannous chloride, tetralin, and hydrogen at high pressure and temperature. The pre-edge peaks are at 4.5 ± 0.2 eV, in good agreement with V(IV) coordinated to oxygen, as in V_2O_4. The heavy fraction vanadium content seems to be made up of V_2O_3 with some entrainment of V_2O_4.

ELEMENTS OF BIOMEDICAL INTEREST. The trace elements that have important beneficial roles in human processes, in addition to the first row transition elements mentioned previously, include As, F, I, Mo, Se, Si, and Sn. Other trace elements that can be toxic and that are commonly found in the environment include Be, Cd, Hg, Pb, Pd, and Tl (1). A majority of the essential elements become toxic at levels greatly exceeding the ideal levels. An extensive literature of EXAFS and XANES studies on biologically interesting compounds has accumulated over the past fifteen years. These studies concerned structural and bonding information of metal interactions in enzymes and proteins. This literature can be part of a data base of model compounds for XANES speciation studies. There is an effort to gather this literature into a centralized and computerized data bank. The compounds of these elements with the more common inorganic ligands have been reported and need not be further discussed here.

A brief mention of classes of structures that have been studied for iron and copper follows and is representative of studies with a majority of the beneficial elements. Iron is one of the most studied elements using EXAFS, in part because of many studies performed by other techniques, including Mossbauer spectroscopy (20). The studies of Fe-S cluster sites include rubredoxin, ferredoxin, aconitase, and the Fe-Mo cofactor. There is an extensive literature on hemoproteins, where the iron is bound to four planar nitrogen atoms and to a variety of axial ligands (20). These include c-type cytochromes and related cytochrome oxidases. Ferritins and iron-tyrosinate proteins have had structures elucidated in the Fe vicinity by EXAFS (20).

Copper proteins and enzymes have also been studied, among them the "blue" copper proteins including azurin, stellacyanin, and plastocyanin (26). There is considerable controversy over the structure and oxidation state of Cu in cytochrome oxidase preparations. Studies of superoxide dismutase enzymes, which catalyze the disproportionation of the superoxide radical, show the existence of Cu(II) and Zn(II). Again, Cu(I) with a filled 3d shell would not show a 1s → 3d transition. Metallothioneins of copper and other metals have also been studied. They function as regulators and storage media for metals and can also be used as detoxification agents for metals (20).

Studies have been made on metal compounds used as therapeutic agents, among them cis-Pt(NH$_3$)$_2$Cl$_2$, an anticancer drug (21); GaCl$_3$, which decreases calcium resorption in bone cancer (22); and gold-based drugs in the treatment of arthritis (23).

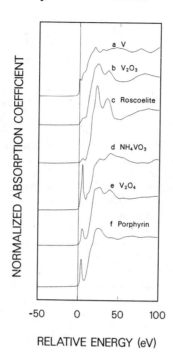

Figure 9.   Absorption spectra of V in fractions resulting from coal processing. (Reproduced with permission from Ref. 19.   Copyright 1981 AAAS.)

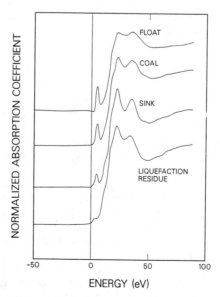

Figure 10.   Absorption spectra of V model compounds for analysis of coal fractions. (Reproduced with permission from Ref. 19.   Copyright 1981 AAAS.)

In this latter case it is interesting to note that Au(0) and Au(I) have filled 5d shells. Therefore, in L-shell XANES, the large sharp peak for the allowed $2p \rightarrow 5d$ transition seen in Au(II) and Au(III) compounds is absent in Au(0) and Au(I) compounds.

FUTURE DIRECTIONS. Improvements in multi-segment solid state detectors will make it possible to accept extremely high count rates expected with next-generation SR sources now under construction. Also new designs for "barrel" monochromators using multilayer mirrors are expected to permit EXAFS experiments at concentration levels as low as 0.1 mg/kg at milliprobe spatial resolutions (20).

There are many unknown factors in the application of XANES and EXAFS to chemical speciation. The application is just now beginning. The possibility of tying speciation to the x-ray microprobe for analysis of unstained tissue samples is exciting to many researchers. XANES and EXAFS have matured rapidly since the availability of dedicated storage rings at Daresbury in England, the NSLS at Brookhaven National Laboratory, and the Photon Factory in Japan. In addition there has been some dedicated time at synchrotron radiation facilities such as SSRL at Stanford, CHESS at Cornell, LURE in France, and HASYLAB in Germany. Other dedicated rings are being built in China, Brazil, and Italy. Even more powerful storage rings are now being built at the Argonne National Laboratory, U.S.A. and at Grenoble, France. There are eight dedicated EXAFS beamlines at the NSLS and each is available to outside users. All other rings in the United States are operated as user facilities. With these expanded facilities and maturation of the EXAFS technique, one would expect a sharp increase in chemical speciation applications.

ACKNOWLEDGMENTS

This work was supported in part by the US Department of Energy, Office of Basic Energy Sciences, Chemical Sciences Division, Processes and Techniques Branch, under Contract No. DE-AC02-76CH00016 and the National Institutes of Health Research Resource Grant No. P41RR01838.

LITERATURE CITED

[1]　Wolf, W. R. In The Importance of Chemical "Speciation" in Environmental Processes; Bernhard, M.; Brinckman, F. E.; Sadler, P. J. Eds.; Springer-Verlag: Berlin, 1986; pp 39-58.

[2]　Gohshi, Y.; Hukao, Y.; Hori, K. Spectrochim. Acta. 1972, 27B, 135-42.

[3]　Winick, H. In Synchrotron Radiation Research; Winick, H.; Doniach, S. Eds.; Plenum Press: New York, 1980; pp 11-60.

[4]　Warburton, W. K. Nucl. Instrum. Meth. Phys. Res. 1986, A246, 541-44.

[5]　Brown, G. S.; Doniach, S. In Synchrotron Radiation Research; Winick, H.; Doniach, S. Eds.; Plenum Press: New York, 1980; pp 353-85.

[6]　Parratt, L. G. Rev. Mod. Phys. 1959, 31, 616-45.

[7]　Bianconi, A.; Dell'Ariccia, M.; Durham, P. J.; Pendry, J. B. Phys. Rev. 1982, B26, 6502-8.

[8]　McMaster, W. H.; Kerr Del Grande, N.; Mallert, J. H.; Hubbell, J. H. Compilation of X-ray Cross Sections; National Technical Information Service, Springfield, VA, 1969; Report UCRL-50174-Sec. II Rev. I.

[9]　Stern, E. A.; Heald, S. M. Rev. Sci. Instrum. 1979, 50, 1579-82.

[10]   Cramer, S. P.; Tench, O.; Yocum, M.; George, G. N. Nucl. Instrum. Meth. Phys. Res. 1988, A266, 586-91.
[11]   This work.
[12]   Phizackerley, Z. U. R; Stephenson, G. B.; Conradson, S. D.; Hodgson, K. O.; Matsushita, T.; Oyanagi, H. J. Appl. Cryst. 1983, 16, 220-32.
[13]   Hastings, J. B.; Eisenberger, P.; Lengeler, B.; Perlman, M. L. Phys. Rev. Lett. 1979, 43, 1807-10.
[14]   Jones, K. W.; Gordon, B. M. Anal. Chem. 1989, 61, 341A-58A.
[15]   Cramer, S. P.; Eccles, T. K.; Kutzler, F. W.; Hodgson, K. O.; Mortenson, L. E. J. Am. Chem. Soc. 1976, 98, 1287-88.
[16]   Bianconi, A. In X-Ray Absorption; Koningsberger, D. C.; Prins, R., Eds.; Wiley & Sons: New York, 1988; p 648.
[17]   Kutzler, F. W.; Natoli, C. R.; Misemer, D. K.; Doniach, S.; Hodgson, K. O. J. Chem. Phys. 1980, 73, 3274-88.
[18]   Jaklevic, J. M.; Kirby, J. A.; Ramponi, A. J.; Thompson, A. C. Environ. Sci. 1980, 14, 437-41.
[19]   Maylotte, D. H.; Wong, J.; St. Peters, R. L.; Lytle, F. W.; Greegor, R. B. Science 1981, 214, 554-56.
[20]   Cramer, S. P. In X-Ray Absorption; Koningsberger, D. C.; Prins, R., Eds.; Wiley & Sons: New York, 1988; pp 257-320.
[21]   Hitchcock, A. P.; Lock, C. J. L.; Pratt, W. M. C. Inorg. Chim. Acta 1982, 66, L45-L47.
[22]   Bockman, R. S.; Repo, M. A.; Warrell, Jr., R. P.; Pounds, J. G.; Schidlovsky, G.; Gordon, B. M.; Jones, K. W. Proc. Natl. Acad. Sci. 1990, 87, 4149-53.
[23]   Eidsness, M. K.; Elder, R. C. Springer Proc. Phys. 1984, 2, 83-85.
[24]   Iyengar, G. V. Biol. Trace Elem. Res. 1987, 24, 263-95.
[25]   Brown, F. C. In Synchrotron Radiation Research; Winick, H.; Doniach, S. Eds.; Plenum Press: New York, 1980; pp 61-100.
[26]   Tullius, T. D. Ph.D. thesis, Stanford University, 1979.

RECEIVED July 16, 1990

Chapter 24

Comparative Intestinal Absorption and Subsequent Metabolism of Metal Amino Acid Chelates and Inorganic Metal Salts

H. DeWayne Ashmead

Albion Laboratories, Inc., 101 North Main Street, Clearfield, UT 84015

Intrinsic, extrinsic, and luminal conditions all influence mineral bioavailability resulting in different intestinal absorption rates and subsequent body metabolism for inorganic salts versus amino acid chelates. Those differences are demonstrated by *in vitro* comparisons of mucosal cell uptakes of minerals as amino acid chelates, SO_4, O_2, and CO_3. In perfusion studies the differences in rates of transfer of isotopes of chelates and metal salts from the mucosal to serosal solutions are examined. Finally, *in vivo* isotope studies are summarized which compare the tissue uptake of minerals from amino acid chelates and inorganic salts. These investigations, utilizing Fe, Zn, Mn, Ca, Mg, and Cu, demonstrate that intestinal absorption and subsequent metabolism of the above minerals is greater when they are ingested as amino acid chelates compared to inorganic salts of the same metals.

In its most basic terms, nutrition is the intake of protein, carbohydrates, fats, vitamins, minerals, and water for growth and maintenance of body tissues, for energy, and to regulate all of the body processes. Health and well being are dependent upon the correct balance and intake of these nutrients. Minerals function in all of the nutrient roles. This necessitates having an optimum amount of the essential minerals to carry out their multitude of assigned functions. Optimum is the key to efficacy. Too much or too little of a mineral has an equally deteriorative effect. Deficiencies are generally of greater practical concern than excesses.

The fact that a specific mineral is chemically present in food does not guarantee intestinal absorption. Numerous factors will enhance the bioavailability of a particular metal, while others will detract from its usage in the body as summarized in Table I(*1*).

0097–6156/91/0445–0306$06.00/0
© 1991 American Chemical Society

Table I. Factors Affecting Mineral Bioavailability

Intrinsic factors:

1. Animal species (including man) and its genetic makeup
2. Age and sex
3. Monogastric or ruminant (intestinal microflora)
4. Physiological function: growth, maintenance, reproduction
5. Environmental stress and general health
6. Food habits and nutrition status
7. Endogenous ligands to complex metals (chelates)

Extrinsic factors:

1. Mineral status of the soil on which the plants are grown
2. Transfer of minerals from soil to food supply
3. Bioavailability of mineral elements from food
   a. Chemical form of the mineral (inorganic salt or chelate)
   b. Solubility of the mineral complex
   c. Absorption on silicates, calcium phosphates, dietary fiber
   d. Electronic configuration of the element and competitive antagonism
   e. Coordination number
   f. Route of administration, oral or injection
   g. Presence of complexing agents such as chelates
   h. Theoretical (*in vitro*) and effective (*in vivo*) metal binding capacity of the chelate for the element under consideration
   i. Relative amounts of other mineral elements

In the lumen:

1. Interactions with naturally occurring ligands
   a. Proteins, peptides, amino acids
   b. Carbohydrates
   c. Lipids
   d. Anionic molecules
   e. Other metals
2. Interactions at and across the intestinal membrane
   a. Competition with metal-transporting ligands
   b. Endogenously mediating ligands
   c. Release to the target cell

A complete review of Table I is outside the scope of this chapter. This present discussion will examine the effect of the chemical form of the mineral on mineral bioavailability. Inorganic metal salts have different absorption levels than chelates(2). Even among chelates, absorption rates will vary according to the ligand, stability constant of the chelate, molecular weight, etc.(3), so the field of chelates must be narrowed further to deal only with molecules resulting from the binding of a polyvalent cation to the alpha-amino and carboxyl moieties of an amino acid to form a five-membered ring. The structure of the ring consists of the metal atom, the active carboxyl oxygen atom, the carbonyl carbon atom, the alpha-carbon atom, and the alpha-nitrogen atom. The bonding is accomplished by both coordinate covalent and ionic bonding. At least two and sometimes three amino acids can be bound to a single metal ion, depending upon its oxidative state, to form bicyclic and/or tricyclic-ringed molecules as seen in Figure 1. Even though the oxidative states of certain cations would allow for a fourth amino acid to be chelated to the cation, the bonding angles and the atomic distances required for chelation would tend to preclude its occurrence.

Several *in vitro* studies have shown intestinal uptake of amino acid chelates to be more rapid and in greater amounts than equivalent quantities of metal salts. Jejunal segments from adult male Sprague-Dawley albino rats beginning 10 cm below the pylorus and continuing for another 20 cm were removed, cut into 2 cm segments, and severed longitudinally along the mesenteric line. After randomization, washing, and incubation in Krebs-Ringer buffer solution, they were then exposed for 2 min to 50 μg of Cu, Fe, Mg, or Zn as CO_3, O_2, or SO_4 or as amino acid chelates, all of which had been dissolved in simulated gastric solutions. Following exposure period, the segments were washed and assayed by atomic absorption spectrophotometry. The summarized results are shown in Table II(2). Clearly, under controlled conditions, the uptake of the amino acid chelates was significantly greater than any of the metal salts tested.

Table II. Mean Jejunal Uptake of Different Mineral Forms (ppm)

	Chelate	SO_4	O_2[a]	CO_3	Control
Copper	35	8	11	6	Trace
Magnesium	94	36	23	51	7
Iron	298	78	61	53	23
Zinc	191	84	66	87	14

[a]oxide

Differences in absorption rates can be demonstrated visually as well. Fasted experimental animals were fed Fe as CO_3 or the amino acid chelate. Sections of their mucosal tissue were excised and examined by scanning electron microscopy. Figure 2a shows unabsorbed $FeCO_3$ on the microvilli whereas Figure 2b shows iron amino acid chelate being absorbed. The difference is

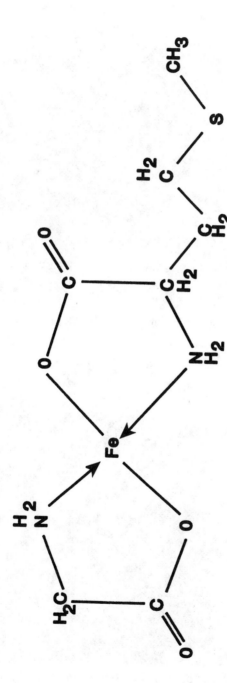

Figure 1. A two dimensional drawing of a bicyclic chelate of iron with glycine and methionine as amino acid ligands.

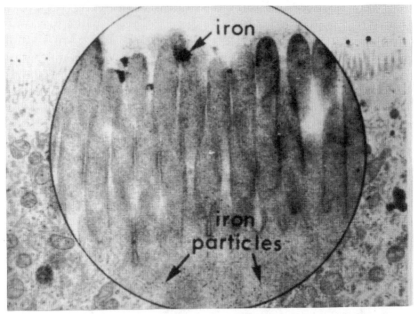

Figure 2a. The absorption of Fe from $FeCO_3$ on the microvilli.

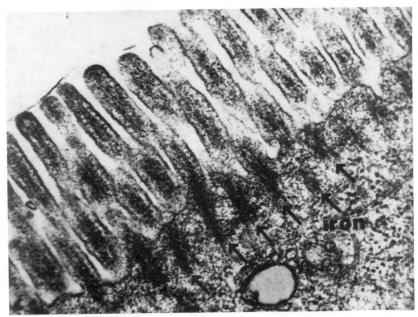

Figure 2b. The absorption of Fe from Fe amino acid chelate on the microvilli.

dramatic. Fe absorption from the amino acid chelate is not only greater, but more rapid.

For absorption of a mineral from a salt to occur, it must be presented to the mucosa as a cation. Numerous studies have demonstrated that, after ingestion, metal salts are generally ionized in the stomach, providing they are soluble. If no interfering chemical reactions occur, the cations enter the intestine where they are bonded to amino acids from the chyme or to the carrier proteins embedded in the luminal membranes of the mucosal cells and transported to the cytoplasm by passive diffusion or active transport. This absorption can occur anywhere in the small intestine, but generally takes place in the duodenum where the metal ions retain their solubility due to a lower pH(*2*).

Conversely, the amino acid chelate is not ionized before absorption. It is not affected by different precipitating anions because the metal ion in the molecule is chemically inert due to the coordinate covalent and ionic bonding by the amino acid ligands. It is not affected by the acid pH of the stomach and survives as an intact molecule particularly after the chelate has been stabilized through a particular buffering process(*4*). Fats and fibers do not interfere with its absorption due to the high formation constant (or the low dissociation constant). And finally, the transport of the amino acid chelate across the mucosal cell does not require vitamin intervention as is the case with some metal ions. Instead, its absorption follows the pathway of a low molecular weight peptide. When an amino acid chelate is formed, it assumes many of the characteristics of a dipeptide or tripeptide. Due to their stabilities, the chemical structures of amino acid chelates are not altered throughout the digestive process. Therefore, as an amino acid chelate, the metal can easily travel through the intestine in the guise of a dipeptide-like or tripeptide-like molecule. The low molecular weight of this molecule (200 to 400 daltons) allows intact absorption into the mucosal cell without further luminal hydrolysis.

Stability of amino acid chelates is of great concern to its absorption. When only one amino acid is used to chelate the metal ion, it may not form a chelate bond that is sufficiently strong to resist dissociation in the gut. Two chelating amino acids provide four bonds (three amino acids - 6 bonds) into a single metal atom. The combination of the bonds projecting out at tetrahedral angles (as in the case of the bicyclic species) and the steric hindrance of the chelating amino acid rings impede competing electrophilic molecules or atoms from disrupting the chelate bond. When only one amino acid chelates to a metal atom, the result is an open side where competitive electrophilic species may attack. Thus, two- and three-ringed amino acid chelates are more stable. The weaker single amino acid chelates, or amino acid complexes, can dissociate chemically in the stomach and intestines, thus freeing the metal ion to the chyme and allowing it to become susceptible to any of the various kinds of competing reactions which are adverse to absorption. An amino acid chelate will also resist the action of the peptidases that break internal peptide bonds, due to the presence of the metal atom in the molecule as well as a result of a higher stability constant. As noted above, luminal changes in pH do not alter the amino acid chelate. Consequently, the amino acid chelate molecule can be absorbed intact through the intestinal mucosa, pulling the metal along with it. As illustrated in Figure

3, when the intact amino acid chelate is absorbed into the mucosal cells, it is believed that a coordinating link is formed through one of its amino acids with the gamma-glutamyl moiety of glutathione, a tripeptide found in the mucosal cell membrane. After the formation of the chelate-glutathione molecule, an enzymatic breakdown of this molecule would occur, resulting in the active transport of the amino acid chelate from the luminal side of the mucosa to the cytoplasm within the cell(2).

After traversing the mucosal cell membrane, the amino acid chelate is released at the terminal web due to a pH change which breaks the linkage of the transport molecule with the chelate. The amino acid chelate can then migrate across the mucosal cell to the basement membrane and from there directly into the plasma, as an intact molecule. No intracellular carrier is required.

While a certain amount of cytoplasmic peptide hydrolysis can take place, intracellular hydrolysis of the amino acid chelates occurs less frequently than the hydrolysis of dipeptides without a metal, due to the steric hindrance of the ring structure inherent in the chelate molecule, as noted above, in connection with restrained luminal hydrolysis. The actual separation of the metal from its amino acid ligand molecules generally takes place at the sites of usage. Most metals function in the body as parts of amino acid chelates and complexes and as components in enzyme systems. When hydrolysis occurs, the coordinating bonds break, possibly due to enzymatic action together with site-specific pH changes that can lower the stability constants of the amino acid chelates, favoring the release of the cation to a cytoplasmic ligand with a higher stability constant.

Previously cited work has demonstrated that mucosal uptake of the amino acid chelate is more rapid than that of metal salts. Perfusion studies have demonstrated that the transfer of the amino acid chelate to the serosa is also more rapid. In these experiments, rat small intestines were excised, washed, and everted in an oxygenated (95% O_2 and 5% CO_2) buffer solution. Fifteen-centimeter (15 cm) segments were cut, tied at the ends, and run in tandem in separate vessels at 37° C which contained 1×10^{-4} ^{65}Zn as $ZnCl_2$ or as Zn amino acid chelate in the mucosal solution. Each hour, for a period of eight hours, intestines were removed, rinsed and digested prior to scintillation counting to determine absorbed radioactivity in the tissues. Each hour, for a period of five hours, samples of the serosal solution were removed and analyzed for ^{65}Zn transferred from the mucosa to the serosa. Figures 4a and b show the moles of Zn per gram of tissue in the duodenum and in the jejunum, while Figure 5 shows the quantity of ^{65}Zn transferred to the serosal solution(5).

The transfer of Zn from the mucosa to the serosa in the form of an amino acid chelate is not only more rapid than Zn as $ZnCl_2$, but the quantity of Zn absorbed and transferred is significantly higher. When chelated to amino acids, approximately four times as much Zn was transported to the serosa leaving less in the intestinal tissues. When the mucosal cell is dependent upon a cytoplasmic generated carrier (metallothionein) to transport the Zn to the serosa, as in the case of $ZnCl_2$, the movement of the Zn appears hindered, or even blocked due to lack of carrier molecules. This results in higher concentrations of Zn from $ZnCl_2$ in the mucosal cells but less in the serosal solution.

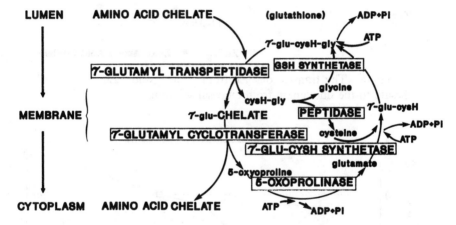

Figure 3.   The membrane transport of an amino acid chelate illustrating steps involved.

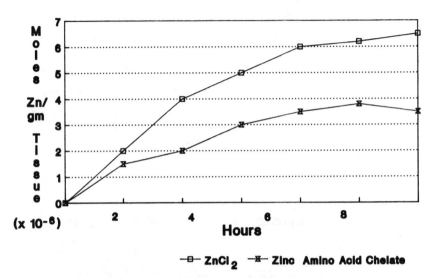

Figure 4a. The transfer of ^{65}Zn from ^{65}ZnCl$_2$ and ^{65}Zn amino acid chelate to the duodenum from mucosal solution.

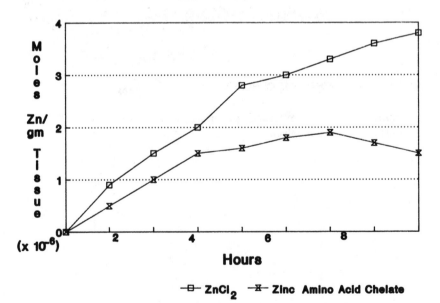

Figure 4b. The transfer of ^{65}Zn from ^{65}ZnCl$_2$ and ^{65}Zn amino acid chelate to the jejunum from mucosal solution.

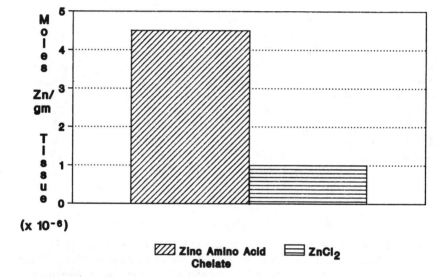

Figure 5. The transfer of ^{65}Zn from ^{65}ZnCl$_2$ and ^{65}Zn amino acid chelate to the serosal solution from the mucosal solution.

Greater mucosal absorption and transfer of amino acid chelates results in greater body tissue levels of the minerals from those chelates. Several studies presented below illustrate this.

In the first example, Sprague-Dawley albino male adult rats were fasted for 24 hours before being divided into two groups of eight animals each. While under light ether anesthesia each animal was given an intraperitoneal dose of 5.0 μg of ^{65}Zn as an amino acid chelate or as $ZnCl_2$. Four hours after dosing, all of the animals were sacrificed and their tissues assayed for ^{65}Zn with a gamma-ray spectrometer. Table III presents the amount of ^{65}Zn found in the assayed tissues of the animals(2). While the levels differed from tissue to tissue, the mean increase of Zn in the tissues was almost 19% when the amino acid chelate was ingested.

Table III. Mean ^{65}Zinc Levels in the Tissues (cc/m/gm)

Tissue	Zn amino acid chelate	$ZnCl_2$	% Increase
Muscle	186	153	22
Heart	1,457	1,433	2
Liver	10,250	7,529	36
Kidney	8,629	7,797	11
Brain	541	444	22

In a second study, which was designed similarly to the above Zn study, ^{54}Mn amino acid chelate was compared to $^{54}MnCl_2$. After being fasted for 24 hours, two groups of adult male Sprague-Dawley albino rats were orally administered 32 μg of ^{54}Mn as either the chelate or chloride salt. Fourteen days following dosing, the animals were sacrificed and their tissues assayed for ^{54}Mn. The delay in sacrificing was to allow for homeostasis. When the body is loaded with a stable Mn there is a rapid excretion of old tissue Mn via the bile and a redistribution of new Mn within the tissues(6). The 14 days between dosing and tissue assay allowed the distribution of ^{54}Mn to occur. The results are shown in Table IV(2).

When Mn was chelated to the amino acids, its metabolism within body tissues was much greater ($\overline{X}$=82.3%) than as an inorganic salt. Fecal recovery of $^{54}MnCl_2$ was 33% more than from ^{54}Mn amino acid chelate. The remainder of the ^{54}Mn from $MnCl_2$ was excreted via the urine after absorption, indicating lower metabolism of Mn from $MnCl_2$.

Table IV. Mean ^{54}Manganese Levels in Tissue (cc/m/gm)

Tissue	Mn amino acid chelate	MnCl$_2$	% Increase
Heart	107	36	197
Liver	106	52	104
Kidney	97	80	21
Spleen	397	190	109
Lung	56	54	4
Sm Intestine	141	89	58
Muscle	28	22	27
Bone	266	112	138

A third comparative study was designed using ^{45}Ca. Adult male Sprague-Dawley albino rats were divided into two groups. After fasting for 24 hours each animal received 1 mg of ^{45}Ca orally as either Ca amino acid chelate or as CaCl$_2$. One week following the dosing, all of the animals were sacrificed and their tissues assayed for ^{45}Ca. The delay in sacrificing the animals was to allow for hormonal production of a vitamin D dependent carrier to transfer the Ca from CaCl$_2$ through the mucosal cell cytoplasm. The Ca amino acid chelate does not require this carrier molecule in order to be transferred to the plasma. Table V presents the results(2).

Table V. Mean ^{45}Calcium Levels in the Tissues (cc/m/gm)

Tissue	Ca amino acid chelate	CaCl$_2$	% Increase[a]
Bone	5,772	3,682	57
Muscle	1,206	614	96
Heart	932	642	45
Liver	742	664	12
R. Cerebrum	804	698	15
Kidney	730	686	6
Serum	31	8	288
Erythrocytes	13	19	(-32)
Whole Blood	44	27	63

[a]With the exception of the erythrocyte compartment.

The chelate group excreted 76% less ^{45}Ca in their feces than the Cl$_2$ group, indicating greater absorption from the chelate. Furthermore, the mean absorption and transfer of the ^{45}Ca from the amino acid chelate was 61.1% greater.

In a fourth series of studies, 4.4 μc of ^{59}Fe were administered in a single oral dose as either the amino acid chelate or as $FeCl_3$ to fasted pregnant Sprague-Dawley female albino rats 4 days before expected parturition. Approximately 72 hours after dosing, the 15 animals in each group were sacrificed, and their tissues and fetuses assayed for ^{59}Fe. The delay between dosing and assaying was to allow placental transfer of ^{59}Fe. Table VI presents the results(2).

Table VI. Mean ^{59}Fe Levels in the Tissues of Pregnant Rats and Fetuses (cc/m/gm)

Tissue	Fe amino acid chelate	$FeCl_2$	% Increase
Uterus	4,925	3,333	48
Liver	9,675	8,167	18
Kidney	950	567	68
Spleen	325	134	143
Heart	1,425	333	328
Lung	2,925	1,367	114
Fetus ($\overline{X}$/Fetus)	46	16	188

The data in Table VI demonstrate that when Fe is chelated to amino acids, its uptake, and certainly, its utilization ($\overline{X}$=129.6%), are significantly greater than when equivalent amounts of Fe in the chloride form are ingested. An added finding in this study is the placental transport of the Fe as the amino acid chelate. Traditionally, in the latter stages of gestation, transplacental Fe transport has been difficult. Other studies have indicated that increased placental transport of Fe amino acid chelate is primarily due to the molecular weight of the chelate in the plasma (approximately 300 daltons) versus that of iron transferrin in the blood (approximately 90,000 daltons)(7). This research further demonstrated that the amino acid chelate is absorbed into the blood intact in the same molecular form as when it was ingested. The smaller molecule can then be more easily transferred through the so-called "placental barrier" than Fe attached to the much larger transferrin molecule.

In conclusion, it has been shown that there are numerous factors which will affect mineral bioavailability. One of these factors is the chelation of the mineral with amino acids. The above data demonstrate that chelation with amino acids result in greater absorption and metabolism of the mineral than in the form of soluble salts.

Literature Cited

1. Kratzer, F.; Vohra, P. In *Chelates in Nutrition*; CRC Press: Boca Raton, Fl., 1986; pp 35.

2. Ashmead, H. D., Burton, S. A.; Peterson, R. V. In *Intestinal Absorption of Metal Ions and Chelates*; Charles C Thomas: Springfield, Il., 1986.
3. Miller, R. In *Chelating agents in poultry nutrition*; Proc. Delmarva Nutr. Short Course, 1968.
4. Jensen, N. *Biological Assimilation of Metals*, U.S. Patent No. 4,167,546, Washington D. C., Sept. 11, 1979.
5. Fang, S.; Burton, S. A.; Peterson, R. V., *et al.* In *Chelated Mineral Nutrition in Plants, Animals, and Man*; Ashmead, D., Ed.; Charles C Thomas: Springfield, Il., 1982; pp 137-151.
6. Underwood, E. In *Trace Elements in Human and Animal Nutrition*; Academic Press: New York, N.Y., 1977; pp 174-176.
7. Ashmead, D., Graff, D. In *Placental transport of chelated iron*, Proc. IPVS, Mexico, 1982; pp 207.

RECEIVED August 1, 1990

Chapter 25

Mineral Supplementation in Plants Via Amino Acid Chelation

Robert B. Jeppsen

Albion Laboratories, Inc., 101 North Main Street, Clearfield, UT 84105

Plants must obtain their vital nutrients by absorption from
air, water, and/or soil. Improvements in bioavailability and
assimilation can be gained through the use of amino acid
chelates of the required minerals. Research presented compares
metal amino acid chelates with complementary EDTA or
inorganic sources. Data for corn, tomatoes, apples, potatoes and
wheat indicate improvement through the supplementation of
amino acid chelates of various metals. When increased
bioavailability and anabolic usage are coupled with a multiple
ratio assessment of all nutrients compared to each other, as
determined by plant assays from crop to crop, optimal amounts
and proportions of the amino acid chelates can be supplemented
and significantly improved crop yields can result.

While chelation of metals is well known, physiological sequellae for mineral
bioavailability in plant and animal nutrition have been less understood. A
common misconception has been that all chelates of a particular metal are the
same, with identical or very similar absorptive and metabolic properties.
However, each type of ligand molecule has its own unique properties and,
therefore, each type of chelate resulting from a particular class of ligands has
its own properties. Chelation of a metal occurs when two or more sites from
the ligand molecule bond into the same metal atom. This results in one or
more cyclic ring structures which give the new molecule unique properties
different from those that the metal would exhibit if it were in an inorganic or
ionic form. More ligands than one may join the atom and form rings in
addition to the first ligand molecule. The metal atom encased in the center of
four or more bonds and the outward projecting backbones of the ligand rings
is subsequently protected from chemical reactions or attractions which could
interfere with its absorption.

The amino acids form chelates between the terminal carboxyl and the adjacent alpha-amino group. This results in a ring structure around a metal of five members comprising bonds between the metal, carboxyl oxygen, carbonyl carbon, alpha-carbon, alpha-nitrogen and then back into the metal with a coordinate covalent bond from the nitrogen atom. The resulting chelate bonds are compatable to animal and plant metabolic systems in that they maintain their integrity prior to their metabolic usage and yet are capable of rupture, allowing the organism to benefit from the mineral and amino acid nutrition. This may be contrasted with a powerful synthetic chelate such as EDTA which forms strong bonds which are not as readily broken for metabolic usage of the chelated metal by plants and animals. Since a historical tendency in agriculture and animal husbandry has been to lump all of the chelates together, the properties of EDTA metal chelates have been presumed to represent those of other kinds of chelates. This assumption is unwarranted. The amino acid chelates have proven to be well-adapted vehicles for increasing epidermal absorption of metals through the surfaces of leaves, stems, and fruits.

High Bioavailability of Amino Acid Chelates

In a greenhouse experiment administered by Albion Laboratories(_1_), the effects of different sources of iron were combined with different levels of nitrogen supplementation. Corn plants were grown in pots in a greenhouse with four plants per pot and each treatment was repeated three times. The pH was 8.1 in the pots. The extractable iron was determined to be five parts per million in all soils, including the control soils. Nitrogen was supplemented as null plus two levels (0, 500, and 1000 ppm) by a urea and ammonium nitrate mixture at 50:50 proportions. The controls received nitrogen as above, but no iron source. Iron was applied as a foliar spray at 400 ppm for each of Iron Amino Acid Chelate, ferrous sulfate, and Iron-EDTA. The first spraying was done when the corn plants had acheived growth of ten to twelve inches and a second spraying at the same rates followed one week later. The plants were harvested three weeks after the second spraying. Following harvest, the foliage was washed in 1% hydrochloric acid, followed by a distilled deonized water rinsing, then dried for 24 hours at 75°C. Dry matter yield was determined, thereafter. This was followed by dry ashing, extraction, and measurement of the iron and manganese by DCP (direct current plasma spectrophotometry). The results are shown in Table I.

The highest dry matter yields and, therefore, the greatest vegetative growths were obtained from the Iron Amino Acid Chelate foliar treatments. Duncan's multiple range test (DMRT) was applied to determine significance of the various data within the table. Data which are similar ($P < 0.05$) are followed by a common letter. Administration of nitrogen at 1000 ppm generally yielded less beneficial results than the 500 ppm rates. In terms of dry matter, the Iron Amino Acid Chelate groupings stand unique with a lower end exception of inorganic iron sulfate at 500 ppm nitrogen. The Iron-EDTA was inconsequential in its production of dry matter, regardless of nitrogen levels.

Table I. Comparison of Above-Ground Corn Yields From Three Iron
Sources with Concomitant Nitrogen Treatments

Treatment Fe-source	N (ppm)	Dry Matter (g/pot)	Fe (ppm)	Mn (ppm)
None	0	14.32 cd[1]	28 e	35 c
None	500	15.02 bcd	31 de	37c
None	1000	13.86 cd	35 cde	39 c
Fe-AACH[2]	0	17.62 a	102 a	71 b
Fe-AACH	500	16.73 ab	114 a	90 a
Fe-AACH	1000	16.85 ab	119 a	72 b
$FeSO_4$	0	13.23 de	48 bcde	32 c
$FeSO_4$	500	15.73 abc	53 bcd	42 c
$FeSO_4$	1000	14.82 bcd	63 b	40 c
Fe-EDTA	0	14.12 cd	47 bcde	38 c
Fe-EDTA	500	15.08 bcd	56 bc	37 c
Fe-EDTA	1000	11.24 e	69 b	31 c

SOURCE: Adapted from ref. 1.
[1]Numbers in a column not followed by a common letter differ significantly   at
$P < 0.05$, as determined by DMRT.
[2]Fe-Amino Acid Chelate.

All of the iron foliar spray sources promoted greater iron uptakes, but results from the amino acid chelate were highest with close to four times those of the control. The EDTA source of iron did not differ significantly from the inorganic iron, although both were higher than the control amounts. Other elements can be affected by the ready absorption of a particular ion. Manganese was absorbed in greater levels in conjunction with the Iron Amino Acid Chelate foliar spray. By contrast, the manganese absorption from the Iron-EDTA source was not appreciably different from the control values.

Table II shows data collected from a field crop of corn which had no apparent deficiencies(2). Iron, zinc and manganese were administered singularly as Amino Acid Chelate foliar sprays upon the visually healthy crop. Bushels per acre yields increased in all cases. Also of significance was the increase in absorption of other metals. Zinc Amino Acid Chelate generated a notable increase of both iron and manganese, as well as of zinc.

Table II. Effects of Foliar Application of Iron, Zinc, and
Manganese Amino Acid Chelates on the Yield
and Nutrient Content of Corn

Treatment	Yield (Bu/A)[1]	N	P	K	Fe	Zn	Mn
			%			ppm	
Control	126.5	2.83	0.38	3.31	218	31	62
Fe-AACH[2]	132.2	2.92	0.34	3.84	257	35	67
Zn-AACH[2]	134.6	2.84	0.37	3.24	277	76	71
Mn-AACH[2]	132.9	2.85	0.37	3.42	207	30	82

SOURCE: Adapted from ref. 2.
[1]Yield in bushels per acre.
[2]Metal-Amino Acid Chelate.

Isotope Studies of Absorption and Translocation. In order to more fully understand the absorption of the Iron Amino Acid Chelates, radioactive ^{59}iron was applied to tomato leaves as a foliar spray of either Iron Amino Acid Chelate, Iron-EDTA or iron sulfate(2). All of the iron sources were applied at the same molar concentrations of iron. The data which are shown in Table III indicate that significantly more iron was absorbed through the epidermis of the tomato leaf if in the Iron Amino Acid Chelate form.

Table III. Replicate Absorption Measurements of ^{59}Fe, Applied as a Foliar
Spray to Selected Attached Tomato Leaves

	Fe-Amino Acid Chelate	Fe-EDTA	FeSO$_4$
Same Leaf	43.1[1]	26.6	29.4
Adjacent Leaf	0.20	0.10	0.20
Same Stem	0.30	0.03	0.14
Same Leaf	37.58	23.93	21.08
Adjacent Leaf	0.37	0.13	0.14
Same Stem	0.07	0.04	0.14

SOURCE: Adapted from ref. 2.
[1]Units are corrected counts per minute per milligram.

In further experiments on greenhouse corn plants, nitrogen was administered twice a week from a nutritive solution(2). The results are shown in Table IV. Sprays containing the same amounts of radioactive iron were given to plants which were 60 centimeters tall. Analyses were made five days after spraying. Each of the three treatments was replicated three times and the values shown are the resulting means.

Table IV. Absorption and Distribution of ^{59}Fe Applied as a Foliar Spray to Selected Attached Corn Leaves

Plant Part	Iron Amino Acid Chelate	Iron Sulfate
Point of Application	227[1]	68
Point of Application + 1 cm	0.54	0.17
Opposite Leaf	0.20	0.13
Root	0.13	0.03

SOURCE: Adapted from ref. 2.
[1]Units are corrected counts per minute per milligram.

In each case of either Iron Amino Acid Chelate or iron sulfate, the highest corrected counts per minute of absorbed radioactive iron were obtained at the point of application. Points of application could be considered to be sources or reservoirs of iron that could be drawn from on an as needed basis. Within the same timeframe, nearly four times as much iron had been translocated to the root from the Iron Amino Acid Chelate as compared to the inorganic control source.

Effects of Foliar Applications of Amino Acid Chelates on Flowering Fruits. The data in Table V were generated from experimentation in San Salvador on tomato plants(3). In this case, a foliar spray containing several minerals was administered to field tomatoes of the Santa Cruz Kada variety. Nitrogen and phosphorus fertilizers as well as standard herbicides and pesticides were applied to the soil at the same rate in all treatments including the control. Each of the three rates of spraying of Multimineral Amino Acid Chelate were done ten days before first flowering. The study was a complete randomized block design with four treatments and five replicates. Each plot was 10 meters by 1 meter, as these were row crops.

The number of buds increased progressively according to the rate of foliar application of Multimineral Amino Acid Chelate. The most dramatic increase occurred in the number of fruits obtained from the highest application of the Multimineral. This is especially significant in that the first spraying achieved an average number of fruits that were 2.5% above controls, the second yielded 6% above the first, however, the third yielded 26% above the second, and 33% above the first. The amounts of increases in foliar applications were 15% greater from the first spraying to the second and, 13% greater from the second to the third comprising a 30% increase from the first to the third sprayings.

Table V. Effects of Foliar Applications of Multimineral Amino Acid Chelate on Tomato (*Licopersicum esculentum* L.) Yield

Treatment[1] (ml/ha)[2]	Number of Buds (average)	Number of Fruits (average)
0	28.2	19.8
464.28	30.5	20.3
535.71	31.5	21.5
607.14	33.2	27.0

SOURCE: Adapted from ref. 3.
[1]Multimineral Amino Acid Chelate
[2]Units are milliliters per hectare.

Effects of Mineral Amino Acid Chelate Application Directly on the Surface of Fruits. It has been found that there is a direct correlation between bitter pit and the amount of calcium in apple fruits. Calcium is important in cell wall development in plants with the integrity of cells and cell walls completely dependent on the availability of sufficient calcium. Results in Table VI were obtained when calcium was administered as a post harvest fruit dip (Hymas, T., unpublished research report).

Table VI. Calcium Concentrations in Granny Smith Apples Receiving Various Postharvest Dip Treatments

Treatment	Dilution	Ca in Peel (ppm)
Control	----	241
$CaCl_2$	1:74	219
Ca-AACH[1]	1:120	708
Ca-AACH	1:60	723
Ca-AACH	1:30	670
$CaCl_2$ + Ca-AACH	1:148 + 1:60	568

[1]Calcium Amino Acid Chelate

The calcium from Albion Laboratories was Calcium Amino Acid Chelate at 5% calcium. The commercial dip was calcium chloride at 12% calcium. Granny Smith apples were used and there were eight apples per replication in each treatment. It had previously been determined that better results could be obtained by additional treatments with DPA (diphenylamine) administered to decrease incidence of scalding, and administration of a sucrose ester that had been shown to slow down maturation of fruits. These two products were administered to all apples with the exception of the controls. Prior to calcium measurement, all of the apples were washed in deionized distilled water. The skins were separated from the rest of the flesh for analysis. These were dried and analyzed by atomic absorption spectrophotometry.

The data indicate the amount of calcium in the peel. The control received no dips. In all cases of Amino Acid Chelate administration, the amounts of calcium in the peel were high above the control amounts. A combination of calcium chloride and Calcium Amino Acid Chelate was not as effective, being at 21% less than the highest chelate alone, and 15% less than the lowest chelate alone. The difference between the calcium chloride and highest Calcium Amino Acid Chelate represented a 330% in increase in calcium absorption from the Amino Acid Chelate source, even though the chelated source contained only 42% of the calcium present in the inorganic dip.

Improvements of Yield Through TEAM Evaluation of Albion Amino Acid Chelates

In recognizing that they had suceeded in creating a series of mineral chelates which were highly bioavailable, researchers at Albion Laboratories, Inc. decided to further expand the utilization of these products by determining a way to properly balance minerals within particular crop plants as a way to maximize yields under any prevalent environmental or edaphic conditions. This work, which incorporated several years of experimentation, resulted in what has been termed the TEAM report (Technical Evaluation of Albion's Minerals). It utilizes the mass action proportional relationship of every element pitted against every other element of eleven of the elemental nutrients that are necessary for plant growth. Critical levels were also involved in the calculation parameters. The report is based on the nutritive balances of minerals supplied by amino acid chelates in foliar sprays which will yield the highest amounts of any particular crop. The calculations and subsequent recommendations for kinds and amounts of Albion Laboratories Amino Acid Chelates to be administered by foliar spraying are handed through computer software. The result has been the ability to prescribe the nutritive needs of a particular crop on a particular field for maximized yields. Since nutritive requirements differ from plant-type to plant-type, the evaluation takes these differences into account by selecting the crop-type at the time of data entry of the assayed mineral levels of the plant tissues.

Computer Assisted Recommendations for Amino Acid Chelate Applications for Increased Potato Crop Yields.

To illustrate the advantages of TEAM evaluation, some data are shown in Table VII. The eleven nutrients analyzed by the TEAM report are shown in the left column. Actual nutrient concentrations are indicated next and these are followed by nutrient indices which are assigned by the TEAM program. In addition to these data, the critical levels obtained from other agronomic sources and publications are shown in the rightmost column(4).

**Table VII. Nutrient Analyses for a Potato Crop Showing Actual
Concentrations, TEAM Report Balanced Nutrient Indicies,
and Critical Level Values**

Nutrient	Nutrient Concentration	Nutrient Indices	Critical Level
Ca	0.85 %	-17	0.49 %
S	0.26 %	-16	0.18 %
Mg	0.24 %	-15	0.24 %
P	0.36 %	- 5	0.36 %
Fe	203 ppm	- 1	75 ppm
K	4.32 %	2	3.50 %
Cu	10 ppm	2	6 ppm
B	29 ppm	5	9 ppm
N	5.36 %	8	3.1 %
Mn	105 ppm	11	7 ppm
Zn	36 ppm	26	19 ppm

SOURCE: Reproduced with permission from ref. 4. Copyright 1986 Noyes Publications.

The nutrient indices should be considered to show mineral balance proportions, either as a decriment or adequacy for any particular nutrient that was analyzed from the plant tissues that were submitted for analysis. The most needful nutrients (or the ones most severely out of balance) are listed first, with decreasing needfulness as the series moves downward towards the nutrient measured in the highest abundance and present in sufficient quantities for growth and yield.

The nutrient which is most out of balance and is listed first in this example is calcium. By referring to the assayed nutrient concentration and comparing it with its critical level, calcium would appear to be present in sufficient quantities. Similar arguements could be made for sulfur and iron which are additionally listed as having some degree of imbalance for highest yields of potato crops. In order to assess the results of following the customary farm management practice of determining mineral supplementation needs by comparing tissue analyses with known critical levels, the professional recommendations of an experienced and independent farm manager were used. Both recommendations were followed on separate portions of the field. When assessing the nutrient concentrations indicated in Table VII, the recommendation of the farm manager was to add phosphorus and let the other minerals go untouched, since they appeared to be in sufficient concentrations as compared to critical levels. The TEAM recommendation, however, was to repair an imbalance seen through its evaluation as indicated by the differences of nutrient indices.

Results of following the two divergent recommendations are shown in Table VIII. The control was considered to be the hundred percent yielding value. The recommendation of the farm manager was considered as the critical treatment and the TEAM report results are also indicated. The yield in hundred-weight bags per acre of potatoes was increased by 23% by following the TEAM evaluation recommendations and supplying the correct mineral balances by foliar spraying the appropriate Amino Acid Chelates on the crops. The critical evaluation suggested by the farm manager was able to effect some improvements in yields resulting in a 10% increase over the control, but this was less than half of the increase that was obtainable through following the TEAM recommendation.

Table VIII. Yield of Potato Crops Given Nutrient Supplements Based on TEAM Analysis Versus Standard Farm Management (Critical Levels)

Treatment	Yield (CWT/A)[1]	% Difference
Control	323	100
Critical	354	110
TEAM	398	123

[1]Yield in hundred-weight bags per acre.

<u>Computer Assisted Recommendations for Amino Acid Chelate Application for Increased Wheat Yields</u>. Improvement in wheat yield through use of the TEAM evaluation is shown in Tables IX and X(4). These data are particularly depictive of the capabilities of the TEAM program. Zinc was listed as the most out of balance mineral, and yet had a nutrient concentration within the plant leaves of nearly twice the suggested critical level. Manganese, which was also shown as being deficient or, at least, out of balance, was seventeen times greater than the critical level. A comparison of Tables IX and X, show that the TEAM program is not just a blanket measure and ranking of nutrient concentrations. The proper balance of nutrients form a crucial part of the analysis. Mineral needs vary from crop to crop and what is in balance for one crop may be out of balance for another.

When the farm manager was allowed to assess the nutrient concentrations in Table IX, his choice was to do nothing. All of the balances seemed to be appropriate and were all sufficiently above the critical levels. Again, the recommendation of the TEAM report was to repair a measured empirical imbalance for that particular wheat crop on that particular field.

Table IX. Nutrient Analyses for a Wheat Crop Showing Actual
Concentrations, Critical Level Values and TEAM Report
Balanced Nutrient Indices

Nutrient	Nutrient Concentration	Nutrient Indices	Critical Level
Zn	28 ppm	- 9	15 ppm[1]
K	4.36 %	- 7	1.25 %
Ca	0.30 %	- 7	0.20 %[1]
P	0.50 %	- 5	0.15 %
Mn	85 ppm	- 3	5 ppm
S	0.43 %	- 2	0.15 %[1]
Cu	10 ppm	2	5 ppm
Mg	0.29 %	3	0.15 %[1]
N	5.42 %	4	1.5 %
Fe	215 ppm	5	50 ppm
B	31 ppm	19	5 ppm

SOURCE: Reprinted with permission from ref. 4. Copyright 1986.
[1]Sufficiency level.

The results of these prescribed treatments are shown in Table X. In this case, the critical level treatment was the same as the control treatment, since the farm manager opted to do nothing. There was a slight decriment in bushels per acre of yield from the critical treatment over the control, but this would be considered a random difference. The TEAM recommendation, however, was able to increase the yield of bushels per acre of wheat by 20% over the control, which represented a substantial increase.

Table X. Yield of Wheat Crops Given Nutrient Supplements Based on
TEAM Analysis Versus Standard Farm Management (Critical Levels)

Treatment	Yield (Bu/A)[1]	% Difference
Control	103	100
Critical	101	98
TEAM	124	120

[1]Yield in bushels per acre.

Effects of TEAM Recommendations of Amino Acid Chelates on Plants Exibiting Chronic Pathogenicity. In addition to helping increase yields in crops that are either deficient in mineral nutrients or out of balance in nutrients,

Amino Acid Chelates manufactured by Albion Laboratories can also benefit plants that have routine pathological problems. A field of potatoes that was commonly infected with early blight disease was given foliar spraying of Albion Amino Acid Chelates according to the TEAM recommendation(5). Early blight characteristics included wilt and death with symptoms commensing in mid-season. The plant tissue analyses indicated that three minerals, zinc, iron, and manganese, were low. The TEAM computer evaluation included three additional elements, sulfur, boron, and copper as being out of balance sufficiently with the first three mentioned to justify taking further corrective action. Most of the field was sprayed with the recommended chelates, but a portion along the fence was set aside as the control portion. The soil had been examined through different parts of the field and was found to have no differences between the field in general and the control area.

At mid-season, plants in the control area showed the effects of early blight disease with its necrosis and wilt. Some deaths were apparent. Plants in the treated area showed improved appearances over previous years. The treated area was able to yield 332 hundred-weight bags per acre while the control area yielded 294 hundred-weight bags per acre. This represented a 13% increase for the plants that were sprayed with Amino Acid Chelates according to the TEAM computer evaluation in comparison to the controls. This was a significant improvement in plants that were normally prone to early blight infection.

Value of TEAM Computer Assisted Recommendations of Amino Acid Chelates. It can readily be seen how crucial a proper balance of available minerals is to high-yielding crops (Tables VII, VIII, IX, and X). Even though most all elements were present within the plant tissue in sufficient quantities according to critical levels, there were important characteristics of balance which were not in alignment and required correction for yield improvement. Often, classical agronomic practice cannot assess the effects of trace mineral insufficiencies until the end of the season, and only then, if classical symptoms of mineral deficiencies become apparent. Plans may then be made to remedy the situation before the next season. However, the crop yields for that particular season are lost as to what they could have been, had the imbalance or deficiency been known in time for corrective actions to be taken. The advantage of TEAM is to allow making that evaluation early in the season. The plant sampling materials can be collected soon after they arise from the ground or emerge from trees or shrubs. The TEAM program can then assess mineral analyses and evaluate the balance on a crop-by-crop basis. This results in a recommendation for administering Albion Laboratories Amino Acid Chelated Minerals at particular concentrations to promote an appropriate balance of minerals in the plant and maximize yields for that particular crop in the same season of growth.

Conclusions

Economic uses of plants can include fruits, stocks, vegetables, fibers, wood, and chemicals. With the exceptions of some chemicals that require stress on normal plant physiology for maximal production, most of the uses of plant crops benefit from maximal health and growth. Requirements for optimal growth include sufficient carbon dioxide, water, light, temperature, and minerals. Carbon dioxide is relatively constant in most all arable lands. Presuming that the crop is grown in an appropriate climate with sufficient water, light and temperature, the only remaining hurdles to surpass in maximizing growth and yields become the issues of mineral balances and adequacies. The use of Albion Laboratories Amino Acid Chelated Minerals represents a highly bioavailable source of minerals that can be utilized by plants for both vegetative growth and fruit production by taking advantage of the rapid acquisition possible through foliar or fruit surface application. This allows the needs of a particular crop to be assessed, prescribed and fulfilled within the same season. The data that have been presented have shown how seemingly healthy plants may also benefit from foliar application of these chelated minerals. While the plants may appear to be healthy, they may not be maximized as to the balance of minerals required for optimal performance and growth. By using the highly bioavailable Amino Acid Chelates, and also taking advantage of the computer mass action rationing capability of the TEAM evaluation report, optimal crop-by-crop performance can be maximized through applying optimal mineral balances for the particular crop.

Literature Cited

1. Hsu, H. H.; Ashmead, H. D. In *Foliar Feeding of Plants with Amino Acid Chelates*; Ashmead, H. D.; Ashmead, H. H.; Miller, G. W.; Hsu, H. H., Eds.; Noyes Publications: Park Ridge, N.J. 1986; pp 273-280.
2. Hsu, H. H.; Graff, D. In *A New Era in Plant Nutrition*; Ashmead, D., Ed.; Albion Laboratories, Inc.: Clearfield, Utah, 1982, pp 37-51.
3. Reyes, N.; Escobar, C. In *A New Era in Plant Nutrition*; Ashmead, D., Ed.; Albion Laboratories, Inc.: Clearfield, Utah, 1982, pp 84-90.
4. Hsu, H. H. In *Foliar Feeding of Plants with Amino Acid Chelates*; Ashmead, H. D.; Ashmead, H. H.; Miller, G. W., Hsu, H. H., Eds.; Noyes Publications: Park Ridge, N.J. 1986; pp 183-200.
5. Moss, E. In *A New Era in Plant Nutrition*; Ashmead, D., Ed.; Albion Laboratories, Inc.: Clearfield, Utah, 1982, pp 109-119.

RECEIVED August 1, 1990

Chapter 26

Trace Element-Induced Toxicity in Cultured Heart Cells

Craig C. Freudenrich, Shi Liu, and Melvyn Lieberman

Department of Cell Biology, Division of Physiology, Duke University
Medical Center, Durham, NC 27710

We have undertaken a multidisciplinary approach to study the
toxic effects of several trace elements (e.g. Sb, Cd, Co, Pb, Hg,
Zn) on the structure and function of cardiac cells. Although the
basic mechanisms underlying these toxic effects remain unknown,
altered membrane ion transport has been proposed as a major
determinant in the onset of cell injury. By applying an array of
techniques including light microscopy, atomic absorption
spectroscopy, electrophysiology, and microspectrofluorometry, we
show that cultured embryonic chick heart cells can serve as a
useful model to study how trace elements affect the transport
mechanisms that maintain intracellular ionic homeostasis. We
present examples of how this model can be used to characterize
the basic effects on ion homeostasis, to identify the mode of
uptake of trace elements, and to quantitate the effects of trace
elements on identified ion transport mechanisms. With this
approach it should be feasible to elucidate the basic mechanisms
underlying trace element-induced cardiotoxicity.

Several trace elements (e.g. Sb, Cd, Co, Pb, Hg, Zn) have been shown to induce a
variety of cardiotoxic effects such as arrhythmias, conduction abnormalities, reduced
contractility, cardiac swelling, myofibril degeneration, and sudden death (1-6), but
the mechanisms underlying these effects remain unknown. One hypothesis suggests
that alterations of ion transport mechanisms are implicated in the initiation and
maintenance of cell injury (7-10). However, this hypothesis has been difficult to
evaluate for the following reasons: (a) studies involving animal models have
revealed trace element-induced tissue damage, but have not detected the early
physiological and biochemical disturbances in heart muscle (11) and (b) studies with
tissue culture systems have often been used to screen toxic agents based on
observations of morphologic changes and/or reduced viability (12,13), but have not
provided data that permit conclusions as to the cellular mechanisms of injury. Thus,

0097–6156/91/0445–0332$06.00/0
© 1991 American Chemical Society

the progress in this area can be adequately summarized by Iyengar's comment that "the lack of a multidisciplinary approach has been the Achilles heel of biological trace element research" (14). We agree that such an approach is necessary to elucidate the mechanisms of trace element-induced cardiotoxicity.

Studies in our laboratory over the past 20 years have established that cultured embryonic chick heart cells can provide an ideal model to study mechanisms of ion transport (15). The cells lack significant diffusional barriers, contract spontaneously, and have well-defined voltage-dependent ion channels as well as co- and counter-transport mechanisms (e.g. Na/K pump, Na/H exchange, Na/Ca exchange, Na-dependent Cl/HCO$_3$ exchange, Na+K+2Cl cotransport). Furthermore, the cells can be grown in a variety of configurations suitable for study with biochemical, microscopic, spectroscopic, and electrophysiological techniques.

In this chapter, we will present examples of how the cultured chick heart cell model can be used to characterize the toxic effects of trace elements on ion transport, to identify the mode of uptake of trace elements, and to quantitate the effects of trace elements on specific ion transport mechanisms. This information will be necessary to assess the role of membrane transport in trace element-induced cardiotoxicity.

Characterization of Cardiotoxic Effects of Trace Elements on Ion Transport

Hearts from 11 day old chick embryos were disaggregated and the cells were cultured in antibiotic-free medium (15) either as monolayers, aggregates, or polystrands. The cells were noted to contract spontaneously within 1-2 days of culture. All experiments were conducted at 37 °C and cells were incubated or superfused with a balanced salt solution (HTBSS), pH 7.4 composed of the following (in mM): NaCl, 142.2; KCl, 5.4; MgSO$_4$, 0.8; CaCl$_2$, 1.0; NaH$_2$PO$_4$, 1.0; glucose, 5.6; Hepes + Tris, 10; bovine serum albumin, 1 g/l; in some cases, the phosphate and sulfate were removed to prevent precipitation of any of the trace elements.

Heart cell monolayers in culture for 3 days were incubated for 1-4 h in a HTBSS containing either 100 μM CoCl$_2$, 25 μM PbCl$_2$, 100 μM CdCl$_2$, or 500 μM ZnSO$_4$. Phase-contrast microscopic observations of contractile activity at 1 and 4 h revealed that cells exposed to Co and Pb became arrhythmic, whereas those exposed to Cd and Zn were quiescent. These contractile changes were accompanied by morphologic changes as shown in Figure 1. In all cases, intercellular spaces were increased, indicative of cell shrinkage. Also, cells exposed to Cd and Zn became rounded and contained vacuoles, densities, and membrane blebs, indicative of cell damage.

Cell monolayers exposed to Co, Pb, Cd, and Zn for 1-4 h were rinsed first with ice cold, isotonic choline chloride solution to remove extracellular ions and then processed for K and Na analyses by atomic absorption spectroscopy (16). In most cases, cells exposed to the trace elements lost K at 1 and 4 h (Table I). At 1 h, cells exposed to Co, Pb, and Cd lost Na; whereas, cells exposed to Zn gained Na by 4-fold. By 4 h, cells exposed to all of the trace elements, except Co, gained Na. These data indicate that changes in cell volume and in the regulatory processes for K and Na (e.g. Na and K conductances, Na/K pump) occurred in the presence of trace elements.

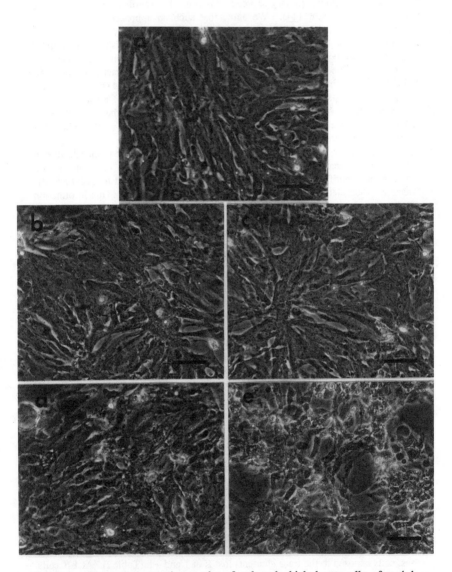

Figure 1.   Phase-contrast micrographs of cultured chick heart cells after 4 h incubation at 37 °C in medium ± trace elements: (a) Control (b) 100 µM CoCl$_2$ (c) 25 µM PbCl$_2$ (d) 100 µM CdCl$_2$ (e) 500 µM ZnSO$_4$. The horizontal bar = 50 µm.

Table I. Effects of Trace Elements on K and Na Content in Cultured Heart Cells

Trace Element	(μM)	Ion Content (nmol/mg protein)	
		K	Na
		1 hour	
Control		1312 ± 131	139 ± 7
CoCl₂	100	1172 ± 9	114 ± 1
PbCl₂	25	1153 ± 27	123 ± 12
CdCl₂	100	1214 ± 60	105 ± 2
ZnSO₄	500	534 ± 89	526 ± 89
		4 hours	
Control		1056 ± 13	118 ± 5
CoCl₂	100	1102 ± 35	115 ± 3
PbCl₂	25	1078 ± 58	141 ± 19
CdCl₂	100	816 ± 55	631 ± 39
ZnSO₄	500	38 ± 25	992 ± 68

Data are mean ± SE (3)

Transmembrane potentials were recorded from aggregates of spontaneously contracting cells according to Liu et al (17) and the cells were exposed to 100 μM concentrations of $CdCl_2$, $CoCl_2$, $ZnSO_4$, $PbCl_2$, Sb (tartar emetic), and $HgCl_2$ as shown in Figure 2. In all cases, except Co, the beating rate of the cells was increased prior to becoming quiescent (depolarized to between -35 and -40 mV); these events usually occurred within 3 min. Examination of the action potentials prior to the cessation of beating revealed that the amplitude and plateau phase were decreased and that the configuration of the action potential was altered. These results can be related to the changes in contractile activity described previously and suggested that the basic ionic processes underlying generation of the cardiac action potential were altered by acute exposure to the aforementioned trace elements.

We have used ion-selective microelectrodes (ISME) to measure intracellular activities of Na, K, Cl, and H in cultured chick heart cells (17). Although trace elements could possibly interfere with the ISME resins, we found no evidence of interference when Na-, K-, Cl-, and pH-ISMEs were calibrated in the presence of 100 μM Sb, Cd, Co, Pb, Hg, or Zn. Therefore, it is feasible to use these ISMEs in experiments involving trace elements.

In preliminary experiments to determine the effects of trace elements on ion regulation, we monitored intracellular pH in polystrands by pH-ISME that were made with the Fluka proton cocktail in a manner similar to that described previously (17). The cells were exposed to either 100 μM $PbCl_2$ or 100 μM $ZnSO_4$. As shown in Figure 3, exposure to Pb increased intracellular pH by about 0.1 units within 5 min. In contrast, exposure to Zn decreased intracellular pH by approximately 0.2 units. These results indicated that Pb and Zn can alter the regulation of H in heart

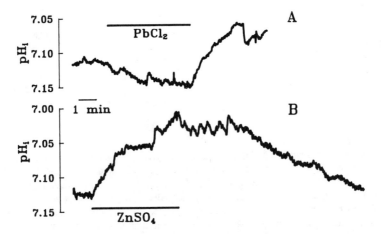

Figure 2. Effects of trace elements on cardiac action potentials (AP). Left: 100 μM trace element was added (up arrow) and removed (down arrow); horizontal bar = 1 min. Right: AP (dashed line, element; solid line, control); horizontal bar = 50 msec and is located at 0 mV; vertical bar = 50 mV for both sides.

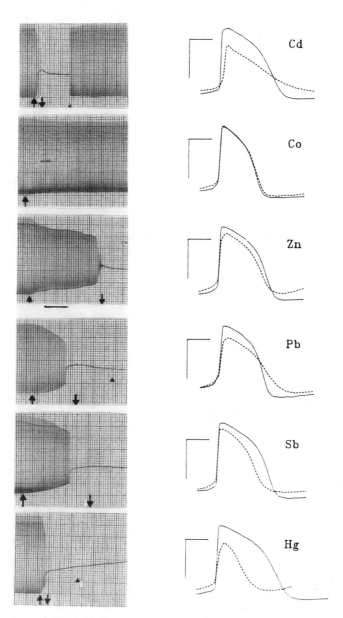

Figure 3. Effects of 100 μM PbCl$_2$ (A) and 100 μM ZnSO$_4$ (B) on intracellular pH as measured using a pH-ISME.

cells by affecting one or more pH regulatory mechanisms (e.g. Na/H exchange, Na-dependent Cl/HCO$_3$ exchange, pH buffering capacity, Ca/H interactions). Future studies will be directed toward assessing the effects of trace elements on specific pH regulatory mechanisms and toward characterizing their effects on the regulation of other ions using ISME techniques.

Identification of the Mode of Uptake of Cardiotoxic Trace Elements

To resolve mechanisms associated with trace element-induced cardiotoxicity, it will be necessary to determine whether the trace elements can act from the outer sarcolemmal surface and/or the cytoplasm. The uptake of trace elements by cells can be measured by conventional methods (i.e. radioisotope tracer flux, atomic absorption spectroscopy) that are quantitative and sensitive to low concentrations; however, these methods are limited because they (a) cannot distinguish between trace elements bound to the cell surface from those that have permeated the sarcolemma (b) may have variable responses due to heterogeneity of the cell population (c) require processing of samples and data and, therefore, do not provide information in real time. Although radioisotopic tracer flux and atomic absorption spectroscopic techniques can be applied to the cultured chick heart cells (15), we have developed a new approach to detect trace element uptake in single, living cells in real time by using the Ca-sensitive fluorescent dye, fura2 (18).

The method takes advantage of the fact that many of the trace elements can alter, by direct binding, the fluorescent properties of fura2. To demonstrate these effects, solutions of 10 μM fura2 were prepared with a 140 mM KCl, 10 mM Mops buffer (pH 7.2) that had been passed through a Chelex (Bio-Rad) column to remove all divalent cations. The solutions were placed in a chamber on the heated (37 °C) stage of a microscope attached to a Spex spectrofluorometer (19). Fluorescence excitation spectra (emission = 510 nm) were acquired before and after the addition of various trace elements (3-1000 μM final concentrations). Figure 4 shows the spectra for Cd, Co, Pb, Mn, Hg, and Zn (100 μM) normalized to the respective trace element-free spectrum. Note that all trace elements, except Pb, had maximal effects at concentrations $\leq$ 100 μM. The type of changes in the fluorescent properties of fura2 evoked by trace elements were variable. Cd and Zn enhanced fluorescence at all wavelengths; whereas, Hg, Co, Mn, and Pb quenched fluorescence (Pb predominantly quenched the fluorescence, but at higher concentrations). By measuring the fluorescence at the Ca-insensitive excitation wavelength (360 nm), we determined the half-maximal concentrations (μM): Cd, 4.7; Zn, 23; Hg, 31; Co, 2.8; Mn, 3.6; and Pb, 447.

Although the effects on fura2 fluorescence would interfere with measurements of cytosolic Ca^{2+}, they provide a convenient means to detect the permeation of trace elements into single, living cells. Cell monolayers were grown on glass coverslips, single cells were microinjected with fura2, and the cells were perfused with HTBSS on the stage of the microspectrofluorometer (19). Fluorescence (excitation, 350 and 360 nm; emission, 510 nm) was measured from single cells as shown in Figure 5. The fluorescence of cells exposed to 0.5 mM Cd increased; whereas, the fluorescence of those exposed to 1 mM Co decreased. In these examples, the fura2 fluorescence changed as predicted from the spectra (Figure 4) and confirmed that both Cd and Zn had permeated the sarcolemma.

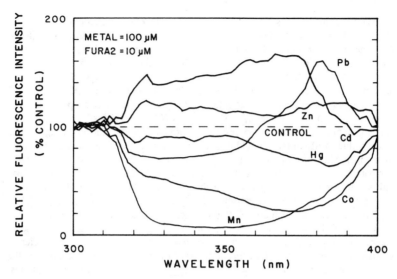

Figure 4. Effects of trace elements (Cd, Co, Pb, Mn, Hg, Zn) on the fluorescence excitation spectra of fura2.

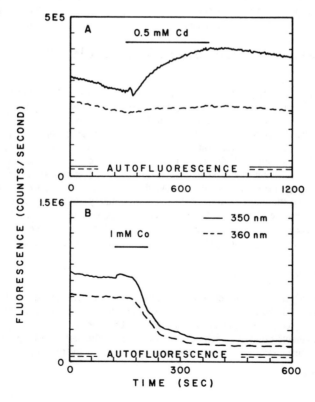

Figure 5. Alterations of intracellular fura2 fluorescence by Cd (A) and Co (B).

We have utilized this method to follow Mn influx in cultured chick heart cells (20). We found that Mn quenched the intracellular fura2 fluorescence as predicted from the spectra in Figure 4 and that the rate of quenching, which is related to the rate of influx, was proportional to the extracellular Mn concentration. The quenched signal evoked by 50 μM Mn was not affected by perfusing the cells with impermeant metal chelators, 3 mM ethylene-diaminetetraacetic acid (EDTA) and 0.2 mM diethylenetriaminepentaacetic acid (DTPA), but was reversed by perfusion with a permeant metal chelator, 20 μM N,N,N',N'-tetrakis(2-pyridylmethyl)-ethylenediamine (TPEN); these results confirmed that the quenched signal was actually due to Mn entry. In addition, the rate of quenching of intracellular fura2 by 0.1 mM Mn was reduced by about 90% when the cells were treated with 10 μM verapamil and 10 μM nifedipine; these results indicate that a major pathway of Mn influx is through voltage-dependent Ca channels. This approach has also been used to detect Cd influx in smooth muscle cells (21) and La influx in heart cells (22).

The advantages of using fura2 fluorescence to monitor trace element influx into cells are as follows: (a) trace elements in the cytosol can be detected because of complexation with fura2 that was microinjected into the cell (b) fura2 is sufficiently sensitive to detect micromolar concentrations of trace elements in the cytosol (c) the method provides data in real time in living single cells and, thus, can be used to test many experimental conditions. The major disadvantage of this fluorescence assay is that it does not allow for quantitation of the absolute amount of trace element uptake. However, when used in conjunction with conventional radioisotopic or atomic absorption methods, the fluorescence method can provide useful information about rates, amounts, and mechanisms of toxic trace element uptake in heart cells.

Quantitation of Toxic Effects of Trace Elements on Defined Ion Transport Mechanisms

As mentioned earler, cultured chick heart cells have many defined ion transport mechanisms that can be characterized by a variety of electrophysiological techniques. For example, Stimers et al (23) showed that Na/K pump current can be quantitated in aggregates of cultured chick heart cells under voltage-clamp conditions by exposing the cells to K-free solution or to 1 mM ouabain; this has also been accomplished in single heart cells using the whole-cell patch voltage-clamp technique (24). Using this approach, we examined the effect of 100 μM $ZnSO_4$ on the Na/K pump current of a single cell. As shown in Figure 6, exposure to Zn for 1 min reduced the Na/K pump current by approximately 16%; these results were consistent with the changes in Na and K content upon exposure to Zn (Table I). This preliminary result demonstrates the feasibility of this approach to determine the effects of trace elements on specific electrogenic transport mechanisms.

Summary

The cultured chick heart cell model offers several advantages to study trace element-induced cardiotoxicity: (a) the cells are in a physiological state (e.g. spontaneously contracting) and lack any diffusional barriers so that the direct effects of trace elements on cardiac cells can be assessed (b) the cells can be grown in a variety of configurations that are well suited to characterize the toxic effects of trace

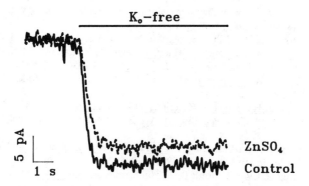

Figure 6. Effect of 100 μM $ZnSO_4$ on the Na/K pump current in a single heart cell.

elements on cell structure and function by various microscopic, biochemical and biophysical methods (c) the cells are suitable for determining trace element uptake by radioisotope tracer flux, atomic absorption spectroscopy and a novel fluorescence method (d) the cells have many well-defined ion transport mechanisms that can be quantitated using biophysical methods so that the effects of toxic trace elements on specific ion transport mechanisms can be assessed. The types of information that can be obtained from this model will be necessary to elucidate the mechanisms underlying trace element-induced cardiotoxicity.

Acknowledgments

We thank Kathleen Mitchell and Meei Liu for their work in preparing the cell cultures and performing the ion content experiments. This work was supported in part by National Institutes of Health Grants HL-07101, HL-17670, and HL-27105 to M.L. and C.C.F.

Literature Cited

1.   Van Stee, E. W. In Cardiovascular Toxicology; Van Stee, E. W., Ed.; Raven Press: New York, 1982; p 1.
2.   Carson, B. L.; Ellis, H. V.; McCann, J. L. Toxicology and Biological Monitoring of Elements in Humans; Lewis Publishers, Inc.: Michigan, 1986.
3.   Van Fleet, J. F.; Ferrans, V. J. Am. J. Pathol 1986, 124, 98.
4.   Hurst, J.W. The Heart; McGraw-Hill Book Co: New York, 1986, Sixth Edition.
5.   Kopp, J. Handbook of Experimental Pharmacology 1986, 80, 195.
6.   Braunwald, E. Heart Disease; W. B. Saunders Co.: Philadelphia, 1988, Third Edition.
7.   Trump, B. F.; Berezesky, I. Current Topics in Membranes and Transport; Academic Press, Inc.: New York, 1985; Vol. 25, p 279.
8.   Bridges, J.W.; Benford, D. J.; Hubbard, S. A. Ann. New York Acad. Sci. 1983, 407, 42.
9.   Buja, L. M.; Hagler, K. K.; Willerson, J.T. Cell Calcium 1988, 9, 205.
10.   Ferrans, V.J. In Physiology and Pathophysiology of the Heart; Sperelakis, N., Ed.; Kluwer Acad. Publ., 1989, Second Edition, p 691.
11.   Zbinden, G. Cardiac Toxicology; CRC Press: Boca Raton, FL, 1981; Vol. 3, p 8.
12.   Thelstam, M.; Mollby, R. Toxicology 1980, 17, 189.
13.   Wenzel, D. G.; Cosma, G. N. Toxicology 1984, 33, 117.
14.   Iyengar, G. V. Elemental Analysis of Biological Systems; CRC Press: Boca Raton, FL, 1989; Vol. 1, p 1.
15.   Horres, C. R.; Wheeler, D. M.; Piwnica-Worms, D.; Lieberman, M. In The Heart Cell in Culture; Pinson, A., Ed.; CRC Press: Boca Raton, FL;1987, Vol. 1, p 77.
16.   Murphy, E.; Aiton, J. F.; Horres, C.R.; Lieberman, M. Am. J. Physiol. 1983, 245, C316.
17.   Liu, S.; Jacob, R.; Piwnica-Worms, D.; Lieberman, M. Am. J. Physiol. 1987, 253, C721.

18. Grynkiewicz, G.; Poenie, M.; Tsien, R.Y. J. Biol. Chem. 1985, 260, 3440.
19. Freudenrich, C. C.; Lieberman, M. Spex Biomedical Application Note B-3; Spex Industries, Inc.: Edison, NJ, 08820, 1989.
20. Freudenrich, C. C.; Lieberman, M. Biophysical J., 1989, 47a.
21. Smith, J. B.; Dwyer, S. D.; Smith, L. J. Biol. Chem., 1989, 264, 7115.
22. Peeters, G. A.; Kohmoto, O.; Barry, W. H. Am. J. Physiol., 1989, 256, C351.
23. Stimers, J. R.; Shigeto, N.; Lieberman, M. J. Gen. Physiol., 1990, 95, 61.
24. Stimers, J. R.; Liu, S.; Lieberman, M. FASEB J., 1990, 4, A295.

RECEIVED July 16, 1990

Chapter 27

Stable Isotope Dilution for Zinc Analysis in Low Birth Weight Infants

J. P. Van Wouwe,[1] and R. Rodrigues Pereira

Interfaculty Reactor Institute, Delft University of Technology, Mekelweg 15, 2628 JB Delft, Netherlands and Department of Pediatrics, St. Clara Hospital, Olympiaweg 350, 3078 HT Rotterdam, Netherlands

We investigated the feasibility of destructive neutron activation analysis for determining the isotopic ratios of zinc. An equation was derived to relate the $^{69m}Zn/^{65}Zn$ photopeak ratios measured in blood-plasma to indicate the degree of dilution of the enriched dietary zinc over the time interval with zinc that was present in the body. The sources of error in determining the $^{69m}Zn/^{65}Zn$ photopeak ratio, and thus in the dilution factor are discussed. Experiments were performed on infants fed 60 mL of formula (containing 0.9 μmol zinc and supplemented with 3.7 μmol of 98.8% enriched ^{68}Zn). One hour later blood was collected. In 1 mL of plasma the $^{69m}Zn/^{65}Zn$ photopeak ratio was determined to be higher than in controls by a factor of 1.069 ± 0.009. This gives a dilution factor of 23 ± 2 for the formula zinc if absorption is 31%. Within a period of 1 h, the dietary zinc was diluted with approximately 90 μmol of body zinc which was only 13% of the assumed total body zinc.

Naturally occurring zinc is a mixture of five stable isotopes, ^{64}Zn, ^{66}Zn, ^{67}Zn, ^{68}Zn and ^{70}Zn, accounting for 48.89, 27.71, 4.11, 18.57 and 0.62 mol% of the total zinc, respectively (Table I). In principle, stable isotope mixtures, enriched in respect to each of the five isotopes, can be used in vivo as tracers to study the zinc status by measuring the dilution factors using isotope-analysis techniques. Also, the radioactive zinc tracers, ^{65}Zn and ^{69m}Zn are

[1]Current address: Rijnsburgerweg 102, 2333 AE Leiden, Netherlands

commercialy available. The stable zinc isotopes which can be used in combination with destructive neutron activation analysis (DNAA) are ⁶⁸Zn and ⁷⁰Zn. Procedures for the performance of an oral loading test by application of the stable ⁶⁸Zn to evaluate aspects of the zinc metabolism have been reported (1). In this paper, the reliabilty of the DNAA measurements of ⁶⁸Zn was tested and the means for prediction of the limits of confidence were given. For example, the calculations were applied to samples from a study of infant formula supplemented with ⁶⁸Zn. The formula was fed to low birth weight infants, known to be at risk to develop zinc deficiency (2-4).

Table I. Composition of Natural, Tracer and Formula Zn Isotopes.

Isotope	Mole Fraction (%)		
	Natural	Tracer	Formula
⁶⁴Zn	48.89	0.56	10.25
⁶⁶Zn	27.71	0.43	5.90
⁶⁷Zn	4.11	0.19	0.98
⁶⁸Zn	18.57	98.80	82.72
⁷⁰Zn	0.62	0.02	0.14

The aim of the present study is to show how, from the measured enrichment (R) of a stable nuclide in blood-plasma, the dilution factor (D) of the administered nuclide can be calculated. From the degree of dilution of the enriched dietary zinc, the (diluted) zinc in the body can be estimated, assuming the absorption of dietary (infant formula) zinc to be 31% (5).

Theory
The aim is to calculate the dilution factor D ($1 <= D <= \infty$) of the zinc tracer in infant blood-plasma for a fixed time interval after the administration of a known amount of the tracer. This requires the measurement of the isotope ratio of the diluted tracer against that of a reference isotope. A factor R, representing the quotient of both of the isotope ratios in plasma after uptake of the tracer following the administration of the dose, can be measured if the absorption of the dose is known. Assuming 100% absorption (e.g., after intravenous injection):

$$R = \frac{([trac]/[ref] \text{ after dilution})}{([trac]/[ref] \text{ natural})} \quad \{1\}$$

where [trac] and [ref] are ⁶⁸Zn and ⁶⁴Zn, respectivelly. Knowing the mole fraction of the [trac] and [ref] in the natural mixture and in the dose (e.g., α, β, in the natural mixture and γ, δ in the dose, respectively), equation {1} can be changed into equation {2}, in which the dilution factor D is expressed as function of the mole fractions and R:

$$D = 1 + \frac{(\alpha^{-1} \times \gamma) - (\beta^{-1} \times \delta) \times R}{R - 1} \quad \{2\}$$

The values for α, β, γ and δ are contained in Table I. Substitution of these values in equation {2} yields:

$$D = 1 + \frac{4.4545 - 0.2096 \times R}{R - 1} \qquad \{3\}$$

For R <5, but > 1.1, the term $0.2096 \times R$ can be neglected and equation {3} can be written:

$$D = 1 + \frac{4.5}{R - 1} \qquad \{4\}$$

For R <1.1, the formula futher simplifies to:

$$D = \frac{4.5}{R - 1} \qquad \{5\}$$

so:

$$\log D = 0.65 - \log (R - 1) \qquad \{6\}$$

The value for D needs only to be corrected for the absorption of the orally given dose (for infant formula 31%). Equation {6} clearly shows that one can calculate D from the measured value of R. The influence of the variation of R on the value of D can be shown when R varies between $R - 2\sigma$ and $R + 2\sigma$, where σ is a fraction of $(R - 1)$. Thus, when R = 1.01 and the fraction varies from 1 to 0.125 times the value of $(R - 1)$, then:

$\sigma/(R-1)$	$\rightarrow R \pm 2\sigma$	D for R$-$/$+$ 2σ
1	$\rightarrow 1.01 \pm 0.01$	$\infty/177$
0.5	$\rightarrow 1.01 \pm 0.005$	$\infty/266$
0.25	$\rightarrow 1.01 \pm 0.0025$	1064/356
0.125	$\rightarrow 1.01 \pm 0.00125$	710/427

For R >1.01, the fraction should be <= $0.25 \times [R - 1]$, in order to get realistic values for D. The variation in R, the lower margin for R to be still determined, the acceptable relative variation in R, and the upper limit of the detectability of D are given in Table II.

METHODS AND MATERIALS
Reagents and Samples
The natural zinc standard used was 99.99% zinc oxide #8842-00 (Koch-Light Lab Ltd, Coinbrook, Bucks, UK). ^{68}Zn (98.8% enriched), purchased from Rohstoff-Einfuhr GmbH (Dusseldorf, FRG) as ZnO was dissolved in 1 M HCl and diluted with 0.1 M Na-acetate buffer of pH 5.6. The formula under study was Frisolac (CCF, Leeuwarden, NL). It was supplemented with ^{68}Zn instead of natural zinc as the normal procedure calls for in Leeuwarden. A 61 µmol solution of ^{68}Zn in 1 mL was added to 1 L of unsupplemented formula which already contained 15 µM zinc after pasteurization. The formula contained per feeding of 60 mL (= 1 dose): 0.06×76 µmol= 4.59 µmol Zn, which was composed of 0.92 µmol natural zinc (0.45 µmol ^{68}Zn plus 0.17 µmol^{68}Zn from the raw formula), and 3.67 µmol from the tracer zinc dose containing 98.8% ^{68}Zn (= 3.63 µmol ^{68}Zn) and 0.56% ^{64}Zn (= 0.02 µmol ^{64}Zn). Thus the total dose in one feeding of 60 mL amounts to 3.80 µmol ^{68}Zn and 0.47 µmol ^{64}Zn (Table I). Plasma was collected and stored at -20°C in zinc free tubes (Venoject 100HL, Terumo, Belgium). All other reagents and solutions used were of the highest purity available.

Table II. The Relation Between the Variation in R (sd), its Lower Limit Detectability, its Acceptable Relative Variation and the Upper Limit Detectability of D.

Var R (sd*)	Lower Limit Detectibility R	Acceptable Variation R (%)	Upper Limit Detectability D
0.001	1.004	0.1	1328
0.002	1.008	0.2	664
0.003	1.012	0.3	443
0.005	1.02	0.5	266
0.01	1.04	1.0	133
0.02	1.08	1.9	67
0.03	1.12	2.7	45
0.05	1.2	4.2	28
0.10	1.4	7.1	14
0.20	1.8	11	7.6
0.30	2.2	14	5.4
0.50	3.0	17	3.7

sd* = standard deviation.

Procedure

Plasma samples were collected 1 hour after administration of the dose. The samples were subjected to a pre-irradiation chemical separation to remove Na and to separate radio-Zn from the other isotopes produced (2). Prior to irradiation, a 1 mL sample was digested overnight with 1 mL HNO_3 at 140° C in a closed Teflon digestion bomb. After cooling, the digest was transferred to a centrifuge tube; the pH was adjusted to 2.0 with NH_4OH; 0.4 mL of Pb-acetate carrier was added; zinc was co-precipitated using ammonium pyrrolidinedithiocarbamate (APDC); the precipitate was dissolved in 1 mL HNO_3; transferred to a quartz irradiation vial (diameter 8 mm, length 70 mm); the HNO_3 was evaporated under an infrared lamp; the vial was sealed and irradiated in a flux ($10^{13}/cm^2/s$) of thermal neutrons for 12 h in the Delft Research Reactor.

The contents of the vial were taken up in 5 mL H_2SO_4 and digested at 250° C with H_2O_2 till clear. The H_2SO_4 was evaporated; the residue was taken up in 8M HCl; and passed through a Biorad AG 2-X8 column (diameter 1.2 cm) of 4 g. The column was washed with 20 mL of 0.5 M HCl and the zinc was eluted with 20 mL 0.02 M HBr. A solution containing 20 mg $ZnSO_4.7H_2O$, 4.5 g NH_4Cl and 0.5 g Na-acetate was added and the zinc was precipitated with 5 mL of 10% $(NH_4)_2HPO_4$ at boiling temperature. The precipitate was allowed to stand for 30 min and then was filtered through a Millipore filter. The filter was counted for 3 h on a Ge(Li) semiconductor detector of 2 keV WHM resolution at 1332 keV. The γ-spectrum of the treated plasma samples showed only two main zinc peaks, viz., at 0.44 MeV (^{69m}Zn) and 1.12 MeV (^{65}Zn). Corrections were made for background and decay.

Stable isotope dilution was tested on four infants. They were born after 32-39 gestational weeks. They were bottle-fed the formula under study. When these infants were adequately growing on the regular formula under study they were considered in steady metabolic state. When a bodyweight of 2500 g was attained, they were offered a

single dose of 60 mL of ^{68}Zn supplemented milk. One hour after the
meal, 2 mL whole blood was drawn by venepuncture. Each ^{68}Zn and ^{64}Zn
determination in patient plasma was accompanied by a control plasma
sample of a volunteer and a natural zinc standard. The study was
approved by the Clara Hospital Ethical Committee.

RESULTS AND DISCUSSION
The counts per minute (cpm) and calculated µmoles of the control and
patient plasma samples and their molar ^{68}Zn/^{64}Zn ratios are given in
Table III. The mean molar ^{68}Zn/ ^{64}Zn ratios in the controls were
0.377 ± 0.010, well in accordance to the expected value of 0.380
(Table I). The mean molar ratios in the infants studied were 0.403 ±
0.003, and showed a significant difference compared to the controls:
p= 0.003 based on analysis of variance (ANOVA). The corresponding
value for D would be 65 ± 9 {eq. 6}, yielding after correction for
absorption a value of 20 ± 3. The body zinc available for dilution
within 1 hour was estimated to be approximately 90 µmol. Since the
estimated total body zinc is 700 µmol (7), it is clear that about
13% of the body zinc is diluted with the enriched stable zinc. If
the absorption of the dose varies between 20 and 40%, the value of D
would range between 13 and 26, and the the available body zinc for
dilution between 60 and 120 µmol (8.5 - 17%). In our four patients
the average value amounted 93 ± 11 µmol, corresponding to 13 ± 2%
(Table IV).

Table III. ^{68}Zn/^{64}Zn Mole Ratios in Plasma with only Natural Zinc
(Ct) and in Patient Plasma Samples[a] (Pt).

Sample	^{69m}Zn cpm	^{68}Zn µmol	^{65}Zn cpm	^{64}Zn µmol	Mole Ratio
Ct 1	1914	0.0041	33.98	0.0110	0.372
Ct 2	1701	0.0037	29.77	0.0096	0.378
Ct 3	1700	0.0037	28.84	0.0093	0.390
Ct 4	1114	0.0024	20.11	0.0065	0.366
Average					0.377

Sample	^{69m}Zn cpm	^{68}Zn µmol	^{65}Zn cpm	^{64}Zn µmol	Mole Ratio
Pt 1	1465.9	0.0032	24.22	0.0078	0.400
Pt 2	961.9	0.0021	15.80	0.0051	0.402
Pt 3	931.5	0.0020	15.11	0.0049	0.408
Pt 4	828.3	0.0018	13.59	0.0044	0.403
Average					0.403

a: counts per minute (cpm) and µmol correspond to 1 mL of sample.
   The technique used was DNAA.

Table IV. Values for R, D, and the Estimated Amount of
Body Zinc (B-Zn) Available for Dilution during a 1 hour
period.

Patient	R	D	B-Zn, μmol
1	1.061	73	96
2	1.066	67	104
3	1.082	54	93
4	1.069	65	78
Mean ± s.d.	1.069	65	93 ± 11

Literature Cited
1. Van den Hamer, C.J.A.; Cornelissen, C.; Hoogenraad, T.U.; Van
   Wouwe, J.P. In Trace Elements in Man and Animals-V; Mills, C.F.;
   Bremner, I.; Chesters, J.K., Eds.; Commonwealth Agricultural
   Bureaux, Slough, UK, 1985, pp. 689-91.
2. Janghorbani, M.; Ting, B.T.G.; Istfan, N.W.; Young, V.R. Am. J.
   Clin. Nutr. 1981, 34, 581-91.
3. Kumar, S.P.; Anday, E.K. Pediatrics 1984, 73, 327-29.
4. Sann, L.; Bienvenu, G.G.F.; Bourgeois, J. Pediatr. Res. 1980,
   14, 1040-46.
5. Sandström, B.; Cederblad, Å.; Lönnerdal, B. Am. J. Dis. Child.
   1983, 137, 726-29.
6. Walravens, P.A. Clin Chem. 1980, 26, 185-89.
7. Van Wouwe, J.P.; Van den Hamer, C.J.A.; Van Tricht, J.B. Eur. J.
   Pediatr. 1986, 144, 598-99.

RECEIVED September 4, 1990

INDEXES

Author Index

Affiliation Index

Subject Index

Other ACS Books

Chemical Structure Software for Personal Computers
Edited by Daniel E. Meyer, Wendy A. Warr, and Richard A. Love
ACS Professional Reference Book; 107 pp;
clothbound, ISBN 0–8412–1538–3; paperback, ISBN 0–8412–1539–1

Personal Computers for Scientists: A Byte at a Time
By Glenn I. Ouchi
276 pp; clothbound, ISBN 0–8412–1000–4; paperback, ISBN 0–8412–1001–2

Biotechnology and Materials Science: Chemistry for the Future
Edited by Mary L. Good
160 pp; clothbound, ISBN 0–8412–1472–7; paperback, ISBN 0–8412–1473–5

Polymeric Materials: Chemistry for the Future
By Joseph Alper and Gordon L. Nelson
110 pp; clothbound, ISBN 0–8412–1622–3; paperback, ISBN 0–8412–1613–4

The Language of Biotechnology: A Dictionary of Terms
By John M. Walker and Michael Cox
ACS Professional Reference Book; 256 pp;
clothbound, ISBN 0–8412–1489–1; paperback, ISBN 0–8412–1490–5

Cancer: The Outlaw Cell, Second Edition
Edited by Richard E. LaFond
274 pp; clothbound, ISBN 0–8412–1419–0; paperback, ISBN 0–8412–1420–4

Practical Statistics for the Physical Sciences
By Larry L. Havlicek
ACS Professional Reference Book; 198 pp; clothbound; ISBN 0–8412–1453–0

The Basics of Technical Communicating
By B. Edward Cain
ACS Professional Reference Book; 198 pp;
clothbound, ISBN 0–8412–1451–4; paperback, ISBN 0–8412–1452–2

The ACS Style Guide: A Manual for Authors and Editors
Edited by Janet S. Dodd
264 pp; clothbound, ISBN 0–8412–0917–0; paperback, ISBN 0–8412–0943–X

Chemistry and Crime: From Sherlock Holmes to Today's Courtroom
Edited by Samuel M. Gerber
135 pp; clothbound, ISBN 0–8412–0784–4; paperback, ISBN 0–8412–0785–2

For further information and a free catalog of ACS books, contact:
American Chemical Society
Distribution Office, Department 225
1155 16th Street, NW, Washington, DC 20036
Telephone 800–227–5558